Thermodynamics for Chemists, Physicists and Engineers

Robert Hołyst • Andrzej Poniewierski

Thermodynamics for Chemists, Physicists and Engineers

Robert Hołyst
Institute of Physical Chemistry
Polish Academy of Sciences
Warsaw, Poland

Andrzej Poniewierski
Institute of Physical Chemistry
Polish Academy of Sciences
Warsaw, Poland

ISBN 978-94-017-8304-0 ISBN 978-94-007-2999-5 (eBook)
DOI 10.1007/978-94-007-2999-5
Springer Dordrecht Heidelberg New York London

Printed on acid-free paper

Springer is part of Springer Science+Business Media (www.springer.com)

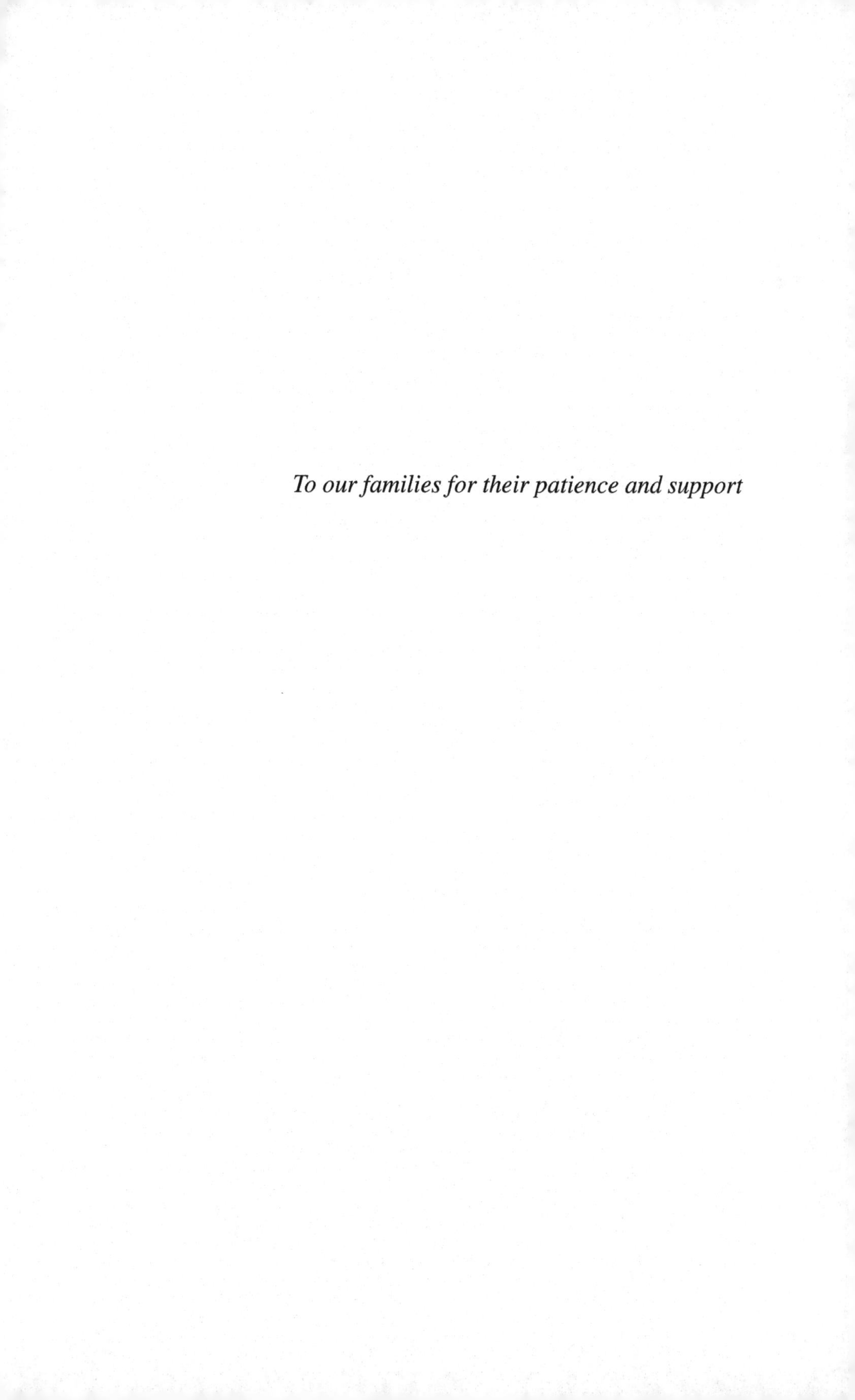

To our families for their patience and support

Preface

The present work is based on our two previous books published in Polish and entitled (in English translation): *Thermodynamics for chemists, physicists and engineers* [6] and *Thermodynamics by exercises* [7]. The first one, besides the part devoted to the fundamentals of phenomenological thermodynamics and its application to phase transitions and chemical reactions, contains also an introduction to statistical thermodynamics written by Alina Ciach. The second book is a collection of exercises on thermodynamics together with their solutions, which correspond to the material presented in [6]. The motivation for writing of these books was the lecture on thermodynamics with the elements of statistical mechanics, given by us at the College of Science, which was a part of the physical chemistry course for the 2nd year undergraduate students. Presently the College of Science forms the department of mathematics and science at the University of Cardinal Stefan Wyszyński in Warsaw, but originally it was established due to the initiative of a few research institutes of the Polish Academy of Sciences, including the Institute of Physical Chemistry where we are employed, and still benefits from their scientific and research potential. Because of large diversity of research carried out in the institutes of the Polish Academy of Sciences, the studies in the College of Science are of interdisciplinary character. Therefore the course of thermodynamics differs from traditional courses of this subject at the physics or chemistry departments. In spite of many excellent textbooks in this field it was difficult to find one, at rather an elementary level, whose scope would correspond to the material lectured by us. This fact inclined us to write a textbook adapted to our needs. However, in the course of writing, we decided that if we extended somewhat the scope of the book, it could also be useful for the Ph.D. students in our institute, who after the second year of studies are obliged to pass an examination on physical chemistry, which is roughly at the level of P.W. Atkins' book [1].

The present book is not simply a compilation of [6] and [7], since we have introduced many significant changes and improvements. Moreover, as we did not want the book to grow in size too much, we decided to limit its scope to phenomenological thermodynamics. To facilitate its use, we have highlighted in the text the postulates and laws of thermodynamics, as well as the most important definitions and conclusions. Mathematical digressions are included in the main text, instead of appendices, as we think that the formalism used in thermodynamics should be treated as its integral part. We pay special attention to the compatibility of definitions, terms, units and notation used in our book with the recommendations of the International Union of Pure and Applied Chemistry (IUPAC) [4].

The book is divided into three parts. At the end of each chapter, there are exercises whose solutions are given at the end of the book. The first part, consisting of five chapters, contains the postulates and laws of phenomenological thermodynamics together with examples of their application. In Chaps. 2 and 3, we introduce and discuss the basic concepts and quantities, such as the equilibrium state, parameters and functions of state, thermodynamic process, pressure, temperature, internal energy, heat and chemical potential, relying mainly on the reader intuition. Writing Chaps. 4 and 5, we were inspired with Callen's book [3]. Chapter 4 is mainly devoted to entropy and the second law of thermodynamics and to the conclusions following from that law. In Chap. 5, we discuss the thermodynamic potentials and natural variables, and also the conditions of intrinsic stability for a pure substance. Less advanced readers can skip the last point.

The second part is devoted to application of thermodynamics to phase transitions in pure substances (Chap. 6) and in mixtures (Chaps. 8 and 9); Chap. 7 is an introduction to thermodynamics of multicomponent systems. In Chap. 6, we give general classification of phase transitions and a few examples of first order and continuous transitions. In the rest of the book, however, we restrict ourselves to first order transitions. The concept of ideal mixture is introduced in Sect. 7.5. Less advanced readers can skip Sects. 7.2 and 7.6. The discussion of phase transitions in mixtures is limited to the case of two-component systems. In Chap. 8, we discuss the phenomena that can be explained by the model of ideal mixture. Non ideal mixtures are considered in Chap. 9. In this case, we study the simplest extension of the ideal mixture model called the simple solution. To understand the whole material presented in this chapter the reader who skipped Sects. 7.2 and 7.6 should return to them. However, less advanced students can skip the formal part of Chap. 9 and concentrate on the phase diagrams presented.

In part three, we consider thermodynamic systems in which chemical reactions occur. Chapter 10 concerns reactions between electrically neutral compounds. The law of mass action, which follows from the condition of chemical equilibrium, is derived for a mixture of ideal gases. Therefore the material presented in this chapter should be understood also by less advanced students. Chapter 11 concerns electrochemical systems, in which chemical reactions occur between ions. Our main aim was to show that due to a chemical reaction a system can perform work other than the mechanical one, which in the framework of thermodynamics can be explained by means of a reversible cell. This chapter is mainly for more advanced graduate students.

We know from our own experience that for the second year students the concepts of the differential and differential form are rather difficult. Since these concepts are crucial for the whole course of thermodynamics, we have tried to explain them in a simple way without going into mathematical details. The second crucial mathematical concept, which is used to introduce the thermodynamic potentials, is the

Legendre transformation. Obviously one can define enthalpy or the Helmholtz or Gibbs free energy without any reference to that concept. On the other hand, we think that it is easier to understand properly the meaning of natural variables of a thermodynamic potential in terms of the Legendre transformation, which was shown in an elegant way by Callen [3].

The exercises together with solutions are to help the readers to evaluate their understanding of the material learned. We believe that our book can be useful both for the students of physics, especially for those who want to extend their knowledge in the direction of physical chemistry, and for the students of chemistry who can treat it as a part of the physical chemistry course. Also students of some engineering departments or biology may use it.

Although the subject of our book is phenomenological thermodynamics, in a few places we refer to statistical mechanics. To the readers who wish to learn more about this important branch of science we recommend the classical books [8] and [14]. From among other books on thermodynamics, physical chemistry and chemistry used by us, we recommend references [9], [15], [13], [1] and [11], and for advanced readers also [5], [10] and [12]. References [16] and [2] can serve as an introduction to the field of phase transitions and critical phenomena.

Warsaw, Poland

Robert Hołyst
Andrzej Poniewierski

Contents

Part I Foundations of Thermodynamics

1 Historical Introduction 3

2 Basic Concepts and Definitions 7
2.1 Concept of Thermodynamic Equilibrium 7
2.1.1 System and Surroundings 9
2.1.2 State Parameters and State Functions 10
2.1.3 Thermodynamic Processes 11
2.1.4 Calculation of Physical Quantities in Quasi-static Processes 14
2.2 Extensive Parameters of State 17
2.2.1 Volume 17
2.2.2 Amount of Substance 18
2.2.3 Internal Energy 19
2.3 Intensive Parameters of State 20
2.3.1 Pressure 21
2.3.2 Temperature 23
2.3.3 Chemical Potential 27
2.4 Equations of State 29
2.4.1 Ideal Gas 29
2.4.2 Van der Waals Gas 31
2.4.3 Photon Gas 31
2.4.4 Equations of State in Terms of Intensive Parameters 32
2.5 Exercises 33

3 Internal Energy, Work and Heat 39
3.1 First Law of Thermodynamics 39
3.2 Isochoric Process 41
3.2.1 Heat Capacity at Constant Volume 42
3.3 Isobaric Process 43
3.3.1 Heat Capacity at Constant Pressure 43
3.4 Adiabatic Process 44

3.4.1 Reversible Adiabatic Process 44
3.4.2 Irreversible Adiabatic Process at Constant Pressure 45
3.5 Isothermal Process 47
3.5.1 Reversible Isothermal Process 47
3.5.2 Irreversible Isothermal Process at Constant Pressure 47
3.6 Evaporation of Liquids 48
3.7 Chemical Reaction 49
3.8 Exercises 51

4 Entropy and Irreversibility of Thermodynamic Processes 57
4.1 Second Law of Thermodynamics 57
4.1.1 Entropy Maximum Principle for Isolated Systems 60
4.2 Conditions of Thermodynamic Equilibrium 61
4.2.1 Thermal Equilibrium 61
4.2.2 Mechanical Equilibrium 65
4.2.3 Equilibrium with Respect to the Matter Flow 65
4.3 Entropy as a Function of State Parameters 67
4.3.1 Fundamental Relation of Thermodynamics 67
4.3.2 Euler Relation 68
4.3.3 Entropy of the Ideal Gas 70
4.3.4 Relation Between Entropy and Heat Capacity 71
4.4 Changes in Entropy in Reversible Processes 72
4.4.1 Isothermal Process 72
4.4.2 Isochoric and Isobaric Processes 73
4.4.3 Evaporation of Liquids 74
4.5 Heat Devices 75
4.5.1 Heat Engine and the Carnot Cycle 75
4.5.2 Efficiency of the Carnot Cycle and Thermodynamic Temperature 77
4.5.3 Refrigerator and the Heat Pump 79
4.5.4 Other Thermodynamic Cycles 81
4.6 Changes in Entropy in Irreversible Processes 82
4.6.1 Irreversible Flow of Heat 82
4.6.2 Free Gas Expansion 84
4.6.3 Irreversible Chemical Reaction 85
4.7 Third Law of Thermodynamics 85
4.8 Exercises 86

5 Thermodynamic Potentials 89
5.1 Legendre Transformation of the Internal Energy and Entropy . . . 89
5.1.1 Definition of the Legendre Transformation 90
5.1.2 Helmholtz Free Energy 92
5.1.3 Enthalpy 93
5.1.4 Gibbs Free Energy 93
5.1.5 Grand Thermodynamic Potential 95
5.1.6 Massieu Functions 96

5.2 Natural Variables 96
5.2.1 Equivalent Representations of the Fundamental Relation . . 97
5.2.2 Thermodynamic Potentials for the Ideal Gas 98
5.3 Free-Energy Minimum Principle 100
5.3.1 Systems at Constant Temperature 100
5.3.2 Systems at Constant Temperature and Pressure 102
5.4 Examples of Application of Thermodynamic Potentials 105
5.4.1 Rules of Calculation of Some Partial Derivatives 105
5.4.2 Maxwell Relations 106
5.4.3 Second Partial Derivatives of the Internal Energy and Thermodynamic Potentials 107
5.4.4 Reversible Adiabatic Process 111
5.4.5 Free Gas Expansion 112
5.4.6 Joule–Thomson Process 113
5.5 Intrinsic Stability of a System 115
5.6 Exercises 117

Part II Phase Transitions

6 Phase Transitions in Pure Substances 123
6.1 Concept of Phase 123
6.2 Classification of Phase Transitions 124
6.2.1 First-Order Phase Transitions 124
6.2.2 Continuous Phase Transitions 125
6.2.3 Ehrenfest Classification 127
6.3 Conditions of Phase Coexistence 127
6.3.1 Two-Phase Coexistence 127
6.3.2 Three-Phase Coexistence 131
6.4 Phase Diagrams 132
6.4.1 Phase Diagram of a Typical Substance 132
6.4.2 Phase Diagram of Water 134
6.4.3 Phase Diagram of ^{4}He 135
6.5 Two-Phase Coexistence Lines 138
6.5.1 Clapeyron Equation 138
6.5.2 Solid–Liquid Coexistence 139
6.5.3 Liquid–Gas Coexistence 140
6.5.4 Solid–Gas Coexistence 140
6.6 Liquid–Vapour Two-Phase Region 140
6.7 Van der Waals Equation of State 143
6.7.1 Maxwell Construction 145
6.7.2 Principle of Corresponding States 146
6.8 Exercises 148

7 Mixtures 151
7.1 Basic Concepts and Relations 151
7.1.1 Definitions 151

7.1.2 Internal Energy . . . 153
7.1.3 Thermodynamic Potentials . . . 154
7.2 Intrinsic Stability of a Mixture . . . 156
7.3 Partial Molar Quantities and Functions of Mixing . . . 160
7.3.1 Partial Molar Quantities . . . 160
7.3.2 Relations Between Partial Molar Quantities . . . 162
7.3.3 Functions of Mixing . . . 163
7.4 Mixture of Ideal Gases . . . 164
7.4.1 Dalton's Law . . . 164
7.4.2 Chemical Potential of a Component . . . 165
7.4.3 Functions of Mixing for Ideal Gases . . . 166
7.5 Ideal Mixture . . . 167
7.6 Real Mixtures . . . 168
7.6.1 Fugacity . . . 168
7.6.2 Activity . . . 170
7.6.3 Dilute Solutions . . . 172
7.6.4 Excess Functions . . . 175
7.7 Phase Rule . . . 176
7.8 Exercises . . . 177

8 Phase Equilibrium in Ideal Mixtures . . . 181
8.1 Liquid–Gas Equilibrium . . . 181
8.1.1 Raoult's Law . . . 181
8.1.2 Liquid–Vapour Phase Diagram at Constant Temperature . . . 183
8.1.3 Lever Rule . . . 185
8.1.4 Liquid–Vapour Phase Diagram at Constant Pressure . . . 185
8.1.5 Boiling Point of a Solution . . . 186
8.1.6 Solubility of Gases in Liquids. Henry's Law . . . 189
8.1.7 Ostwald Absorption Coefficient . . . 190
8.2 Liquid–Solid Equilibrium . . . 191
8.2.1 Freezing Point of a Solution . . . 191
8.2.2 Solubility of Solids in Liquids . . . 193
8.2.3 Simple Eutectic . . . 194
8.3 Osmotic Equilibrium . . . 196
8.4 Colligative Properties . . . 197
8.4.1 Vapour Pressure Depression . . . 197
8.4.2 Boiling Point Elevation . . . 197
8.4.3 Freezing Point Depression . . . 197
8.4.4 Osmotic Pressure . . . 198
8.5 Exercises . . . 198

9 Phase Equilibrium in Real Mixtures . . . 201
9.1 Liquid–Vapour Equilibrium . . . 201
9.1.1 Deviations From Raoult's Law . . . 201
9.1.2 Simple Solutions . . . 202
9.1.3 Zeotropic and Azeotropic Mixtures . . . 204

9.1.4 Zeotropic Mixtures 206
9.1.5 Azeotropic Mixtures 207
9.1.6 Derivation of Equations for the Bubble Point and Dew Point Isotherms and Isobars 208
9.2 Liquid Solutions with Miscibility Gap 210
9.2.1 Miscibility Curve and Critical Temperatures 210
9.2.2 Miscibility Gap in Simple Solutions 213
9.3 Liquid–Vapour Equilibrium in Presence of Miscibility Gap 216
9.4 Liquid–Solid Equilibrium and Solid Solutions 218
9.5 Exercises 220

Part III Chemical Thermodynamics

10 Systems with Chemical Reactions 225
10.1 Condition of Chemical Equilibrium 225
10.1.1 Enthalpy of Reaction 227
10.2 Effect of External Perturbation on Chemical Equilibrium 228
10.2.1 Effect of Temperature 228
10.2.2 Effect of Pressure 229
10.2.3 Le Chatelier–Braun Principle 230
10.3 Law of Mass Action for Ideal Gases 231
10.3.1 Effect of Temperature on the Equilibrium Constant 233
10.3.2 Effect of Pressure on the Equilibrium Constant 233
10.4 Thermochemistry 234
10.4.1 Hess' Law 234
10.4.2 Standard Enthalpy of Formation 235
10.4.3 Kirchhoff Equation 237
10.5 Phase Rule for Chemical Systems 238
10.6 Exercises 240

11 Electrochemical Systems 245
11.1 Electrolyte Solutions 245
11.1.1 Dissociation 245
11.1.2 Chemical Potential of the Electrolyte 246
11.1.3 Debye–Hückel Limiting Law 248
11.2 Aqueous Solutions of Acids and Bases 250
11.2.1 Brønsted–Lowry Theory of Acids and Bases 250
11.2.2 pH of a Solution 251
11.2.3 Dissociation Constant 252
11.3 Electrochemical Cells 253
11.3.1 Daniell Cell 253
11.3.2 Galvanic and Electrolytic Cells 255
11.4 Reversible Cell 256
11.4.1 Work of Chemical Reaction 256
11.4.2 Nernst Equation 257

11.4.3 Half-Cell Potential 259
11.4.4 Standard Hydrogen Electrode 260
11.5 Exercises 261

Solutions 265
Exercises of Chapter 2 265
Exercises of Chapter 3 280
Exercises of Chapter 4 289
Exercises of Chapter 5 297
Exercises of Chapter 6 303
Exercises of Chapter 7 310
Exercises of Chapter 8 315
Exercises of Chapter 9 320
Exercises of Chapter 10 323
Exercises of Chapter 11 331

References 337

Index 339

Part I
Foundations of Thermodynamics

Chapter 1
Historical Introduction

Thermodynamics used to be, above all, a science about heat. Therefore, qualitative and quantitative studies of phenomena related to emanation or absorption of heat determined the historical development of this branch of science. Seemingly no relation exists between thermodynamics and chemistry, geology, mathematics, biology or material sciences. However, if no such relation existed, nowadays thermodynamics would probably be a forgotten branch of science. In the second half of the 19th century, the relation between the second law of thermodynamics and spontaneous chemical reactions was discovered, which led, for instance, to the efficient synthesis of ammonia from nitrogen and hydrogen on an industrial scale. The synthesis of ammonia on a mass scale provided a basis to production of artificial fertilizers, whose application allowed to feed additional two billion people. Nowadays thermodynamics goes far beyond the scope associated originally with its name. It has become a practical science concerned with the states in which matter consisting of a very large number of atoms or molecules can exist, which of these states are preferred in given conditions and how they can be reached.

In everyday life, we use terms such as: *warm, cold, hot*, which are based on our senses. The evidence of how illusive these feelings can be is a sensation of the "heat" experienced both by someone who gets burnt by a hot pot at a temperature of 50 °C and by someone who at a temperature of −50 °C touches a piece of metal with a naked hand. This example is to realize only that various terms used rather freely in colloquial language have precise meaning in thermodynamical terminology.

Nowadays we know that many phenomena in the domain of thermodynamics can be relatively easily explained if they are looked at from a microscopic point of view, which was not known to the creators of thermodynamics in the 18th and 19th century, however. The atomic theory of matter developed only at the beginning of the 20th century due to the theory of Brownian motion elaborated by Smoluchowski and Einstein, Perrin's experiment on sedimentation of colloids and Rutherford's experiment on scattering of the α particles on a thin golden foil.

It was already known in ancient times that the volume of air increases with temperature, which was scrupulously used in temples to open doors after the ignition of the holly fire. Nevertheless, only in 1592 was probably the first thermometer

R. Hołyst, A. Poniewierski, *Thermodynamics for Chemists, Physicists and Engineers*,
DOI 10.1007/978-94-007-2999-5_1, © Springer Science+Business Media Dordrecht 2012

constructed by Galileo, who used the phenomenon of thermal expansion of air. His thermometer had an arbitrary scale and measured a joint effect of temperature and pressure. Then it was difficult to separate these two quantities since the barometer, which is used to measure pressure, was constructed for the first time in 1643 by Torricelli, who was a student and follower of Galileo. In the first thermometers, which were made about the same time as the Galileo thermometer, at first water and then alcohol were used as a working substance placed in a thin capillary. Initially the capillary was open, however, it was soon observed that due to evaporation the amount of the working substance decreased. In order to eliminate this effect, closed capillaries were started to be used. At the beginning of the 18th century, Daniel Gabriel Fahrenheit from Gdansk, known as the father of thermometry, invented the mercury thermometer. He introduced a new scale of temperature, which is still in use in some countries. Another temperature scale was introduced by Celsius, a Swedish physicists and astronomer. Both scales are linear and the temperature of 100 degrees on the Celsius scale corresponds to 180 degrees on the Fahrenheit scale. The temperature of $0\ ^\circ\mathrm{F} \approx -18\ ^\circ\mathrm{C}$ was introduced in an arbitrary way. It was the lowest temperature achieved by Fahrenheit in his laboratory and it corresponds to the freezing point of an aqueous solution of sal-ammoniac (NH_4Cl). The problem of the establishment of a reliable temperature scale was partly solved only when it was observed that melting of ice and boiling of water occur always at the same temperatures at atmospheric pressure. These two points were accepted as a basis of the temperature scale in 1694. On the Celsius scale, the boiling point of water corresponds to the temperature of 100 °C, and the melting point of ice corresponds to 0 °C.

One of the first problems encountered by the creators of thermodynamics was to establish a relation between heat and temperature, which involved application of two devices: the thermometer and calorimeter. The question was whether temperature and heat are the same physical quantity. In the middle of the 18th century, Joseph Black introduced the concept of heat capacity as a proportionality coefficient between the amount of heat and temperature. This seemingly simple idea was similar to that used by Newton in the case of force and acceleration. According to the Black formula, a reading of the thermometer has a meaning of heat per unit mass. Black discovered also the latent heat associated with boiling and freezing.

The emergence of a modern approach to chemistry, based on the invariability of elements and conservation of mass and heat (caloric)[1] in chemical reactions, resulted in a new conceptual framework to be used in many chemical problems. These laws were set forth by Lavoisier in 1789, a few years before his decapitation by the French revolutionists. Nowadays we know that no separate law of heat conservation exists but a more general law of energy conservation, which at that time was only at the stage of formation. Since in chemical reactions occurring in liquids or solids a change in the volume of a system is small, it was not accidental that the theory of caloric conservation worked quite well in those cases. However, the theory

[1]The concept of *caloric*, i.e., a weightless, invisible fluid that flows from hotter bodies to colder bodies, was introduced by Lavoisier.

failed completely in confrontation with experiments in which a great amount of hit was given off due to large friction. For instance, such a phenomenon was observed by count Rumford during the boring of gun-barrels. His discovery that mechanical work can be converted into heat, and is actually an inexhaustible source of heat, was not given a sympathetic reception at that time, because no new conservation law was proposed instead of the caloric conservation.

James Watt—the inventor of the steam engine and a student of Joseph Black—must have realized the possibility of conversion of work into heat and vice versa. However, it was Sadi Carnot—a French engineer—who described this conversion in terms of the energy conservation principle, in the case of an ideal heat engine. Moreover, Carnot showed that the engine efficiency, i.e., the ratio of the work done to the heat supplied, is a universal quantity. It depends only on the ratio of the temperature of the heat reservoir and the temperature of the radiator, but it is independent of the working substance used in the engine. Carnot used the concepts of caloric and heat in his work. Later, caloric originated in the concept of entropy, whereas the analysis of heat led to the energy conservation principle in the form of the first law of thermodynamics. Both concepts were developed independently by Clausius and Kelvin, and the mathematical formalism of differential forms was applied to the Carnot work by Clapeyron.

The empirical definition of temperature given by Celsius and Fahrenheit was not suitable for theoretical consideration of Carnot. In 1849, after Carnot's death (1832), William Thomson—the later Lord Kelvin—noticed that the observation made by Carnot eliminated the dependence of the temperature scale on properties of the substance used in a thermometer or heat engine. Following Carnot, he noticed that the efficiency of a heat engine does not depend on the working agent. This fact indicated the existence of an absolute temperature scale, named later the Kelvin scale, which is independent of the properties of any substance. Due to his discovery thermodynamics went beyond the stage of description of empirical facts and entered upon the path of search for fundamental laws of nature.

For a long time during the development of thermodynamics, work and heat were considered separately from each other. Therefore, it is not surprising that the calorie is still in use as a unit of heat, apart from the joule—a unit of work and energy. The first experiments which allowed to convert the unit of heat into the unit of work were carried out by Mayer and Joule in the middle of the 19th century. Thus, the route to the formulation of the energy conservation principle, which is the essence of the first law of thermodynamics, was relatively long. It originated from the conservation of mechanical energy, i.e., the sum of the kinetic energy and potential energy, which was formulated by Leibnitz at the end of the 17th century. The crowning achievement was the work published in 1847 by Helmholtz, who extended the range of the energy conservation principle to all branches of physics. Due to his publication, Helmholtz could leave the Prussian army, where he served as a doctor, to work out the student grant received from the Prussian government. The army commanding staff decided that such an outstanding man would render a service to the country better if he worked at the university instead of army. In 1905, on the basis of the energy conservation principle, Albert Einstein proposed mass–energy equivalence and in 1932, Enrico Fermi postulated existence of the neutrino.

The second law of thermodynamics, as the absolute temperature scale, came into existence in the middle of the 19th century. Thorough analysis of the Carnot engine led to a new quantity in place of caloric,[2] which was given the name *entropy*. This term, introduced by Clausius, comes from the Greek words `en-`, *in*, and `trope`, *transformation*. Entropy, as energy, is a function of the thermodynamic state. The third law of thermodynamics, related to the unattainability of the absolute zero temperature, was formulated by Nernst in 1906. Finally, in 1909, Carathéodory formulated the last thermodynamic law, which was named the zeroth law. It was the time when the great German mathematician David Hilbert tried to put all branches of physics in axiomatic form. The zeroth law introduces an equivalence relation between two systems that are in thermal equilibrium with each other, and defines the empirical temperature as the equivalence class of that equivalence relation. At the end of the 19th century, statistical physics was initiated mainly due to the works of Boltzmann, Maxwell and Gibbs. Statistical physics and thermodynamics are complementary branches of science. Thermodynamics introduces various concepts, e.g. entropy, which are used in macroscopic description of systems consisting of a great number of entities (atoms, molecules, etc.), whereas statistical physics provides microscopic interpretation of these concepts.

We have not mentioned here many other scientists who contributed to the development of thermodynamics, such as Boyle, Mariotte or Guy-Lusac. Among other things, their investigations of dilute gases contributed to the discovery of the absolute temperature scale.

At the beginning of the 20th century, thermodynamics was already a mature science, which was additionally strengthened by the formation of statistical physics. Despite this fact some circulating, simplified and often incorrect interpretations of the laws of thermodynamics or its concepts still linger on, for instance, the interpretation of entropy as a measure of disorder or the law of increasing entropy of the Universe and its final thermal death. According to statistical physics, entropy is a measure of all possible ways to arrange the constituents of a given system without changing its macroscopic parameters. This definition of entropy does not have much in common with the order of the constituents of the system, understood colloquially as orderliness nice for the eye. Concerning the second example, thermodynamics does not apply to systems in which gravitation dominates, therefore, extension of its laws to the whole Universe is not well-founded.

The readers who wish to study the formal structure of thermodynamics in more detail are referred to the book by H.B. Callen [3]. We also recommend the book by P.W. Atkins [1], which contains many examples of application of thermodynamics to practical problems in chemistry and physics.

[2]Carnot distinguished heat from caloric in his work.

Chapter 2
Basic Concepts and Definitions

Thermodynamics originated from observation of phenomena which occur on the earth in macroscopic systems consisting of a great number of atoms or molecules. Many concepts used in thermodynamics, such as pressure and temperature, are familiar to us from everyday experience. For instance, everybody knows that when two bodies are brought into contact with each other and then isolated from the surroundings, their temperature settles down after some time, which can be easily verified by putting something warm into the fridge. In this chapter, we discuss the concepts of temperature, pressure and thermodynamic equilibrium in a more formal way, and also give definitions of other basic concepts and quantities used in thermodynamics.

2.1 Concept of Thermodynamic Equilibrium

The concept of thermodynamic equilibrium is fundamental to thermodynamics.

Definition 2.1 *Thermodynamic equilibrium* refers to particular states of a macroscopic system, called equilibrium states, which are independent of time (stationary states) and in which no macroscopic flow of any physical quantity exists.

Note that a stationary state does not have to be an equilibrium state because a steady macroscopic flow of heat or matter, or another physical quantity, may exist in the system. For instance, if we keep the two ends of a wire at different and constant temperatures then heat flows through the wire but the temperature gradient along it does not depend on time. Thus, it is a stationary state but not an equilibrium state because of the heat flow.

R. Hołyst, A. Poniewierski, *Thermodynamics for Chemists, Physicists and Engineers*,
DOI 10.1007/978-94-007-2999-5_2,

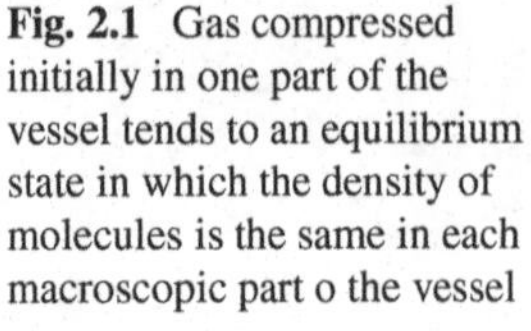

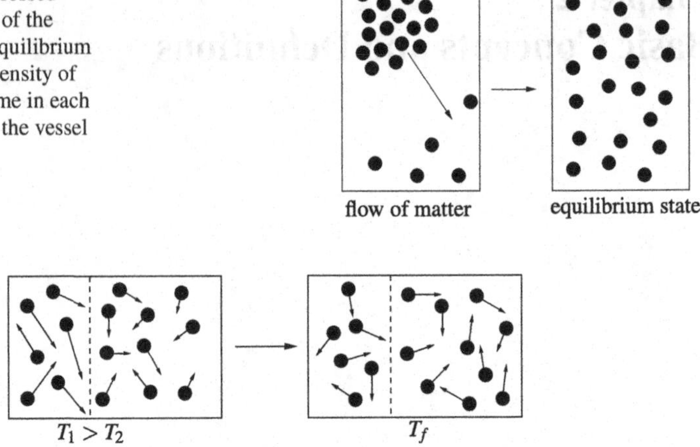

Fig. 2.1 Gas compressed initially in one part of the vessel tends to an equilibrium state in which the density of molecules is the same in each macroscopic part o the vessel

Fig. 2.2 Vessel with a gas is divided with a stiff wall permeable to heat (*broken line*). The *arrows* represent the velocities of molecules. The system tends to the equilibrium state due to the flow of heat from the part of higher temperature T_1 to the part of lower temperature T_2. In equilibrium, the temperature in both parts has the same value T_f

The fundamental postulate of thermodynamics
Every isolated macroscopic system reaches eventually thermodynamic equilibrium independently of the initial state of the system.

This postulate applies to systems observed in the scale of Earth and nothing indicates that it can also be applied to systems in the cosmic scale and to the whole Universe, in particular.

Two examples how systems reach their equilibrium states are shown in Figs. 2.1 and 2.2. In Fig. 2.1 (on the left), the gas has been compressed in one part of the vessel, so its density in that part is larger than in the rest of the vessel. When the dividing wall is removed the gas is not in an equilibrium state because of a macroscopic flow of matter from the region of higher density to the region of lower density. After some time the densities equalize and the system reaches the equilibrium state (on the right).

Figure 2.2 shows a gas in a vessel divided into two parts with a stiff wall permeable to heat. Initially the temperature in both parts is the same. Then the gas in the left part is heated quickly, so its temperature T_1 is higher than the temperature T_2 in the right part. When the heating stops the system tends to the equilibrium state. The heat flows from the high temperature region to the low temperature region. Since an increase in the temperature means an increase in the kinetic energy of molecules, the flow of heat is associated with a transfer of the kinetic energy from molecules in the left part to molecules in the right part. It occurs through collisions of molecules

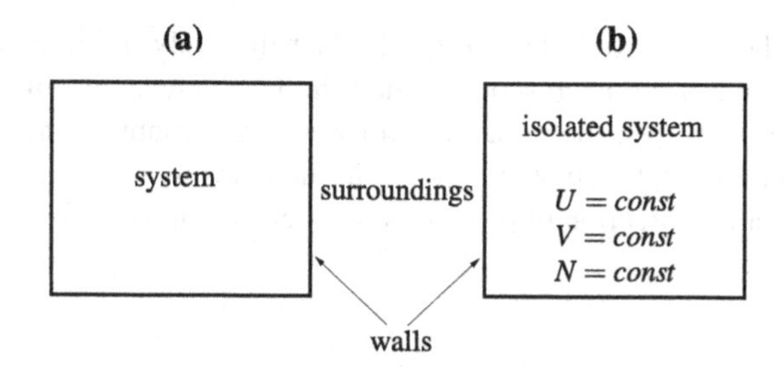

Fig. 2.3 The system is separated by walls from the surroundings. The walls can be movable, can conduct heat or can be permeable to matter. The system interacts with the surroundings through the walls. If the walls are stiff and do not allow neither heat nor matter to pass through them then the system is said to be isolated

with the dividing wall. When the equilibrium state is reached the temperature in both parts is the same.

2.1.1 System and Surroundings

The *system* is a separate part of the world (Fig. 2.3a). Everything that does not belong to the system is called the *surroundings*. The system is separated from the surroundings by *walls*. These concepts are present in discussions of all phenomena described by thermodynamics. The wall is called *adiabatic* if it is impermeable to heat. The wall is called *diathermal* if it permits a heat flow. Finally, the wall can be permeable to matter or not. The system with all walls impermeable to matter is a *closed system*. The system is said to interact with the surroundings if any change in the surroundings causes a measurable effect in the system. A system which does not interact with the surroundings in any way is called an *isolated system*. Thus, the energy, volume and amount of matter do not change in an isolated system.

We consider an isolated macroscopic system, which contains only a pure substance, without any internal walls limiting the motion of molecules, for instance, a pure gas in a vessel. Only a few parameters are needed to completely characterize an equilibrium state of such a simple system (Fig. 2.3b). They are: the *internal energy* U, which is a sum of the kinetic energy and the potential energy of intermolecular interactions of all molecules in the system, the *volume* V and the *number of molecules* N. We accept this empirical fact, confirmed by numerous experiments, as a postulate. Other physical quantities, e.g., pressure, temperature or *entropy* (which will be introduced later on), can be treated as functions of these three parameters. Although energy, volume and number of molecules are also present in classical mechanics, entropy does not have a mechanical counterpart, which means that thermodynamics cannot be reduced to classical mechanics. This is because the number of microscopic variables needed to completely describe the motion of all molecules, i.e., the number of all positions and momenta, is enormous in a macroscopic system. For instance, for 1 mole of a pure substance (18 g in the case of water) it is of the order 10^{24}. Therefore, the fact that only a few parameters can completely characterize a macroscopic system in thermodynamic equilibrium may seem surprising. However, the remaining microscopic variables play a role in the flow of heat (see Fig. 2.2). They are hidden in the entropy, which means that they have influence on the direction of processes in macroscopic systems.

2.1.2 State Parameters and State Functions

Definition 2.2 *State parameters* are physical quantities that characterize a system in thermodynamic equilibrium.

For instance, the volume, number of molecules, pressure, temperature and internal energy are state parameters. Not all state parameters are independent. As we have already mentioned, to completely characterize an isolated system in thermodynamic equilibrium only three parameters are needed: U, V and N. Other quantities, e.g., the pressure p and temperature T, are functions of these three parameters. However, it depends on the actual physical situation which parameters are treated as independent variables. For instance, if energy can be transferred between the system and surroundings in the form of heat it is easier to control the temperature of the system than its internal energy. Then T, V and N are treated as independent variables. In the case of a gas closed in a cylinder with a movable piston, which remains in thermal contact with the surroundings, the pressure and temperature of the system can be easily controlled.

Definition 2.3 *State function* is a physical quantity which assumes a definite value for each equilibrium state of a system, independent of the way that state is reached.

It follows from the definition of state parameters that they are also state functions because their values depend only on the state of the system. In thermodynamics, quantities which are not state functions are also considered. They depend on the way a given equilibrium state is reached. The best known examples of such quantities are work and heat.

Intensive and Extensive State Parameters We consider a system in thermodynamic equilibrium. Each macroscopic part of the system is called a *subsystem*. If the system is in equilibrium the same applies to its all subsystems. In Fig. 2.4a, the system is divided into several subsystems but the boundaries between the subsystems are only imaginary. An *intensive parameter* is a physical quantity which has the same value in each subsystem, therefore, it cannot depend on the mass of a given subsystem. In consequence, intensive parameters do not depend on the mass of the whole system. For instance, temperature, pressure and *chemical potential* (discussed later on) are intensive parameters.

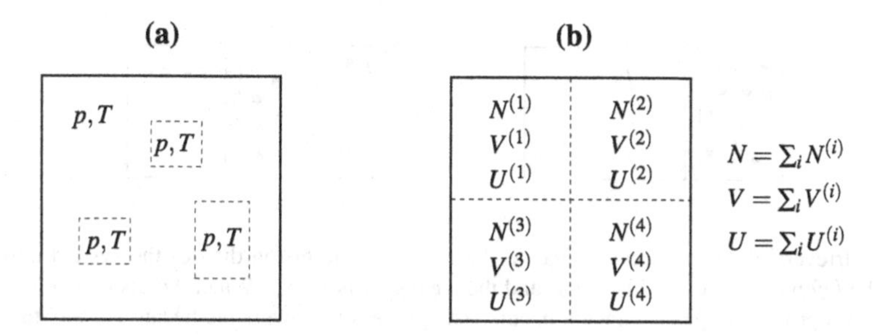

Fig. 2.4 (**a**) System in thermodynamic equilibrium is divided into several subsystems. An intensive parameter, e.g., the pressure p or temperature T, has the same value in all subsystems. (**b**) In a composite system, the value of an extensive parameter, e.g., the number of molecules N, volume V and internal energy U, is the sum of contributions from all subsystems

Imagine now that we form a composite system made up of several identical systems, which become subsystems of the composite system (Fig. 2.4b). Then the volume of the composite system is the sum of the volumes of its subsystems and the same concerns the number of molecules. Also the internal energy and entropy have this property. In general, a physical quantity whose value for the composite system is the sum of its values in individual subsystems is called an *extensive parameter*. This means that for an extensive parameter X we have

$$X = \sum_i X^{(i)}, \tag{2.1}$$

where the index i numbers the subsystems and $X^{(i)}$ is the value of X in the ith subsystem. It follows from the above definition that extensive parameters are proportional to the mass of the system.

2.1.3 Thermodynamic Processes

Definition 2.4 *Thermodynamic process* is a change occurring in a system between the initial and final equilibrium states.

The process is called *adiabatic* if there is no flow of heat between the system and surroundings. The process is said to be: *isochoric*, if it occurs at constant volume, *isothermal*, if it occurs at constant temperature, and *isobaric*, if it occurs at constant pressure.

According to Definition 2.4, the initial and final states of the system are equilibrium states. However, the intermediate states are not equilibrium states, in general. Real processes proceed at a finite rate and the system cannot reach thermodynamic equilibrium during the process. However, we feel intuitively that the slower the process is, the more time the system has to approach thermodynamic equilibrium at each successive stage of the process. In the framework of thermodynamics, only

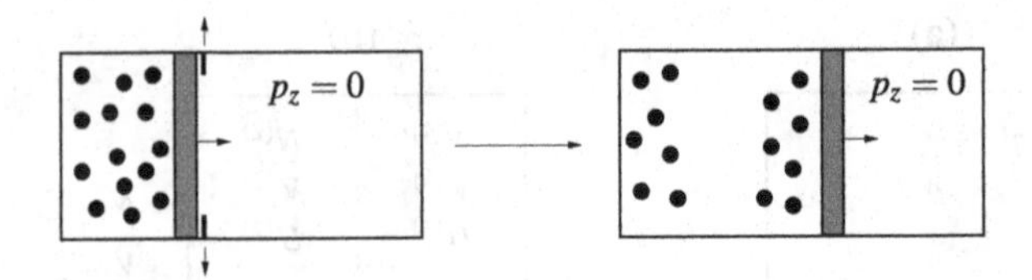

Fig. 2.5 Irreversible expansion of a gas to the vacuum. The piston divides the system into two parts. The *left part* is occupied by a gas and the *right part* is empty. When the piston is released a spontaneous process of gas expansion occurs. The process is irreversible because at any moment the system is not in thermodynamic equilibrium and the gas pressure is not well defined during the process. Inhomogeneities in the molecular density may form during the process

equilibrium states are considered. Any description of non-equilibrium states would involve many more parameters than in the case of equilibrium states. Therefore, it is convenient to consider an idealized process which proceeds infinitely slowly and can be treated as an approximation of real slow processes. Such an infinitely slow process is called a *quasi-static* process.

Definition 2.5 *Quasi-static* process is a sequence of successive equilibrium states of a system.

Differently from real processes, time does not appear in a quasi-static process. The latter proceeds simply from one equilibrium state of the system to another equilibrium state in a continuous manner. A real process can be approximated by a quasi-static process if it is sufficiently slow. In practice, the process is said to be slow if it lasts much longer than the longest characteristic time of a given system. Imagine, for instance, that we move the piston to compress a gas in a vessel. Then we should compare the time of compression with the time the sound needs to reach the wall opposite to the piston. For a system of a length of 3 m and for the speed of sound of 332 $\mathrm{m\,s^{-1}}$, we find that the longest characteristic time of the system is of the order 10^{-2} s. If we moved the piston with the speed of sound then regions of a higher and lower density would form in the system, and the energy of the sound wave would be dissipated. At any moment of the process, the system would not be in thermodynamic equilibrium (cf. Fig. 2.5).

Mathematically a quasi-static process is represented by a curve in the space of equilibrium states. The dimension of this space is equal to the number of independent state parameters needed to define an equilibrium state of a given system. If X denotes a state parameter then a change in X in a thermodynamic process is denoted by ΔX, where

$$\Delta X = X_f - X_i, \tag{2.2}$$

and X_i and X_f are the values of X in the initial and final state, respectively. In calculations, instead of finite changes of parameters, we often consider *infinitesimal*, i.e., infinitely small changes. For the parameter X, an infinitesimal change of the parameter is denoted by the symbol $\mathrm{d}X$. Such a procedure is justified, since according to

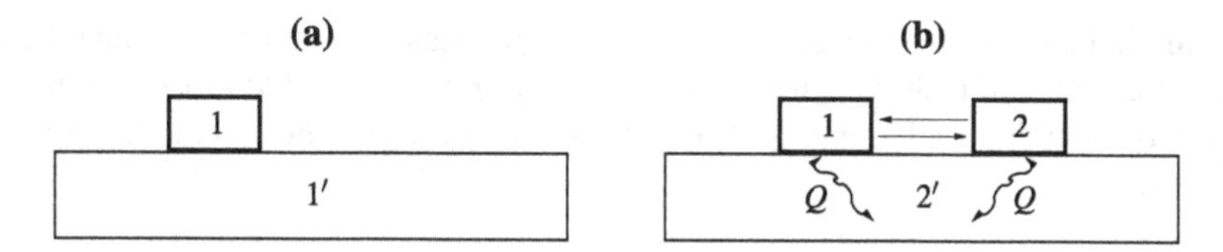

Fig. 2.6 Example of an irreversible process. **(a)** Initially the block is in the state 1 and the surroundings, i.e., the substrate is in the state 1′. **(b)** The block moves on a rough surface towards the state 2 and then back from 2 to 1. In both cases, heat is produced because of the friction. The heat is transferred to the surroundings, which causes that the surroundings do not return to the original state even though the block does

Definition 2.5 we can consider processes between two arbitrarily close equilibrium states.

Definition 2.6 *Reversible process* is a process to which a reverse process exists that restores the original states of both the system and surroundings.

In other words, if in a given process the systems goes from the state 1 to the state 2 and the surroundings go from the state 1′ to the state 2′ then the process is reversible, provided that there exists a process that simultaneously brings the system from 2 to 1 and the surroundings from 2′ to 1′. A process that is not reversible is called an *irreversible* process. Irreversibility is often associated with dissipation of energy in the form of heat. All real processes are irreversible. Two examples are shown in Figs. 2.5 and 2.6. In Fig. 2.6 a block moves on a rough surface. Because of friction heat is produced during this process. The block can be moved back to its original position but due to the friction some amount of heat is transferred to the surroundings independently of the direction of the motion, which means that the surroundings do not return to their initial state.

In order a given process could be considered reversible, it must proceed without friction (dissipation of energy) and at a vanishing rate. Thus, any reversible process is also a quasi-static process, which means that during the process the system passes over successive equilibrium states. However, it does not follow from Definition 2.5 that the reverse statement is also true. Only the second law of thermodynamics, which postulates the existence of entropy as an additional function of state (see Chap. 4), provides a deeper understanding of reversible and irreversible processes. In particular, in Sect. 4.2.1, we derive an expression for an infinitesimal flow of heat in reversible processes (see (4.24)). Since a quasi-static process is an idealization of a real process, and not only a mathematical concept, it must be in agreement with the formula mentioned above. In Chap. 4, we show that quasi-static processes are also reversible processes.

Even though reversible processes are only idealizations of real processes, they actually enable us to calculate changes in state functions in real processes. If a given equilibrium state is reached as a result of an irreversible process then, in general, we cannot calculate directly how state functions have changed in the process. However, according to Definition 2.3, state functions do not depend on the process. Thus, if we can find a reversible process from a given initial state to a given final state then we can calculate the change in any state function in the reversible process instead of the irreversible one. Examples of such calculations are presented in the following chapters.

2.1.4 Calculation of Physical Quantities in Quasi-static Processes

A quasi-static process is defined by a curve in the space of independent state parameters of the system. The curve starts from the initial state i and ends in the final state f. During the process the system is all the time in thermodynamic equilibrium, and only its state parameters change. Together with changes in the state parameters also the state functions change. To determine how a state function changes during the whole process, we have to know how it changes if two equilibrium states are infinitesimally close to each other.

Let us assume, for instance, that the independent state parameters are: the internal energy U, volume V, and amount of substance n, where the unit of the latter is the *mole*. The definition of the mole is given in Sect. 2.2.2. Here, it is sufficient to say that the number of molecules N is proportional to n, and the proportionality coefficient is a very large number, thus, in practice, we can treat n as a continuous variable. As a function of state we can take the temperature, $T = T(U, V, n)$, for instance. A change in T in any process amounts to

$$\Delta T = T_f - T_i = T(U_f, V_f, n_f) - T(U_i, V_i, n_i). \tag{2.3}$$

If the process is quasi-static we can express ΔT as a sum of successive small steps, which in the limit of infinitesimal steps, $\mathrm{d}T$, reduces to the integral, i.e.,

$$\Delta T = \int_i^f \mathrm{d}T, \tag{2.4}$$

where the integration goes from the initial state to the final state. An infinitesimal change in the temperature is associated with infinitesimal changes in the state parameters, i.e.,

$$\mathrm{d}T = T(U + \mathrm{d}U, V + \mathrm{d}V, n + \mathrm{d}n) - T(U, V, n). \tag{2.5}$$

Expanding the first term on the right-hand side in a Taylor series and leaving only the terms linear in the infinitesimal increments of the state parameters, we get

$$\mathrm{d}T = \left(\frac{\partial T}{\partial U}\right)_{V,n} \mathrm{d}U + \left(\frac{\partial T}{\partial V}\right)_{U,n} \mathrm{d}V + \left(\frac{\partial T}{\partial n}\right)_{U,V} \mathrm{d}n. \tag{2.6}$$

Since the choice of independent variables can be different, e.g., U, p and n, to avoid confusion, the constant parameters at which the differentiation is performed are shown explicitly. For instance,

$$\left(\frac{\partial T}{\partial U}\right)_{V,n} \neq \left(\frac{\partial T}{\partial U}\right)_{p,n}, \tag{2.7}$$

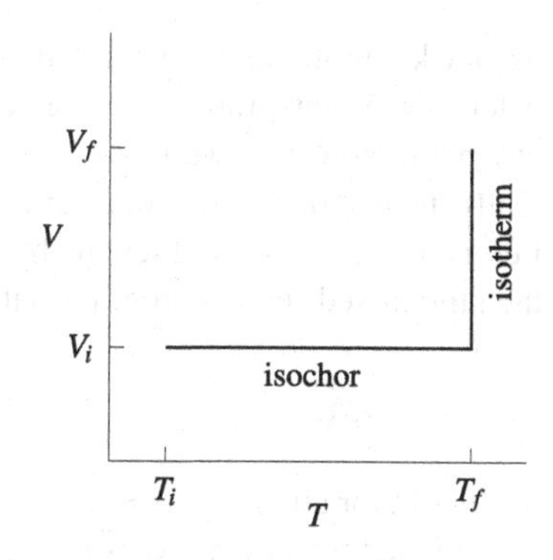

Fig. 2.7 Example of a quasi-static process in the TV plane, which consists of an isochoric and isothermal process. First, the temperature changes from T_i to T_f at constant volume $V = V_i$ (the isochor). Then the volume changes from V_i to V_f at constant temperature $T = T_f$ (the isotherm)

thus, the notation $\partial T/\partial U$ is ambiguous if we do not indicate the other variables.

As we have already mentioned, to define a quasi-static process we have to define a curve (or a path) along which it proceeds in the space of state parameters. For instance, the internal energy U can be treated as a function of T and V if we assume $n = const$. An infinitesimal change in the function $U(T, V)$ amounts to

$$\mathrm{d}U = \left(\frac{\partial U}{\partial T}\right)_V \mathrm{d}T + \left(\frac{\partial U}{\partial V}\right)_T \mathrm{d}V. \tag{2.8}$$

Here the space of independent state parameters is the subset of the plane: $T > 0$, $V > 0$. Often we consider processes in which only one parameter of state varies. In the example considered, it can be an isochoric process, if $V = const$, or an isothermal process, if $T = const$. It can also be a combination of these processes as shown in Fig. 2.7.

Differentials and Differential Forms We consider a state function $Y(X_1, \ldots, X_M)$, where $X_1, \ldots, X_M$ are independent state parameters. An infinitesimal change in Y in a quasi-static process has the following general form:

$$\mathrm{d}Y = \sum_{i=1}^{M} \left(\frac{\partial Y}{\partial X_i}\right)_{X_{j\neq i}} \mathrm{d}X_i, \tag{2.9}$$

where the notation $X_{j\neq i}$ means that the derivative with respect to X_i is calculated at constant X_j for $j \neq i$. In mathematics $\mathrm{d}Y$ is called the *differential* of Y.[1] To calculate a finite change in Y in the quasi-static process, we integrate $\mathrm{d}Y$ along the path that defines the process in the parameter space:

$$\Delta Y = Y_f - Y_i = \int_i^f \mathrm{d}Y. \tag{2.10}$$

An important property of the differential is that the above integral does not depend on the path of integration. Otherwise Y would not be a state function because its value in the final state would depend on the process leading to that state. If we

[1] The term *total differential* is also used.

do not know the explicit form of Y, but we know all its first derivatives, we can determine Y, integrating $\mathrm{d}Y$ from a selected initial state to a given final state along any path. However, we have to be sure that the coefficients at $\mathrm{d}X_1, \ldots, \mathrm{d}X_M$ are really the first derivatives of a certain function Y with respect to the state parameters. Let us denote these coefficients by $Y_1, \ldots, Y_M$. If they are the first derivatives of Y then the mixed second order partial derivatives must be equal, thus, the identities

$$\frac{\partial Y_i}{\partial X_j} = \frac{\partial Y_j}{\partial X_i}, \tag{2.11}$$

must hold for all pairs $i, j = 1, \ldots, M$. It can also be shown that if conditions (2.11) are satisfied then there exists a function Y such that $Y_i = \partial Y / \partial X_i$ for $i = 1, \ldots, M$.

Not all quantities used in thermodynamics are state functions. To explain this, we use a mechanical analogy and compare the work against gravitation with the work against friction forces. In the first case, the work is equal to the change in the potential energy of a body in the gravitational field. As we know it depends only on the space points chosen, which is equivalent to saying that the infinitesimal work performed by the gravitational force is equal to the differential of its potential energy. In the second case, the friction force acts always in the direction opposite to the motion of the body. The infinitesimal work is defined as a scalar product of the friction force and an infinitesimal displacement of the body. However, it cannot be a differential of a certain function because the total work depends on the length of the path along which the body moves.

In thermodynamics, the work W done on the system and the heat Q supplied to the system depend on the process, and not only on the initial and final state. This means that they cannot be expressed as changes in some state functions. However, for a quasi-static process, we calculate W and Q by integrating infinitesimal contributions, in a similar way as in the case of a state function (see (2.10)), i.e.,

$$W = \int_i^f đW, \tag{2.12}$$

$$Q = \int_i^f đQ. \tag{2.13}$$

We use the notation $đW$ and $đQ$ for infinitesimal amounts of work and heat, respectively, to distinguish them from differentials, since the result of integration depends on the path in the space of state parameters. In general, an infinitesimal quantity $đ\omega$ has the following form:

$$đ\omega = \sum_{i=1}^{M} \omega_i \, \mathrm{d}X_i, \tag{2.14}$$

where $\omega_i = \omega_i(X_1, \ldots, X_M)$. If $đ\omega$ is not a differential of a function then the coefficients ω_i do not satisfy conditions (2.11). For instance, if $đ\omega = \omega_1(X_1, X_2)\mathrm{d}X_1$, which means that $\omega_2 = 0$, then $\partial\omega_1/\partial X_2 \neq 0$ and $\partial\omega_2/\partial X_1 = 0$. In mathematical terminology, $đω$ is called a *differential form*.[2] The differential of a function is simply a special case of a differential form.

An example of a differential form is the work performed on the system (a gas, for instance) whose volume changes by $\mathrm{d}V$ at the pressure p. The change in the volume is equal to $\mathrm{d}V = A\,\mathrm{d}x$, where A denotes the area of the piston and $\mathrm{d}x$ is its displacement. The force exerted on the piston by the gas amounts to pA. To compress the gas quasi-statically, the external force must balance exactly the force exerted by the gas, hence, the work performed by the external force amounts to

$$đW = -p\,\mathrm{d}V, \tag{2.15}$$

which means that $đW > 0$ if $\mathrm{d}V < 0$. Since V is not the only state parameter (the remaining parameters are T and n, for instance) conditions (2.11) are not satisfied and $đW$ is not a differential of a function. In Chap. 4, we will see that $đQ = T\mathrm{d}S$, where S stands for the entropy.

2.2 Extensive Parameters of State

It took scientists about 300 years, from the times of Galileo to the times of Helmholtz, to establish the parameters needed to describe thermodynamic equilibrium of a system. Nowadays we know that in the case of an isolated system formed by a pure substance, an equilibrium state is defined by three extensive parameters: the volume, internal energy and amount of substance. All other properties of a system in thermodynamic equilibrium can be expressed in terms of these parameters. Below we present a short discussion of these quantities.

2.2.1 Volume

We denote the volume by V. The SI derived unit of volume is the *cubic meter* (m^3) but the volume of gases and liquids is often expressed in litres (L):

$$1\,\mathrm{L} = 1\,\mathrm{dm}^3 = 10^{-3}\,\mathrm{m}^3 = 1000\,\mathrm{cm}^3.$$

The fact that volume is an extensive parameter follows from its definition. If we form a system made up of subsystems then the volume of the composite system equals the sum of the volumes of all subsystems, provided that they do not overlap. It is a well known geometrical property.

---

[2] To be precise, it is a differential form of rank 1 or 1-form.

2.2.2 *Amount of Substance*

The amount of substance is denoted by n. If there are several chemical components (compounds) in the system we add an index to number them, e.g., $n_1, n_2, \ldots$ or n_A, $n_B, \ldots$. The SI base unit for the amount of substance is the *mole* (mol). The amount of substance is also called the *number of moles*.

Here we consider only a pure chemical substance. In everyday live, to specify the amount of a given substance, we usually specify its mass or rather its weight. However, from the stoichiometric point of view, the mass of a substance is not a convenient measure of its amount. In 1811, Amadeo Avogadro formulated a hypothesis that a gas occupying a given volume at a given temperature and pressure always contains the same amount of substance, independently of its kind. *The same amount* was understood in the sense of stoichiometry of chemical reactions. For instance, one portion of H_2 and a half portion of O_2 give one portion of H_2O in the chemical reaction. In the times of Avogadro, *the portion* was a measure of volume, but it was used to determine the number of moles. At the beginning of the 19th century it was not known that matter consists of atoms or molecules, hence, its amount can be expressed as a number of pieces, e.g., 100 atoms of Ar or 200 molecules of H_2O. Despite this fact, the concept of the mole was introduced, which actually allowed to measure the amount of substance in pieces. The number of elementary entities, i.e., atoms, molecules, ions, etc., contained in 1 mol of the substance is called the *Avogadro constant*, N_A. It amounts to

$$N_A = 6.022\,141\,79\,(30) \times 10^{23}\ \mathrm{mol}^{-1}, \tag{2.16}$$

which is an enormous number. Having determined the Avogadro constant, we can express the number of elementary entities in the relation to N_A.

Definition 2.7 *Amount of substance* is the number of elementary entities divided by the Avogadro constant.

Thus, the amount of substance (number of moles) n is proportional to the number of elementary entities in the substance, and the proportionality constant is N_A^{-1}, hence

$$n = \frac{N}{N_A}. \tag{2.17}$$

Since the above relation is the same for all substances, it is necessary to define the elementary entity, e.g., H_2, NaCl, CO_2, etc.

It is needless to say that the modern knowledge of the structure of matter was not available to the inventors of the mole. Their analysis was based mainly on the stoichiometry of chemical reactions. However, the mole as a unit of the amount of substance implicitly contains information about the molecular structure of matter. It is also evident why the mass is not a convenient parameter to measure the amount of substance in chemical reactions. For example, in the reaction of carbon combustion: $C + O_2 \rightarrow CO_2$, 1 mol of carbon and 1 mol of oxygen give one mol of carbon dioxide. Expressing the same amounts of the substances in the units of mass, we have 12 g C and 32 g O_2 which give 44 g CO_2. No simple proportion can be deduced in this case. However, such a proportion becomes obvious if the amount of substance is expressed in moles.

As in the case of volume, it is quite obvious that the amount of substance is also an extensive parameter. If we form a system composed of two or more subsystems, each of which contains a given number of elementary entities of the same substance, the total number of elementary entities in the system is simply the sum of these numbers. The same concerns the number of moles, which follows from relation (2.17).

2.2.3 Internal Energy

The internal energy is denoted by U. The SI derived unit of energy is the *joule* (J):

$$1\,\mathrm{J} = 1\,\mathrm{N\,m} = 1\,\mathrm{kg\,m^2\,s^{-2}}.$$

Another non-SI unit of energy used in thermodynamics is the calorie (cal). In fact, there are several differently defined calories. One of them is defined as the amount of energy required to warm up one gram of water from 14.5 °C to 15.5 °C at standard atmospheric pressure (101325 Pa), which is

$$1\,\mathrm{cal} = 4.1855\,\mathrm{J}.$$

Another one is the thermochemical calorie: $1\,\mathrm{cal_{th}} = 4.184\,\mathrm{J}$. To express the ionization energy or the energy liberated in nuclear reactions, the electronvolt (eV) is used:

$$1\,\mathrm{eV} \approx 1.602 \times 10^{-19}\,\mathrm{J}.$$

It is the amount of energy gained by a single electron in an electric potential difference of one volt (V). The electronvolt is not an SI unit.

Energy is an original concept and neither physics nor chemistry can give us the answer to the question about its nature. They only show how energy can be measured or calculated. We know the kinetic energy of bodies and their gravitational energy. We can calculate the energy of electric charges in an electric field and the energy of magnets in a magnetic field. A few examples of energy calculations are given below. For instance, the kinetic energy of a body of the mass m, moving with the speed v small compared to the speed of light, is given by:

$$E = \frac{1}{2}mv^2.$$

A change in the gravitational energy of the same body elevated to the height h above sea-level amounts to

$$\Delta E = mgh.$$

The energy of a body at rest is given by the famous Einstein equation:

$$E = mc^2,$$

which expresses mass-energy equivalence, where m is the rest mass of the body and c is the speed of light. It was derived by Einstein in the framework of the special theory of relativity.

In thermodynamics, we also know some explicit formulae to calculate changes in the internal energy, for instance, a change due to heating of a system. In general, the internal energy of a system, U, consists of the kinetic energy of molecules and the potential energy of intermolecular interactions. The latter manifests itself, for instance, in the form of heat released during condensation of a gas. We can also include in U the energy of chemical bonds, which shows up as the heat released or absorbed in chemical reactions, and also the energy of electrons in atoms (ionization energy), etc. The internal energy is a sum of all kinds of energy existing in a given system. However, changes in the internal energy, rather than its absolute value, have physical meaning. In practice, we choose a reference state, to which we assign zero of the internal energy, and determine the energies of other states with respect to the reference state. For instance, if there are no chemical reactions in the system and electron excitations can be neglected then we include in the internal energy only the kinetic energy of atoms or molecules and the potential energy of their mutual interactions and the interactions with external fields (e.g., the gravitational energy).

The internal energy is an extensive parameter, as are volume and amount of substance. This statement is less obvious, however, than in the case of the last two quantities. To show this, we consider a system combined of two identical systems. The internal energy of the composite system can be expressed in the following form:

$$U = U^{(1)} + U^{(2)} + U^{(12)},$$

where $U^{(1)}$ and $U^{(2)}$ denote the internal energy of the original systems, and $U^{(12)}$ comes from the interactions of molecules of system (1) with molecules of system (2). Since intermolecular interactions decay quickly with distance, only the molecules close to the surface of contact can contribute to $U^{(12)}$. If L is the linear size of the system then $U^{(12)}$ is proportional to aL^2, where a is a molecular size, whereas $U^{(1)} + U^{(2)}$ is proportional to L^3, hence, the ratio of $U^{(12)}$ to $U^{(1)} + U^{(2)}$ is proportional to a/L. Therefore, all surface effects can be neglected if the system is sufficiently large (macroscopic). In what follows, we always assume that the condition $L \gg a$ is satisfied and the internal energy is an extensive parameter.

2.3 Intensive Parameters of State

In this section, we discuss, in a rather intuitive way, three intensive parameters: the pressure, temperature and chemical potential. Their formal definitions are given in Chaps. 3 and 4. Then, it will turn out that each of these parameters is equal to a derivative of the internal energy with respect to an appropriate extensive parameter. Note that the derivative of one extensive parameter with respect to another extensive parameter must be an intensive parameter because the dependence on the mass of the system cancels out.

We know from everyday experience that if two bodies are brought into thermal contact then a difference between their temperatures causes a flow of heat until the temperature of one body becomes equal to the temperature of the other body. Similarly, a difference in the pressure of a gas on the two sides of a piston causes the action of a force proportional to the piston area and its motion, which stops when the pressure on both sides is the same. This simple observation leads to the conclusion that temperature and pressure are intensive parameters because they have the same values in all subsystems of a given system in thermodynamic equilibrium. Otherwise a macroscopic flow of heat or internal changes in the volume of the subsystems would occur in the system, which would be in contradiction with the definition of thermodynamic equilibrium (see Definition 2.1). A similar argumentation applies to the chemical potential, which we discuss in more detail in Sect. 2.3.3. We will see, however, that the chemical potential of a pure substance is uniquely defined by the temperature and pressure of that substance. Thus, if the temperature and pressure have the same values in all subsystems of a given system then the same is true for the chemical potential. Therefore, it is easier to understand the concept of chemical potential if at least two components are present in the system. For instance, if we add a dye to water and wait for a sufficiently long time we will observe a uniform colour in the whole volume. Even if the temperature and pressure are the same throughout the system we observe a flow of the dye from the regions of higher concentration to the regions of smaller concentration. This process continues until the concentration of the dye becomes uniform. The flow of the dye is caused just by the difference in its chemical potential in various parts of the system.

To summarize, a difference in the temperature is responsible for a flow of heat, a difference in the pressure is responsible for a change in volume (flow of volume), and a difference in the chemical potential of a given component is responsible for a flow of that component. In thermodynamic equilibrium, no flow of any quantity exists, therefore, the temperature, pressure and chemical potential of each component must have the same values in all subsystems of a given system.

2.3.1 Pressure

To define any physical quantity, a method of its measurement must be given. The method can be arbitrary provided that it is reproducible. For instance, Galileo in his studies on the uniformly accelerated motion measured time by means of the amount of water that flowed during the experiment. His measurement of time was reproducible, and the accuracy achieved by Galileo between the 16th and 17th century was comparable with the accuracy of the 19th century watches.

Pressure is defined as the normal force acting on a surface, divided by the area of that surface. For the pressure, symbol p is used. To determine the pressure exerted by a gas on a movable piston of given area, we have to measure the force acting on the piston. The force can be measured with a dynamometer, for instance. The SI derived unit of pressure is the *pascal* (Pa):

$$1\ \mathrm{Pa} = 1\ \mathrm{N\,m^{-2}} = 1\ \mathrm{kg\,m^{-1}\,s^{-2}}.$$

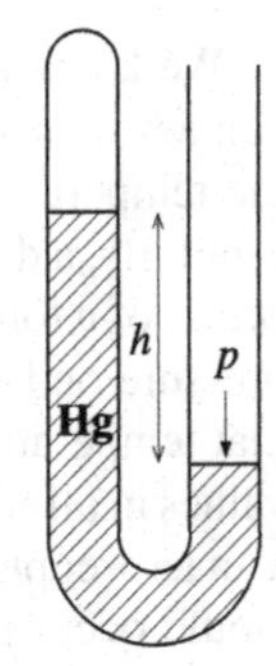

Fig. 2.8 Schematic picture of the mercury barometer used to measure the pressure p. In the closed part of the tube, the pressure is nearly zero. From the balance of the forces acting on the mercury column of the height h, we have: $\rho g h = p$, where ρ is the mercury density and g denotes the gravity of the earth

Other non-SI units of pressure are: the bar (bar),

$$1 \text{ bar} = 10^5 \text{ Pa},$$

the atmosphere (atm),

$$1 \text{ atm} = 101\,325 \text{ Pa},$$

the torr (torr),

$$1 \text{ torr} = \frac{1}{760} \text{ atm} \approx 133.322 \text{ Pa},$$

and the millimeter of mercury (mmHg),

$$1 \text{ mm Hg} \approx 1 \text{ torr}.$$

In the mercury barometer, the force exerted by air is balanced by the weight of the mercury column of the height h in a glass tube (Fig. 2.8). The balance of the forces gives:

$$p = \frac{mg}{A} = \rho g h,$$

where m is the weight of the mercury column, A is the area of its cross-section, ρ is the density of mercury ($\rho \approx 13.6 \text{ g cm}^{-3}$) and $g \approx 9.81 \text{ m s}^{-2}$ denotes the gravity of the earth. In the past, it was assumed that 1 torr = 1 mm Hg. However, the density of mercury varies with temperature and the gravity depends on the place on the earth. Nowadays 1 torr is defined exactly as $\frac{1}{760}$ atm. Why is mercury used in barometers instead of water, for instance? Water evaporates quickly and its density is too low (1 g cm^{-3}) to be used as a barometric liquid. It is easy to calculate the height of the barometer if water or flaxseed oil (its density is about 0.94 g cm^{-3}) was used instead of mercury, to measure atmospheric pressure on the earth surface. On the other hand, in the case of low pressure (air pressure decreases with the altitude), the relative accuracy of the mercury barometer worsens because the height of the mercury column decreases. At an altitude of 20 km, the pressure drops to 0.05 atm, and at 700 km it is of the order 10^{-12} atm (a state of high vacuum). Such low pressures cannot be measured with the mercury barometer.

The microscopic interpretation of pressure follows from classical and statistical mechanics, and is related to a change in the momentum of molecules that collide

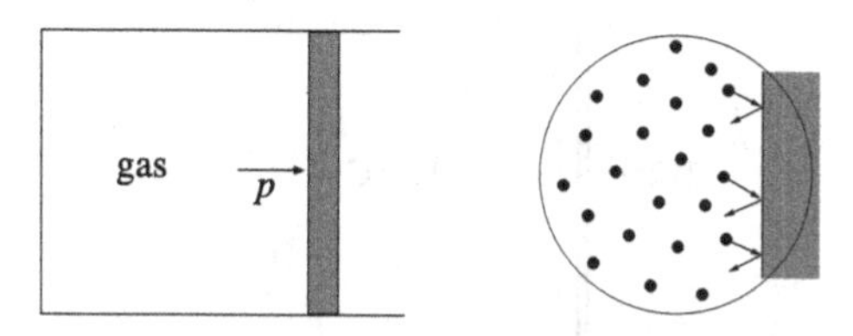

Fig. 2.9 Gas pressure p that is measured (*left picture*) results from a great number of molecular collisions with the piston. The *right picture* presents a great magnification of a small fragment of the gas near the piston

with a wall (Fig. 2.9). The change in the momentum per unit time is equal to the force exerted by the molecules on the wall. Thus, to determine the pressure exerted on the wall one needs to calculate the average number of molecular collisions with the wall per unit time.

2.3.2 Temperature

Zeroth Law of Thermodynamics and the Empirical Temperature The zeroth law of thermodynamics, which leads to the concept of *empirical temperature*, was formulated in 1909, which is more than 200 years after the discovery of the thermometer. In order to formulate this law, the state of *thermal equilibrium* is to be defined first. It is a state of thermodynamic equilibrium of a system restricted by diathermal walls, i.e., walls permeable to heat.

The Zeroth Law of Thermodynamics
If two systems are in thermal equilibrium with a third system they are also in thermal equilibrium with each other

From the point of view of mathematics, thermal equilibrium between two systems is an *equivalence relation*, and the empirical temperature t is defined as the *equivalence class* of that relation. All systems that are in thermal equilibrium with one another belong to the same equivalence class, to which a common value of the temperature t is assigned. Different classes correspond to different values of the empirical temperature. Obviously such an assignment can be done in many ways because it can be based on different reproducible physical phenomena. Thus, it is not surprising that many temperature scales have been in use, for instance, the Celsius and Fahrenheit scales. None of these scales is particularly favoured. For instance, the zero point on the empirical temperature scale is conventional. Only the Kelvin temperature scale is fundamentally different from the empirical scales. It is strictly related to entropy and the second law of thermodynamics, which we discuss in Chap. 4. The temperature on the Kelvin scale is always positive and the zero

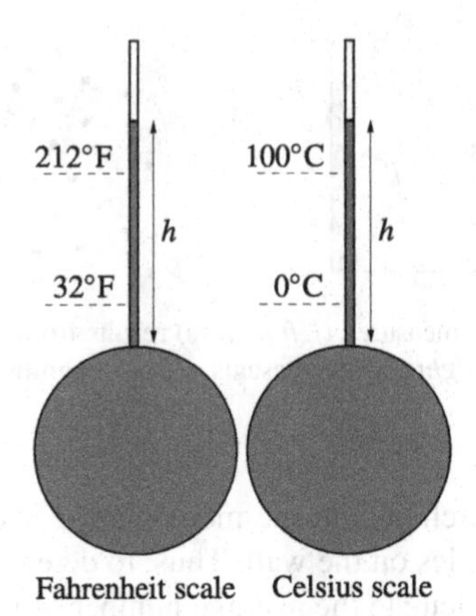

Fig. 2.10 Thermal expansion of liquid is used in both thermometers shown. They are physically identical but one has the Celsius scale and the other has the Fahrenheit scale. The height of the liquid (e.g., mercury) column, h, shows the temperature measured. The freezing point and boiling point of water correspond to the same values of h on both thermometers. However, on the Celsius scale we read 0 °C and 100 °C, whereas on the Fahrenheit scale 32 °F and 212 °F, respectively

point corresponds to the lowest temperature, which is unattainable experimentally, however.

Measurement of Temperature Temperature is one of the intensive parameters that we know very well from everyday life. To measure it we often use mercury thermometers. Nowadays two empirical temperature scales are in common use: the *Celsius scale* in Europe and the *Fahrenheit scale* in the United States.

Celsius chose two reproducible phenomena (which means that they occur always at the same temperatures): the freezing and boiling of water, and assigned to them 0 °C and 100 °C, respectively. Then he divided the scale between these temperatures into 100 equal degrees. Fahrenheit did a similar thing but he chose different phenomena. The temperature 0 °F corresponds to the freezing point of a mixture of water and sal-ammoniac, and the value of 100 °F was assigned to the temperature of his (sick) wife. Fahrenheit also divided his scale into 100 equal degrees. The conversion of one scale to the other is given by the following formula:

$$t_F/°\mathrm{F} = \frac{9}{5}t_C/°\mathrm{C} + 32, \tag{2.18}$$

where t_C and t_F denote the temperatures in the Celsius and Fahrenheit scale, respectively. It follows from the comparison of the two scales that 0 °C corresponds to 32 °F and 100 °C corresponds to 212 °F. Figure 2.10 shows that there is no fundamental difference between these two scales.

In the measurement of temperature, a physical phenomenon sensitive to its variation should be used. Such a phenomenon is, for instance: thermal expansion of liquid (used in household thermometers), electric resistance of metals (platinum thermometer), electric potential difference at the junction of two different metals (thermopile). To measure temperature in the outer space or on the surface of a star, the temperature dependence of the photon–energy distribution is used (pyrometer).

A thermometer based on thermal expansion of liquid is shown in Fig. 2.10. Let us assume that in the range of temperature between t_0 and t, the volume of the liquid used in the thermometer satisfies the following relation:

$$V = V_0\big[1 + \alpha(t - t_0)\big], \tag{2.19}$$

where α denotes the *thermal expansion coefficient* and the volume V_0 corresponds to the temperature t_0. Measuring the height of the liquid column, $h = (V - V_0)/A$, where A is the cross-section area of the glass tube, we determine the empirical temperature difference:

$$t - t_0 = \frac{hA}{V_0\alpha}. \tag{2.20}$$

The larger V_0 and smaller A (narrower tube) are, the more accurate the thermometer is.

In the case of the platinum thermometer, the electric resistance of platinum is measured, which in the range 0–630 °C varies according to the formula:

$$\mathscr{R} = \mathscr{R}_0\big[1 + a(t - t_0) + b(t - t_0)^2\big]. \tag{2.21}$$

In order to determine the constants $\mathscr{R}_0$, a and b, the following three points are used: the freezing point and boiling point of water and the boiling point of sulphur. These three constant points are needed to calibrate the thermometer. It should be added that a good thermometer must satisfy one more condition. In the measurement of temperature, a flow of heat between the thermometer and the system studied always exists, which disturbs the temperature of the system. To minimize that disturbance, the thermometer should be small compared to the system the temperature of which we measure.

In the case of the thermopile, one measures the electric potential difference E between two junctions of two different metals, which are placed at different temperatures. E depends on the temperature difference Δt as follows:

$$E = a + b\Delta t + c(\Delta t)^2. \tag{2.22}$$

One of the junctions can be immersed in a mixture of ice and water, whose temperature is treated as the reference temperature. Using other constant points, it is possible to determine the constants a, b and c.

Kelvin Scale There exists a favoured temperature scale called the *absolute scale* or *Kelvin scale*. The base SI unit of temperature is *kelvin* (K). The temperature that appears in all fundamental laws of nature, called the *absolute temperature* or *thermodynamic temperature*, denoted by T, is always expressed in the Kelvin scale. Thus, if we use a different temperature scale we have to convert it to the Kelvin scale. For instance, the temperature t_C in the Celsius scale which corresponds to the temperature T in the Kelvin scale is obtained from the following formula:

$$T/\mathrm{K} = t_C/^\circ\mathrm{C} + 273.15. \tag{2.23}$$

The temperature $t_C = 0\ °C$ corresponds to $T = 273.15$ K and the lower limit of all temperatures, called the *absolute zero* ($T = 0$ K), corresponds to $t_C = -273.15\ °C$. In Chap. 4, we show that the thermodynamic definition of temperature defines T up to an arbitrary constant factor, which means that the choice of the temperature unit is arbitrary. However, since the Celsius scale appeared before the Kelvin scale, it was convenient to use the same unit, i.e., 1 K and 1 °C mean the same temperature difference. Because of this choice, however, the freezing point of water is not an integer in the Kelvin scale.

It had passed more than one hundred years since the first experiments on dilute gases, performed by Boyle at the end of the 17th century, before all relations between pressure, volume and temperature of a dilute gas were discovered. It was found that these three parameters satisfy, to a good approximation, the following equation:

$$pV = B(t_C + 273.15\ °C), \tag{2.24}$$

where B is constant for a given amount of gas. Moreover, it was shown that

$$B = nR, \tag{2.25}$$

where n is the number of moles of the gas, and

$$R = 8.314472\,(15)\ \mathrm{J\,K^{-1}\,mol^{-1}} \tag{2.26}$$

is called the *gas constant*. It is easy to recognize that the temperature in Eq. (2.24) is actually expressed in the Kelvin scale, hence, (2.24) can be rewritten as follows:

$$pV = nRT. \tag{2.27}$$

Thus, the Kelvin scale can be inferred from the studies of the properties of dilute gases. We notice that the gas constant does not depend on the kind of a gas. This is related to the fact that the amount of substance is expressed in moles. If we used kilograms instead of moles, the value of R would depend on the molecular mass of a given substance. For instance, it would be 16 times larger for hydrogen (H_2) than for oxygen (O_2).

It follows from (2.27) that for a given amount of a dilute gas and at constant volume, we have

$$\frac{p}{p_0} = \frac{T}{T_0}, \tag{2.28}$$

where T_0 and p_0 define a reference state. This equation shows that measuring the gas pressure, we can determine its temperature T. To calibrate such a gaseous thermometer, one constant point is to be chosen, for instance, the freezing point of water at atmospheric pressure.[3]

The absolute temperature also appears in the internal energy of a photon gas in thermodynamic equilibrium with the *perfect blackbody*. According to the Stefan–Boltzmann law

$$U = \gamma V T^4, \tag{2.29}$$

[3]In fact, it should be the *triple point*, i.e., a thermodynamic state in which water vapour, liquid water and ice are in thermodynamic equilibrium (at $T_0 = 273.16$ K).

where γ is a constant. This energy can be determined from the intensity of the electromagnetic radiation emitted by the perfect blackbody, which means that it absorbs photons independently of their energy. This method can be used, for instance, to determine the temperature of stars. For example, the temperature on the sun surface determined in this way amounts to 6000 K. The thermometers based on the Stefan-Boltzmann law are called *pyrometers*. Expression (2.29) can also be applied in measurements of very high temperatures, at which most of solids melt.

In laboratories, we can obtain both very high and very low temperatures. The lowest temperature obtained on the earth amounts to 10^{-8} K (Boulder Colorado, 1995), and the highest temperature amounts to 10^{12} K (CERN, 2000). In both cases, the amount of substance used in the experiment was very small compared to 1 mol.

2.3.3 Chemical Potential

In a closed system, energy can be transferred only in the form of work or heat. In the general case, also matter can flow between the system and surroundings. To describe thermodynamic equilibrium between the system and surroundings if a flow of matter is possible, we have to introduce an additional intensive parameter, apart from pressure and temperature, which is called the *chemical potential*. In the case of a pure substance, it is just one quantity, denoted by μ. In a mixture, each component has its own chemical potential. The SI derived unit of the chemical potential is *joule/mol* (J/mol).

While temperature and pressure are the concepts we are familiar with, it is not the case of the chemical potential. To explain this concept, we recall first that the condition of mechanical equilibrium requires that pressures in all subsystems of a given system are equal. Similarly, thermal equilibrium means that the temperatures of all subsystems are the same. If these conditions are not satisfied simultaneously then heat flows between the subsystems or their volumes change, which means that the system is not in thermodynamic equilibrium.

We consider now a system composed of two subsystems separated by a immobile diathermal wall impermeable to matter. The subsystems are in thermal equilibrium but their pressures can differ because the wall is stiff. We assume that one subsystem is formed by a two-component liquid mixture, in which the number of solvent molecules is much greater than the number of molecules of the other component (the solute). The other subsystem is formed by a pure solvent (Fig. 2.11). We also assume that the temperatures and pressures in both subsystems are initially the same.

What will happen if we replace the wall impermeable to the matter flow with a semi-permeable wall, permeable only to the solvent molecules? Since we have assumed equal temperatures and pressures on both sides of the wall, it may seem that nothing will happen, because the subsystems are in thermal and mechanical equilibrium. Such an answer is wrong, however. In fact, a certain number of the solvent molecules will flow from the subsystem containing the pure solvent to the subsystem containing the mixture. Moreover, this flow causes the pressure of the mixture

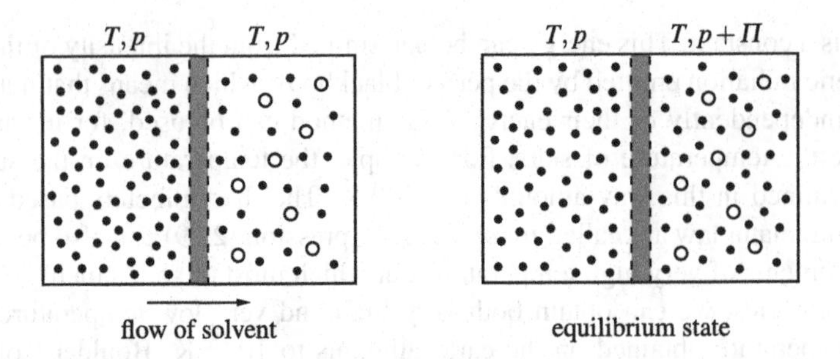

Fig. 2.11 System is divided into two subsystems by an immobile diathermal wall. The *left part* is occupied by a pure solvent (*black circles*), and the *right part* contains a mixture of the solvent with a small amount of a solute (*white circles*). The wall is permeable only to the solvent molecules. Initially the temperatures and pressures in both parts are the same (*left picture*). However, because of the difference in the chemical potential of the solvent, a certain number of the solvent molecules flow to the *right part*. In thermodynamic equilibrium (*right picture*), the chemical potential of the solvent is the same in both subsystems but there is a pressure difference Π, called the *osmotic pressure*

to increase, which is possible because the wall is immobile. The pressure difference between the mixture and the pure solvent is called the *osmotic pressure*.[4] Thus, the flow of the solvent molecules occurs in the direction of increasing pressure, which seems counterintuitive. Apparently *something* is missing in our description of this phenomenon. That what is missing is just the chemical potential. It turns out that the chemical potential of the solvent in a mixture differs from the chemical potential of the pure solvent. In the phenomenon discussed here, the temperature and pressure in the mixture and in the pure solvent are initially the same but the chemical potential of the pure solvent is higher than the chemical potential of the solvent in the mixture, which we explain in Chap. 8. Anyway, the chemical potential difference causes a flow of the solvent in the direction of the lower chemical potential. The flow stops when the chemical potential of the solvent in both subsystems is the same.

If the chemical potential of a given component has the same value in all subsystems then the system is said to be in equilibrium with respect to the flow of that component. This statement concerns only the components whose flow is not restricted by some internal constraints, as in the example above. As we have already mentioned, the chemical potential of a pure substance is a function of temperature and pressure. It depends also on the form in which a given substance exists in given conditions, i.e., as a gas, liquid, or solid. For example, the chemical potentials of liquid water and ice are different functions of temperature and pressure. Therefore, if we ask about the conditions in which liquid water and ice are in thermodynamic equilibrium we have to take into account the equality of chemical potentials, apart from the equality of pressures and temperatures. If the chemical potentials were different then the matter would flow between the liquid and solid, thus, they would

[4]The phenomenon of osmosis is discussed in Chap. 8.

not be in thermodynamic equilibrium. These problems are discussed extensively in Part II, which is devoted to phase transitions.

2.4 Equations of State

Having determined all state parameters for a given system in thermodynamic equilibrium: the temperature T, volume V, pressure p, internal energy U, amount of substance n, etc., we discover that they are not independent of one another. A relation between the state parameters $X_1, X_2, X_3, \ldots$:

$$\mathscr{F}(X_1, X_2, X_3, \ldots) = 0, \tag{2.30}$$

is called an *equation of state*. A system in thermodynamic equilibrium is uniquely defined by a specified number of independent parameters of state. The remaining parameters of state are functions of the independent parameters. Thus, they are state functions, which can be determined if we know the equations of state.

In general, the equations of state depend on a given substance and on the range of the parameters in which we want to describe that substance. For example, the equations of state of a substance in the gaseous state are different from the equations of state of the same substance in the liquid or solid state. Below we discuss three model systems: the ideal gas, the van der Waals gas and the photon gas, which have different equations of state. The parameters of state that appear in these equations of state are: T, V, p, n and U. In the case of the photon gas, there is no dependence on n, but it is an exception rather than a rule.

2.4.1 *Ideal Gas*

The parameters of state that describe a dilute gas in thermodynamic equilibrium satisfy, to a good approximation, the following equations of state:

$$pV = nRT, \tag{2.31}$$

$$U = \frac{f}{2} nRT. \tag{2.32}$$

R denotes the gas constant (see (2.26)), and f specifies the number of *degrees of freedom* of a single molecule. For monatomic molecules, e.g., argon, $f = 3$. For linear molecules, e.g., oxygen, $f = 5$, and for more complex molecules, e.g., methane, $f = 6$.[5] A hypothetical gas for which Eqs. (2.31) and (2.32) are satisfied for all

[5]This follows from the *equipartition theorem*, which is derived in the framework of classical statistical mechanics. According to the equipartition theorem each degree of freedom of the translational or rotational motion of a molecule gives the same contribution to the internal energy, equal to $RT/2$ per 1 mol. A monatomic molecule has three degrees of freedom related to the translational motion of the center of mass, hence, $f = 3$. A linear molecule has, in addition, two independent axes of rotation perpendicular to its symmetry axis, hence, $f = 3 + 2 = 5$. Other molecules have three axes of rotation, which gives $f = 3 + 3 = 6$.

values of the state parameters is called the *ideal gas*, and (2.31) and (2.32) are the equations of state of the ideal gas. The ideal gas can be treated as a limiting case of a real gas when the pressure $p \to 0$. In a very dilute gas, all intermolecular interactions can be neglected and the only contribution to its internal energy comes from the kinetic energy of molecules.

It follows from the equations of state of the ideal gas that we can choose three independent variables from the five state parameters that appear in these equations. For instance, if we substitute

$$T = 298\ \text{K}, \qquad p = 1\ \text{atm}, \qquad n = 1\ \text{mol},$$

into (2.31) and (2.32) we get, for $f = 3$:

$$V = 0.024453\ \text{m}^3, \qquad U = 3716.6\ \text{J}.$$

If we increase the number of moles to $n = 2$, keeping T and p unchanged, then the volume and internal energy also increase twice.

Choosing U, V and n as independent variables, we can transform the equations of state of the ideal gas as follows:

$$p = \frac{2U}{fV}, \tag{2.33}$$

$$T = \frac{2U}{fnR}. \tag{2.34}$$

Here the temperature and pressure are treated as functions of state: $p = p(U, V, n)$ and $T = T(U, V, n)$. We use this form of the equations of state in Chap. 4, to calculate the entropy of the ideal gas.

As independent parameters of state we can also choose T, V and n, and rewrite the equations of state in the following form:

$$p = \frac{nRT}{V}, \tag{2.35}$$

$$U = \frac{f}{2}nRT. \tag{2.36}$$

In this case, the pressure and internal energy are functions of state, which are formally expressed as $p = p(T, V, n)$ and $U = U(T, V, n)$, although U does not depend on V for the ideal gas. Relation (2.35) is used, for instance, to calculate the work done during the isothermal compression ($T = const$) of the ideal gas.

Finally, as independent parameters of state one can choose T, p and n, which gives

$$V = \frac{nRT}{p}, \tag{2.37}$$

$$U = \frac{f}{2}nRT. \tag{2.38}$$

Here, the volume and internal energy are functions of state: $V = V(T, p, n)$ and $U = U(T, p, n)$, where U does not depend on p for the ideal gas. For instance, Eq. (2.37) is used to determine the dependence of the chemical potential of the ideal gas on pressure.

2.4.2 Van der Waals Gas

If the gas is not sufficiently dilute the intermolecular interactions cannot be neglected and equations of state (2.31) and (2.32) are not satisfied. Then Eq. (2.31) can be replaced by the *van der Waals equation of state*:

$$p = \frac{nRT}{V - nb} - \frac{an^2}{V^2}. \tag{2.39}$$

It describes some real gases in an approximate way. Moreover, Eq. (2.39) applies not only to gases but, to some extend, also to liquids. In Chap. 6, we show that it explains qualitatively the change of gas into liquid. The constants a and b in Eq. (2.39) are to be determined experimentally for a given substance. For example, $a = 0.1358\ \mathrm{J\,m^3\,mol^{-2}}$ and $b = 3.85 \times 10^{-5}\ \mathrm{m^3\,mol^{-1}}$ for nitrogen (N_2). For most of simple substances, $b \approx 3 \times 10^{-5}\ \mathrm{m^3\,mol^{-1}}$ and a can vary from $0.003\ \mathrm{J\,m^3\,mol^{-2}}$, for helium, to $1\ \mathrm{J\,m^3\,mol^{-2}}$, for Freon.

For the internal energy of the van der Waals gas, the following formula is assumed:

$$U = \frac{f}{2}nRT - \frac{an^2}{V}. \tag{2.40}$$

Equations (2.39) and (2.40) are empirical but they can also be derived from a molecular theory (see Sect. 6.7). We notice that they reduce to the equations of state of the ideal gas when $a = 0$ and $b = 0$. Since the ideal gas model is based on the assumption of non-interacting molecules, the constants a and b must be related to intermolecular interactions. The constant a stands at the term which takes into account, in an approximate way, attraction between molecules. Thus, it is a measure of the strength of attractive forces. The presence of the constant b stems from the fact that a molecule is not a point object but it occupies some volume. Therefore, it is not possible to compress a given substance into the volume nb taken up by the molecules themselves because such a state would have infinite pressure.

2.4.3 Photon Gas

The photon gas is a system formed by quanta of electromagnetic radiation (photons). As an ordinary gas of molecules, the photon gas has its volume, temperature, internal energy and pressure. However, in contrast to a molecular gas, the number of photons is not conserved. Photons are absorbed and emitted by the walls of a vessel and because of that the number of moles is not a state parameter for the photon gas. The state of thermodynamic equilibrium of the photon gas is called the *blackbody radiation*, and the equations of state have the following form:

$$p = \frac{1}{3}\gamma T^4, \tag{2.41}$$

$$U = \gamma V T^4, \tag{2.42}$$

where $\gamma \approx 7.56 \times 10^{-16}\ \mathrm{J\,m^{-3}\,K^{-4}}$.[6] Thus, the pressure, $p = p(T, V)$, and the internal energy, $U = U(T, V)$, are state functions, and T and V are state parameters. If we treat U and V as state parameters, and $T = T(U, V)$ and $p = p(U, V)$ as state functions then we obtain

$$p = \frac{U}{3V}, \tag{2.43}$$

$$T = \left(\frac{U}{\gamma V}\right)^{1/4}. \tag{2.44}$$

2.4.4 Equations of State in Terms of Intensive Parameters

Dividing the equations of state of the ideal gas (see (2.31) and (2.32)) by the mole number n, we get

$$pv = RT, \tag{2.45}$$

$$u = \frac{f}{2} RT, \tag{2.46}$$

where $v = V/n$ and $u = U/n$ denote the molar volume and the molar internal energy, respectively. Both v and u are intensive parameters, as are temperature and pressure, i.e., they do not depend on the mass of the system. Equations (2.45) and (2.46) provide as much information about properties of the system as Eqs. (2.31) and (2.32) do. This stems from the fact that any macroscopic subsystem of a system in thermodynamic equilibrium have the same physical properties as the whole system. It simplifies the description of thermodynamic systems because the number of independent parameters of state is reduced by one.

Treating in the same way Eqs. (2.39) and (2.40) for the van der Waals gas, we find:

$$p = \frac{RT}{v - b} - \frac{a}{v^2}, \tag{2.47}$$

$$u = \frac{f}{2} RT - \frac{a}{v}. \tag{2.48}$$

In the case of the photon gas, we divide both sides of Eq. (2.42) by V, which gives $U/V = \gamma T^4$. Thus, instead of the molar internal energy, we have the internal energy per unit volume.

The conclusions presented above follow from the definition of extensive and intensive parameters. When we join together m identical systems, to form a new m times larger system, the intensive parameters, such as pressure and temperature, do not change, and the extensive parameters, such as volume and internal energy, are

[6] The constant $\gamma = 8\pi^5 k_B^4/(15c^3h^3)$, where $k_B = R/N_A$ is the Boltzmann constant, and c and h denote the speed of light in the vacuum and the Planck constant, respectively.

multiplied by m. For example, if pressure is treated as a function of three extensive parameters: U, V and n, then

$$p(mU, mV, mn) = p(U, V, n). \tag{2.49}$$

Here $m > 0$ is an integer but it is easy to show that m can be treated as a positive real variable. To show this, we consider a large system in thermodynamic equilibrium, which is divided to $\mathscr{N}$ identical small (but macroscopic) fragments. Then we use $\mathscr{M}$ fragments, to form a subsystem of the original system. Assuming that the extensive parameters U, V and n correspond to the large system, for the subsystem, we have mU, mV and mn, where $m = \mathscr{M}/\mathscr{N}$ is a rational number. If $\mathscr{N}$ is a very large number we can assume that m is practically a continuous variable. Thus, we can substitute $m = 1/n$ in (2.49), hence,

$$p(u, v, 1) = p(U, V, n), \tag{2.50}$$

i.e., p is a state function of two intensive parameters: u and v. A similar relation can be derived for temperature.

Then we consider the internal energy as a state function of T, V and n. Since U is an extensive parameter, the following identity must hold:

$$U(T, mV, mn) = mU(T, V, n). \tag{2.51}$$

Substituting $m = 1/n$, we obtain

$$U(T, V, n) = nU(T, v, 1) = nu(T, v). \tag{2.52}$$

We note finally that relations (2.49)–(2.52) are exact, as they do not refer to any particular form of the equations of state.

2.5 Exercises

2.1 Calculate the kinetic energy of molecules in 22 L of air, assuming that the average speed of molecules is equal to the speed of sound and the density of air is 10^{-3} g cm^{-3}. Ignore the fact that air is a mixture of several gases.

2.2 Calculate the change in the gravitational energy of 1 mol of water transferred from the level $h = 0$ to the height $h = 5$ km.

2.3 Calculate approximately the change in the internal energy of 9 g of water caused by its evaporation. The heat of evaporation amounts to 40 kJ mol^{-1}. Compare it with the change in the gravitational energy found in Exercise 2.2 and with the kinetic energy of air. Draw a conclusion about the energy of interactions of water molecules.

2.4 Calculate approximately the change in the internal energy of 40 g of argon caused by its evaporation. The heat of evaporation amounts to 6 kJ mol^{-1}. Compare it with the energy needed to evaporate water. Draw conclusions about the interaction energies of water molecules and argon atoms.

2.5 Calculate approximately the change in the internal energy of 12 g of carbon caused by its combustion. The heat of combustion amounts to 400 kJ mol^{-1}. Compare it with the energy of intermolecular interaction found in Exercise 2.4. Comparing the changes in the internal energy due to evaporation and combustion, draw a conclusion about the energy of chemical bonds and intermolecular interactions.

2.6 The sun shines due to thermonuclear reactions in which nuclei of light elements combine to form nuclei of heavier elements. As a result of a few nuclear reactions, four hydrogen nuclei (protons) combine to form one nucleus of helium (2 protons and 2 neutrons) and the energy of 26 MeV is released (1 MeV = 10^6 eV, 1 eV $\approx$ 1.6×10^{-19} J). Calculate the energy released during the nuclear reaction in 1 mol of hydrogen nuclei (protons).

2.7 As a result of annihilation, 1 mol of carbon is transformed into the energy of a photon gas. Calculate how much energy has been released. Compare it with the forms of energy discussed previously.

2.8 One mol of water vapour condenses. Find the change in its volume, assuming 1 g cm^{-3} for the density of liquid water and 10^{-3} g cm^{-3} for the density of water vapour.

2.9 In the reaction of ammonia formation, 2 mol of NH_3 are formed from 1 mol of N_2 and 3 mol of H_2. Find the changes in the total amount of substance in the system and in the amounts of individual components: N_2, H_2 and NH_3.

2.10 A mixture of three gases: 3 mol of H_2, 4 mol of H_2O and 1 mol of Ar, occupies a vessel of the volume of 100 L. A second vessel contains 5 mol of H_2, 2 mol of H_2O and 1 mol of Ar, which also occupy 100 L. Then we fuse the vessels. Give the total mole number and the volume of the gases in the fused vessels.

2.11 There are four vessels, each of which contains the same gas of the internal energy U, volume V and mole number n. Then we fuse the vessels. Find the total internal energy, volume and mole number of the gas in the fused vessels.

2.12 A man needs 2000 kcal a day, on average, in order his organism could function properly. Calculate the power used up by a man. It is worth mentioning that the first studies on the heat given off by animals were carried out by Lavoisier and Laplace in the 18th century with the use of calorimeters. Thus, thermodynamics provided a basis for the determination of the human diet from the energetic point of view.

2.13 Why do we use mercury in the barometer instead of water or oil, for instance? What would the height of the barometer have to be if we used flaxseed oil, of the density about 0.94 g cm^{-3}, instead of mercury, to measure atmospheric pressure on the earth surface?

2.14 Is 1 atm a large pressure? What force does air exert on a human body? What would happen to a human body if the pressure of air decreased to the value 10^{-6} atm?

2.15 Assume you drink water using a 20 cm long straw. What pressure difference do you have to apply with your mouth (when you suck in air in the straw you simply produce an underpressure above the liquid surface), to drink up water?

2.16 At a depth of 10 m there is a submarine. The pressure of air inside the submarine amounts to 1 atm. A sailor wants to lift the lid to let a diver out into water. The lid area is equal to 2 m^2. Calculate the force the sailor has to apply to lift the lid. Explain why to let the diver out of the submarine, the sluice-gate must be filled with water first.

2.17 Convert 0 °F, 70 °F and 451 °F to the Celsius scale.

2.18 What temperature in the Kelvin scale corresponds to −273.15 °C?

2.19 Express the temperature of 0 K in the Fahrenheit scale.

2.20 Convert 100 °F to the Celsius scale. How do we know that Fahrenheit's wife was sick when he calibrated the thermometer? Is the human body temperature suitable for the calibration of thermometers?

2.21 How would you calibrate a pyrometer if you know that the temperature of water at the triple point is equal to $T = 273.16$ K? You have a means of measuring the energy emitted by the perfect blackbody at a given temperature. The internal energy U is proportional to the intensity of radiation.

2.22 Prove that the internal energy per mole does not depend on the size of the system, thus, it is an intensive parameter. Assume that you join together m identical systems characterized by the same parameters of state: U, V and n. Calculate the internal energy per mole for each system and for the composite system and compare the results.

2.23 We have two vessels containing water at the same temperature and pressure. The volume of the vessels amounts to $V_1 = 18\ \text{cm}^3$ and $V_2 = 36\ \text{cm}^3$, respectively, and the density of water is $\rho = 1\ \text{g}\,\text{cm}^{-3}$. Calculate the mole number of water and its molar density (the mole number per unit volume) for each vessel. Then we join the vessels together. Calculate the volume, mole number and molar density after the fusion of the vessels. Which of them are extensive parameters and which are intensive parameters? Why is the density ρ an intensive parameter?

2.24 Determine the volume per one molecule of H_2O in liquid water. The density of water $\rho = 1\ \text{g}\,\text{cm}^{-3}$, the Avogadro constant $N_A = 6.022 \times 10^{23}\ \text{mol}^{-1}$ and the molar mass of water $M = 18\ \text{g}\,\text{mol}^{-1}$.

2.25 Calculate an infinitesimal increase in the volume, dV, if the temperature increases by dT, the pressure increases by dp, and the mole number n does not change. Perform the calculations for the ideal gas.

2.26 Calculate an infinitesimal increase in the internal energy, dU, if the temperature increases by dT, the volume increases by dV, and the mole number n does not change. Perform the calculations for the van der Waals gas.

2.27 Calculate an infinitesimal increase in the pressure, dp, for the photon gas if the temperature increases by dT.

2.28 Which of the expressions presented below is a differential of a function of x and y: (1) $đ\omega = 2xy^3 dx + 3x^2y^2 dy$, (2) $đ\omega = xy^4 dx + x^2y^2 dy$. Find this function.

2.29 Show that in the case of the differential form which is not a differential of a function in Exercise 2.28, the integral $\int_i^f đ\omega$ depends on the path of integration. As the initial point (i) and final point (f) in the xy plane assume $(0,0)$ and $(1,1)$, respectively. Perform calculations for two paths: (1) $0 \leq x \leq 1$, $y = 0$ and $x = 1$, $0 \leq y \leq 1$, (2) $0 \leq x \leq 1$, $y = x$. Verify that in the case of the differential df, the integral does not depend on the path of integration and is equal to $\Delta f = f_f - f_i$.

2.30 A system that contains 5 mol of a gas in a vessel of the volume $V = 120$ L and at a temperature of 25 °C is in thermodynamic equilibrium. Then, we let a certain amount of the gas out of the vessel. The pressure of the gas in the new state is equal to $p = 0.5$ atm. What is the amount of the gas that has escaped?

2.31 A vessel of the volume 0.1 L is occupied by 1 mol of N_2 at the temperature $T = 298$ K. Calculate the pressure and internal energy of the gas. Apply the equation of state of the ideal gas.

2.32 A vessel of the volume 0.1 L is occupied by 1 mol of N_2 at the temperature $T = 298$ K. Calculate the pressure and internal energy of the gas. Apply the van der Waals equation of state with the constants: $a = 0.1358$ J m^3 mol^{-2} and $b = 3.85 \times 10^{-5}$ m^3 mol^{-1}. Compare the result with the result of Exercise 2.31.

2.33 A vessel of the volume 1000 L is occupied by 1 mol of N_2 at the temperature $T = 298$ K. Calculate the pressure and internal energy of the gas. Apply the van der Waals equation of state, with the same constants a and b as in Exercise 2.32, and then the equation of state of the ideal gas. Compare the result with the results of Exercises 2.31 and 2.32.

2.34 Calculate the internal energy of a photon gas contained in a vessel of the volume of 1 m^3, for two values of the temperature: 298 K and 400 K. Calculate the pressure exerted by the gas on the walls of the vessel.

2.35 Verify the relations:

$$p(U, V, n) = p(mU, mV, mn) \quad \text{and} \quad U(T, mV, mn) = mU(T, V, n),$$

for the ideal gas and van der Waals gas.

2.36 Calculate $-V^{-1}(\partial V/\partial p)_{T,n}$ for the ideal gas and van der Waals gas. Does this expression make sens for the photon gas? Calculate an analogous quantity for the photon gas at constant internal energy U.

2.37 Calculate $V^{-1}(\partial V/\partial T)_{p,n}$ for the ideal gas and $V^{-1}(\partial V/\partial U)_p$ for the photon gas.

2.38 Calculate $(\partial U/\partial T)_{V,n}$ for the ideal gas, van der Waals gas and photon gas.

2.39 Assuming that most of air is contained in a thin, 10 km in thickness, layer around the earth, the pressure is equal to 1 atm, and the average temperature amounts to 14 °C, calculate the total mole number of gases in the atmosphere. Then assuming that oxygen makes 21 % of air, calculate the amount of oxygen. Living organisms use up 0.5×10^{16} mol of oxygen per year. If the oxygen supply was not renewed in the photosynthesis process, how quickly would it be lacking in the atmosphere?

2.40 The temperature on the sun surface amounts to about 6000 K. The amount of energy radiated by the sun in the form of photons per unit time and per unit area is given by

$$I = \frac{uc}{4},$$

where u is the internal energy of the photon gas per unit volume, and $c = 3 \times 10^8\ \mathrm{m\,s^{-1}}$ is the speed of light. The radius of the sun is equal to 700 000 km. Calculate the energy E radiated by the sun during one second. Consider if the mankind will cope with the energetic crisis after the resources of oil have been exhausted.

Chapter 3
Internal Energy, Work and Heat

3.1 First Law of Thermodynamics

In the previous chapter, we introduced the concept of the internal energy of a system as one of the extensive parameters of state. It follows from the energy conservation principle that the internal energy of an isolated system is a constant quantity. In thermodynamics, however, we are mainly interested in interactions between the system and surroundings, therefore, we have to specify possible ways of energy transfer between them. One of these ways is *mechanical work*. If the system performs work its energy decreases. For example, if we squeeze a spring it can perform work later, decreasing its potential energy. If an external agent (surroundings) performs work on a mechanical system the energy of the system increases.

We already know that thermodynamics deals with macroscopic systems consisting of a great number of atoms or molecules. The internal energy of such a system, as the energy of a mechanical system, can be changed by means of mechanical work. An example of mechanical work is compression of a gas in a vessel with a movable piston. We can determine such work if at any moment we know the force acting on the piston and its displacement, however, we do not need to know how individual molecules interact with the piston. Obviously, it is not the only way to perform work on the system. For instance, when an electric mixer is immersed in a liquid we can perform work by mixing the liquid. Due to the viscosity there exists friction between the liquid and a rotating mixer. Thus, the electric current, which drives the mixer, performs a definite amount of work on the system, which can be easily measured.

In the case of a simple mechanical system, such as the spring, the work done by an external agent is stored in the form of potential energy, which does not change as long as the system does not perform any work. On the other hand, if some work is performed on a thermodynamic system, the energy stored in it can escape to the surroundings in the form of heat. To prevent it, we have to insulate thermally the system with adiabatic walls. Then the increase in the internal energy of the system, ΔU, is exactly equal to the work performed on the system:

$$\Delta U = W. \tag{3.1}$$

R. Hołyst, A. Poniewierski, *Thermodynamics for Chemists, Physicists and Engineers*,
DOI 10.1007/978-94-007-2999-5_3,

If the system performs work on the surroundings its internal energy decreases, hence, $\Delta U < 0$ and also $W < 0$. Note that Eq. (3.1) means that adiabatic walls enable us to measure changes in the internal energy, provided that we can measure the work performed by the system or on the system.

If the system is not adiabatically isolated (3.1) is not satisfied. Then the difference:

$$Q = \Delta U - W, \tag{3.2}$$

is called *heat*, provided that the amount of matter in the system does not change. Heat is simply another form of energy transfer between the system and surroundings. A flow of heat results from chaotic collisions of molecules inside and outside the system with molecules of the walls. Using suitable materials for the walls, we can considerably limit the flow of heat. For example, we can use double glass walls covered with a silver layer, with a very dilute gas in between (thermos is built in that way), which can be considered as a good thermal insulation. Applying better and better materials we can approach the perfect thermal insulation, that is, the adiabatic wall, for which $Q = 0$ by definition.

Apart from work and heat, we can change the internal energy simply by varying the amount of matter in the system. For instance, if we combine two identical systems to form a bigger system then the amount of matter increases twice, compared to the original system, and the same concerns the internal energy. In this case, no heat flows and no work is performed. This form of energy transfer is called sometimes *chemical work*; here we denote it by Z. The possible ways of changes in the internal energy of a thermodynamic system are summarized in the *first law of thermodynamics*. This law expresses the energy conservation principle for macroscopic systems.

The First Law of Thermodynamics

There exists an extensive function of state, called the internal energy U, whose change in a thermodynamic process:

$$\Delta U = Q + W + Z, \tag{3.3}$$

is caused by the flow of heat Q, work performed W, and flow of matter Z.

The first law of thermodynamics postulates existence of the internal energy as a function of state and specifies three possible means of its variation. Note that the same change in the internal energy, ΔU, can be obtained in different processes that begin and end in the same states. For instance, compressing a gas in an adiabatic process ($Q = 0$) from the volume V_i to the volume $V_f < V_i$, we perform the work $W = W_1$, which increases the gas temperature from T_i to $T_f > T_i$. The same final state is reached if we first compress the gas isothermally to the volume V_f, at the temperature $T = T_i$ ($W \neq 0$, $Q \neq 0$), and then heat the gas to the temperature T_f at constant volume $V = V_f$ ($W = 0$, $Q \neq 0$). In both processes $Z = 0$ but the amounts of work and heat are different. In the first process $\Delta U = W_1$, whereas in the second process the work W_2 is done and the heat Q_2 is transferred, hence, $\Delta U = W_2 + Q_2$. This example shows, that we can change the means of energy transfer between the system and surroundings, to obtain the same final state. It also shows that W and Q cannot be functions of state because their values depend on the process linking the initial and final states, and the same concerns Z.

The first law is formulated for finite changes in the internal energy. In general, we can calculate W, Q and Z only for quasi-static processes, which involves the use of infinitesimal quantities. Then Eq. (3.3) adopts the following form:

$$\mathrm{d}U = \text{đ}Q + \text{đ}W + \text{đ}Z, \tag{3.4}$$

where $\text{đ}Q$, $\text{đ}W$ and $\text{đ}Z$ denote infinitesimal amounts of heat, work and chemical work, respectively. In what follows we assume that the system performs work, or work is performed on the system, only if the volume of the system changes.

Chemical work appears in the context of mixtures, phase transitions and chemical reactions. For the time being, we will ignore this contribution to $\mathrm{d}U$, however. That is, we assume that there is no transfer of matter between subsystems of a given system or between the system and surroundings, and that no reaction occurs. Then using (2.15) and (3.4), we get

$$\mathrm{d}U = \text{đ}Q - p\mathrm{d}V. \tag{3.5}$$

The internal energy increases when heat is supplied to the system ($\text{đ}Q > 0$) or work is done on the system ($\text{đ}W = -p\mathrm{d}V > 0$), and it decreases if heat is given off by the system ($\text{đ}Q < 0$) or work is done by the system ($\text{đ}W < 0$). Below we persent a few examples of calculation of work and heat in various processes.

3.2 Isochoric Process

During an isochoric process the volume of the system, V, remains constant, hence

$$\text{đ}W = -p\mathrm{d}V = 0. \tag{3.6}$$

The internal energy can change only if heat can flow between the system and surroundings. We assume that the flow of heat is a quasi-static process, thus

$$\mathrm{d}U = \text{đ}Q. \tag{3.7}$$

It should be emphasized, however, that Eq. (3.7) does not mean that $\text{đ}Q$ is equal to $\mathrm{d}U$ for all values of the state parameters, otherwise Q would be a state function. This equality is satisfied only at constant V. As independent parameters of state we choose the temperature and volume (the number of moles, n, is assumed constant), hence

$$\mathrm{d}U = \left(\frac{\partial U}{\partial T}\right)_V \mathrm{d}T + \left(\frac{\partial U}{\partial V}\right)_T \mathrm{d}V. \tag{3.8}$$

Putting $\mathrm{d}V = 0$ in (3.8) and then comparing with (3.7), we get

$$\text{đ}Q = \left(\frac{\partial U}{\partial T}\right)_V \mathrm{d}T. \tag{3.9}$$

3.2.1 Heat Capacity at Constant Volume

The quantity

$$C_X = \left(\frac{đQ}{\mathrm{d}T}\right)_X, \tag{3.10}$$

is called *heat capacity* of the system at constant parameter X, where for X we can substitute the volume V, pressure p, or another parameter of state. In the isochoric process, heat is transferred at constant V, hence

$$C_V = \left(\frac{đQ}{\mathrm{d}T}\right)_V \tag{3.11}$$

is called the heat capacity at constant volume. Heat capacity specifies the amount of heat $đQ$, to be supplied to the system in a given process, to increase its temperature by $\mathrm{d}T$. The SI derived unit of heat capacity is *joule kelvin*$^{-1}$ ($\mathrm{J\,K^{-1}}$). In general, heat capacity is a function of temperature and other parameters of state of the system. In the case of C_V, we can use relation (3.9), from which we get

$$C_V = \left(\frac{\partial U}{\partial T}\right)_V, \tag{3.12}$$

which means that $C_V = C_V(T, V)$. Heat capacity is an extensive function of state as a derivative of U with respect to an intensive parameter. Therefore, it is convenient to introduce a quantity that does not depend on the amount of substance, i.e., the *molar heat capacity* at constant volume, $c_v = C_V/n$, where $v = V/n$ is the molar volume.

In the case of the ideal gas, the internal energy is given by (2.32), hence, the molar heat capacity at constant volume amounts to

$$c_v = \frac{1}{n}\left(\frac{\partial U}{\partial T}\right)_V = \frac{f}{2}R. \tag{3.13}$$

Note that it depends only on the structure of a single molecule in the gas. For monatomic molecules, $c_v = 3R/2$, for linear molecules, $c_v = 5R/2$, and for non-linear molecules, $c_v = 3R$.

To obtain the heat absorbed by the system during the isochoric process of heating from the initial temperature T_i to the final temperature $T_f > T_i$, we integrate $đQ$:

$$Q = \int_{T_i}^{T_f} đQ = n\int_{T_i}^{T_f} c_v(T, v)\mathrm{d}T. \tag{3.14}$$

If $T_f < T_i$ then $Q < 0$, i.e., heat is given off to the surroundings. If c_v does not depend on temperature, or that dependence can be neglected for the given temperature range, we get

$$Q = nc_v(T_f - T_i). \tag{3.15}$$

The heat Q is equal to the change in the internal energy of the system: $Q = \Delta U$.

3.3 Isobaric Process

The isobaric process occurs at constant pressure. Here we consider a quasi-static process, which means that the pressure in the system is equal to the external pressure during the process. The flow of heat between the system and surroundings in this process causes the volume of the system to change. It is very easy to calculate work in the isobaric process, i.e.,

$$W = \int_{V_i}^{V_f} đW = -p \int_{V_i}^{V_f} \mathrm{d}V = -p(V_f - V_i), \tag{3.16}$$

where V_i and V_f denote the system volume in the initial and final state, respectively. Since the pressure is constant, from expression (3.5) we get

$$đQ = \mathrm{d}U + p\mathrm{d}V = \mathrm{d}(U + pV). \tag{3.17}$$

The quantity

$$H = U + pV \tag{3.18}$$

is called *enthalpy*. Enthalpy, as the internal energy, is a function of state and an extensive quantity, and its unit is *joule* (J). Now relation (3.17) can be expressed as follows:

$$đQ = \mathrm{d}H. \tag{3.19}$$

Comparing (3.19) with (3.7), we see that the infinitesimal amount of heat in the isochoric process is equal to the differential of the internal energy, whereas in the isobaric process, it is equal to the differential of enthalpy.

3.3.1 Heat Capacity at Constant Pressure

The heat capacity at constant pressure, C_p, is defined in a similar way as C_V, i.e.,

$$C_p = \left(\frac{đQ}{\mathrm{d}T}\right)_p. \tag{3.20}$$

In this case, it is natural to take the temperature and pressure as the independent parameters of state. We notice that the equality $C_p > C_V$ must hold because in the isobaric process, only a part of the heat absorbed by the system raises its temperature. The rest of it is used to perform work against the external pressure p when the system expands due to the heating. Treating the enthalpy as a function of T and p, we get

$$\mathrm{d}H = \left(\frac{\partial H}{\partial T}\right)_p \mathrm{d}T + \left(\frac{\partial H}{\partial p}\right)_T \mathrm{d}p. \tag{3.21}$$

In the isobaric process $\mathrm{d}p = 0$, thus, using (3.21) and (3.19), we get

$$đQ = \left(\frac{\partial H}{\partial T}\right)_p \mathrm{d}T. \tag{3.22}$$

Finally, substituting (3.22) into (3.20), we obtain

$$C_p = \left(\frac{\partial H}{\partial T}\right)_p. \tag{3.23}$$

The enthalpy is an extensive state function, hence, the heat capacity $C_p = C_p(T, p)$ is also a state function and an extensive quantity. Therefore, as in the case of C_V, we define the molar heat capacity at constant pressure, $c_p = C_p/n$.

The enthalpy of the ideal gas results from the equations of state: $U = fnRT/2$ and $pV = nRT$, hence

$$H = \left(\frac{f}{2} + 1\right) nRT. \tag{3.24}$$

Substituting (3.24) into (3.23) and dividing by n, we get

$$c_p = \left(\frac{f}{2} + 1\right) R. \tag{3.25}$$

For monatomic molecules, $c_p = 5R/2$, for linear molecules, $c_p = 7R/2$, and for other molecules, $c_p = 4R$. Comparing (3.25) with (3.13), we find the following simple relation between c_v and c_p for the ideal gas:

$$c_p - c_v = R. \tag{3.26}$$

The heat absorbed by the system during the process of heating or cooling at constant pressure p is equal to the change in the enthalpy of the system: $Q = \Delta H$. It can be expressed in terms of c_p as follows:

$$Q = \int_{T_i}^{T_f} đQ = n \int_{T_i}^{T_f} c_p(T, p) \mathrm{d}T, \tag{3.27}$$

where T_i and T_f denote the initial and final temperature, respectively. If c_p does not depend on temperature we get

$$Q = nc_p(T_f - T_i). \tag{3.28}$$

3.4 Adiabatic Process

3.4.1 Reversible Adiabatic Process

In the adiabatic process, there is no transfer of heat between the system and surroundings ($đQ = 0$), and the change in the internal energy is equal to the work performed on the system:

$$\mathrm{d}U = đW. \tag{3.29}$$

The work performed on the system increases its internal energy U as well as its temperature. The parameter of state that we can control in the adiabatic process is the volume V. Then the pressure and temperature are functions of V which we want to determine for the ideal gas.

The ideal gas is defined by the two equations of state: $pV = nRT$ and $U = fnRT/2$. Here, the number of moles n is constant. Note that the term nRT in U can be replaced with the product pV, hence, $U = fpV/2$. Since $đW = -p\mathrm{d}V$, relation (3.29) adopts the following form:

$$\frac{f}{2}\mathrm{d}(pV) = -p\mathrm{d}V. \tag{3.30}$$

To the differential of a product, the usual rules of the differential calculus apply, i.e.,

$$\frac{f}{2}\mathrm{d}(pV) = \frac{f}{2}(p\mathrm{d}V + V\mathrm{d}p) = -p\mathrm{d}V, \tag{3.31}$$

hence

$$(f+2)p\mathrm{d}V + fV\mathrm{d}p = 0. \tag{3.32}$$

Dividing both sides of the last equation by pV, we get

$$(f+2)\frac{\mathrm{d}V}{V} + f\frac{\mathrm{d}p}{p} = \mathrm{d}\ln\left[\left(\frac{V}{V_0}\right)^{f+2}\left(\frac{p}{p_0}\right)^{f}\right] = 0, \tag{3.33}$$

where V_0 and p_0 denote the volume and pressure of a reference state, respectively. When the differential of a function vanishes, the function must be equal to a constant. Thus, the relation between the pressure and volume in a quasi-static (reversible) adiabatic process in the ideal gas has the following form:

$$pV^{1+2/f} = p_0V_0^{1+2/f} = const, \tag{3.34}$$

where the constant can be obtained from the initial values of the pressure and volume. In the case of monatomic molecules, the exponent equals $5/3$. To obtain the relation between the temperature and volume in the adiabatic process, we multiply the equation $pV = nRT$ by $V^{2/f}$, which leads to the following formula:

$$TV^{2/f} = T_0V_0^{2/f} = const, \tag{3.35}$$

where $T_0 = p_0V_0/(nR)$.

3.4.2 Irreversible Adiabatic Process at Constant Pressure

The system we want to consider is shown in Fig. 3.1. A vessel of the total volume V_{tot} is divided into two parts by a movable piston. One part is occupied by a gas and the other part is empty. The whole vessel is isolated adiabatically. We assume that the piston is coupled to a mechanical device which can perform work on the system or receive work from the system without loss of energy in the form of heat. In addition, we assume that processes taking place in the device are always quasi-static, independently of the process occurring in the system coupled to it. We call such a device a *reversible work source*.

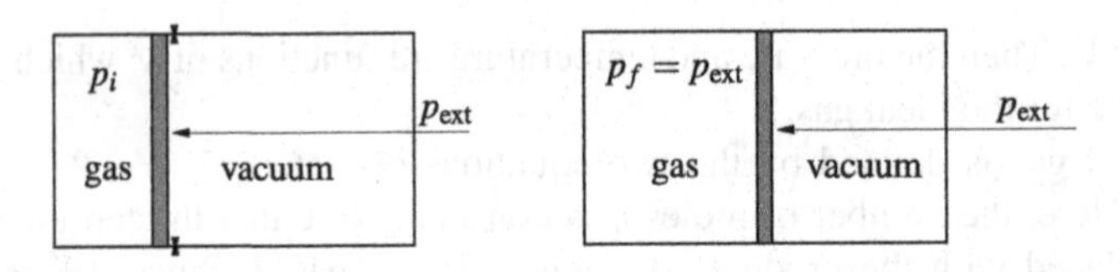

Fig. 3.1 The vessel is isolated adiabatically and the initial gas pressure amounts to p_i. An external constant force acts on the piston, exerting the pressure $p_{ext} < p_i$ on the gas (*left figure*). After the piston is released, it moves with accelerated motion. If the volume of the vessel is sufficiently large then the final gas pressure p_f is equal to p_{ext} (*right figure*). The other possibility is that $p_{ext} < p_f < p_i$ and the gas fills the whole volume of the vessel. The process described is irreversible. The gas pressure is well defined only at the beginning and at the end of the process when the system is in thermodynamic equilibrium

In Fig. 3.1, the reversible work source acts on the piston with a constant external force F_{ext}, hence, the pressure exerted on the gas equals $p_{ext} = F_{ext}/A$, where A is the piston area. An adiabatic process at constant pressure is possible only if the initial gas pressure p_i differs from p_{ext}. Initially the piston is blocked and the gas is in thermodynamic equilibrium in the volume V_i and at the pressure $p_i > p_{ext}$. When the piston is released the gas starts to expand and reaches eventually a new equilibrium state in the volume V_f and at the pressure p_f, where $p_{ext} \leq p_f < p_i$. The equality $p_f = p_{ext}$ holds only if the initial volume of the empty part of the vessel is sufficiently large. If we decreased p_{ext} at constant values of V_{tot}, V_i and p_i, we would get $V_f = V_{tot}$ at a certain value of p_{ext}, i.e., the whole vessel would be filled with the gas.

The parameters of the final equilibrium state, V_f and p_f, are obtained from the first law of thermodynamics:

$$U_f - U_i = U(V_f, p_f) - U(V_i, p_i) = W. \tag{3.36}$$

Note that the process is not quai-static because the gas pressure is not equal to the external pressure p_{ext} during the process, and actually the former is not even well defined. The piston moves with accelerated motion and regions of a larger and smaller density form in the gas. Since the piston gains kinetic energy during its motion, it does not stop when the external force becomes equal to the force exerted by the gas. It can perform a few oscillations damped by internal friction present in the gas. Eventually, the piston stops because its kinetic energy is dissipated. The work done by the gas, to increase the kinetic energy of the piston, is then dissipated into the gas, thus, it does not contribute to the change in the internal energy of the gas. The only contribution comes from the work done by the gas on the reversible work source, which by definition is quasi-static, hence

$$W = -p_{ext}\Delta V = -p_{ext}(V_f - V_i). \tag{3.37}$$

From (3.36) and (3.37), we get

$$U(V_f, p_f) - U(V_i, p_i) = -p_{ext}(V_f - V_i). \tag{3.38}$$

Now we have two possibilities: either $p_f = p_{ext}$ in the final state and V_f is determined from (3.38), or $V_f = V_{tot}$ and we determine $p_f > p_{ext}$ from (3.38). In particular, if $p_{ext} \to 0$ then $V_f = V_{tot}$ for a sufficiently small value of p_{ext}, which means that $W = -p_{ext}(V_{tot} - V_i) \to 0$. In the limiting case $p_{ext} = 0$, the gas expands freely to the vacuum. Since it performs no work and the process is adiabatic, the internal energy of the gas does not change.

3.5 Isothermal Process

3.5.1 Reversible Isothermal Process

The isothermal process occurs at constant temperature. As the second parameter of state we choose the volume. Changing the piston position, we can compress or expand the gas in the vessel. A change in the piston position results in a change in the gas pressure. Here we consider a quasi-static (reversible) process, which means that at any moment the external force acting on the piston is balanced by the force exerted on the piston by the gas. Thus, we can calculate the work W done on the system if we know the equation of state of the gas, i.e., $p = p(T, V)$; the work is given by

$$W = \int_{V_i}^{V_f} đW = -\int_{V_i}^{V_f} p(T, V)\mathrm{d}V. \tag{3.39}$$

In the ideal gas case, $p = nRT/V$, hence

$$W = -nRT \int_{V_i}^{V_f} \frac{\mathrm{d}V}{V} = nRT \ln \frac{V_i}{V_f}. \tag{3.40}$$

If the initial volume V_i is greater than the final volume V_f then $W > 0$. Since the internal energy of the ideal gas depends only on temperature (the mole number is assumed constant), $\Delta U = 0$ in the isothermal process, hence

$$Q = -W = nRT \ln \frac{V_f}{V_i}. \tag{3.41}$$

Thus, the amount of energy absorbed by the system in the form of work is given off in the form of heat. When the system absorbs heat from the surroundings at constant temperature it simultaneously performs the work $W^* = -W$ equal to the amount of heat absorbed. Note that the relation $Q = -W$ is not satisfied, in general, because the internal energy of other systems usually depends on the volume, thus, $\Delta U \neq 0$ in the isothermal process.

3.5.2 Irreversible Isothermal Process at Constant Pressure

Now we assume that the ideal gas expands at constant temperature but the external force acting on the piston is constant and given by the condition

$$p_{ext} = p_f = \frac{nRT}{V_f}. \tag{3.42}$$

During the process the external pressure p_{ext} does not balance the gas pressure. Such a process is not quasi-static, thus, it is not reversible. However, if the piston is coupled to a reversible work source (see Sect. 3.4.2) we can calculate the work done by the gas in this process:

$$W^*_{\text{irr}} = p_{\text{ext}} \Delta V = nRT \left(1 - \frac{V_i}{V_f}\right), \tag{3.43}$$

where the index 'irr' refers to irreversible processes. We can now compare W^*_{irr} with the work done by the system in the reversible isothermal process, i.e.,

$$W^* = nRT \ln \frac{V_f}{V_i}. \tag{3.44}$$

For $V_i < V_f$, we have $x = V_i/V_f < 1$, $W^*/nRT = -\ln x$ and $W^*_{\text{irr}}/nRT = 1 - x$. Since $-\ln x = \int_x^1 \mathrm{d}x'/x'$, $1 - x = \int_x^1 \mathrm{d}x'$ and $1/x' > 1$ for $0 < x' < 1$, the inequality $-\ln x > 1 - x$ holds, hence,

$$W^* > W^*_{\text{irr}}. \tag{3.45}$$

In both processes the initial and final states are the same, hence, the change in the internal energy is also the same, and for the ideal gas it is $\Delta U = 0$ because T is constant. We see, however, that in the reversible isothermal process the work performed by the system is greater than in the irreversible process. Although this result has been obtained for the ideal gas, in Sect. 5.3.1 we show that it is a general conclusion, which follows from the second law of thermodynamics.

3.6 Evaporation of Liquids

An example of a reversible process occurring at constant pressure and constant temperature is evaporation of a liquid. We consider a vessel containing a liquid, which is closed with a movable piston at the top. The temperature of the liquid is equal to the ambient temperature. Then we shift the piston up slowly. The space that forms above the liquid surface is filled with the vapour which is in thermodynamic equilibrium with the liquid; it is called *saturated vapour*. In order to reach thermodynamic equilibrium between the liquid and vapour at the given temperature T, the external pressure p must be equal to the saturated vapour pressure at this temperature. When a small amount of heat is supplied to the system at constant pressure a certain amount of the liquid evaporates, which is shown in Fig. 3.2. The volume of the system increases but its temperature does not change. A new equilibrium state between the vapour and liquid, with a greater amount of the vapour, is established.

We assume that the process of evaporation proceeds quasi-statically. The work done in this process amounts to

$$W = -p(V_g - V_l), \tag{3.46}$$

where V_g denotes the volume of the vapour formed from the liquid that has occupied the volume V_l. The work $W < 0$ because the system increases its volume. At the

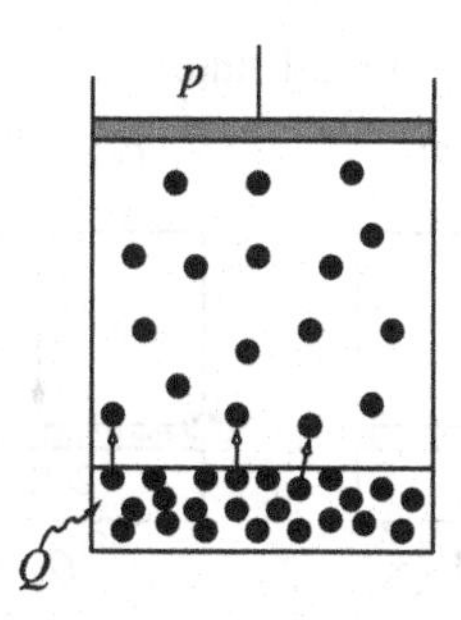

Fig. 3.2 Liquid evaporates at constant external pressure, p, equal to the saturated vapour pressure at a given temperature. When the heat Q is supplied to the system some molecules escape from the liquid surface. Leaving the liquid, they perform the work W against the vapour pressure, due to collisions with molecules present already in the vapour, which is marked with the arrows. An increase in the vapour volume shifts the piston upward. The internal energy increases in the evaporation process because the average distance between molecules grows and, in effect, the intermolecular attraction weakens

same time the system absorbs the heat Q from the surroundings because molecules to leave the surface of the liquid must increase their kinetic energy first. According to the first law of thermodynamics, the change in the internal energy of the system amounts to

$$\Delta U = Q - p(V_g - V_c) \approx Q - pV_g, \tag{3.47}$$

where we have neglected the volume of the liquid since $V_g \gg V_l$.

For example, to evaporate 1 mol of water at the temperature of 25 °C, 44 kJ of heat needs to be supplied. The saturated vapour pressure at this temperature amounts to 3.17 kPa. Assuming that water vapour can be approximated by the ideal gas, we get for 1 mol of the vapour

$$pV_g = RT \approx 2.48 \text{ kJ}.$$

Note that the work done by the system in the evaporation process is much smaller than the heat supplied.

3.7 Chemical Reaction

If a chemical reaction occurs in a liquid or solid then the typical change in the volume caused by the reaction is of the order 0.01 L per mol, and the work performed at the pressure of 1 atm is of the order 1 J per mol. The amount of heat released or absorbed depends on the given reaction and it amounts from a few kilojoules to a few hundred kilojoules. Thus, $\Delta U \approx Q$ in chemical reactions because the energy of chemical bonds is much larger than the work needed to change the volume of a liquid, solid or gas. If gases participates in a reaction which occurs at constant pressure, the ratio of the work performed to the heat released in the reaction amounts to a few percent only.

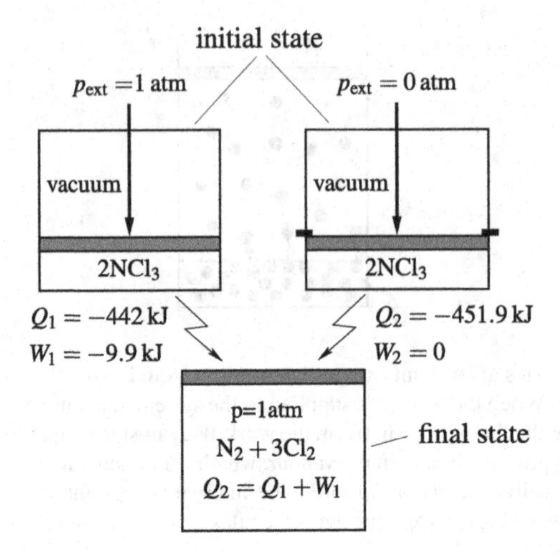

Fig. 3.3 Decomposition of 2 mol of liquid nitrogen trichloride into gaseous products. In the initial state, the pressure in the liquid under the piston is equal to 1 atm, whereas the space above the piston is empty (vacuum). In the final state, the whole volume of the vessel is filled with gaseous products of the reaction at the pressure $p = 1$ atm. In the first process (*left figure*), the reaction proceeds at constant pressure $p_{\text{ext}} = 1$ atm, maintained by the external force acting on the piston. In the second process (*right figure*), the reaction proceeds at $p_{\text{ext}} = 0$ when the piston is released. In both processes, the initial and final states are the same, hence, ΔU must also be the same

We consider the decomposition of 2 mol of liquid nitrogen trichloride into gaseous products (Fig. 3.3):

$$2\text{NCl}_3 \rightarrow \text{N}_2 + 3\text{Cl}_2. \tag{3.48}$$

The liquid reactant and gaseous products occupy the volume of 0.148 L and 97.9 L, respectively, at the pressure of 1 atm and temperature of 25 °C. If we maintain the system at constant pressure, for instance, the vessel containing the reactant is closed with a piston, then the amount of heat $Q_1^* = -Q_1 = 442$ kJ is released and the work $W_1^* = -W_1 = 9.9$ kJ is done by the system. During the reaction the system performs work, shifting the piston, and at the same time, it gives off a certain amount of heat to the surroundings. The motion of the piston is caused by the gases produced in the reaction and we know that the volume of the system at the end of the process amounts to 97.9 L. Using the first law of thermodynamics, we conclude that the internal energy has decreased by

$$\Delta U = Q_1 + W_1 = -451.9 \text{ kJ}. \tag{3.49}$$

We can also determine the change in the internal energy in the case of the reaction at the pressure $p_{\text{ext}} = 0$. Then, comparing with the previous case, we can calculate what part of the change in the internal energy of the system comes from the work performed during the reaction. In order to do so, we prepare a system of the volume 97.9 L at the temperature of 25 °C and pressure of 1 atm. Then we block the piston, pomp out the air from above and then initiate the decomposition of NCl_3 with an electric spark, for instance, releasing the piston at the same time. When the reaction is completed the gaseous products occupy the whole volume of the vessel. The pressure in the vessel amounts to 1 atm, and the temperature is equal to 25 °C. No work has been done in the process since the force acting on the piston was equal to zero. Measuring the heat released to the surroundings, we can determine the change in the internal energy. The heat released in the process amounts to $Q_2^* = -Q_2 = 451.9$ kJ, hence,

$$\Delta U = Q_2 = -451.9 \text{ kJ}. \tag{3.50}$$

The above example is illustrative of the fact that the internal energy is a state function, thus, it does not depend on the process that brings the system to the final state. In the first process, the heat $Q_1^* = 442$ kJ is released to the surroundings and the work $W_1^* = 9.9$ kJ is performed by the system. In the second process, the heat $Q_2^* = 451.9$ kJ is released, and the work $W^* = 0$. However, in both processes the change in the internal energy is the same because the initial and final states are the same.

Chemical reactions are usually studied at constant pressure, therefore, it is rather inconvenient to use the internal energy for this purpose. We return to this point in Part III.

3.8 Exercises

3.1 One mole of a monatomic ideal gas is maintained at constant pressure of 1 bar. How much heat is to be supplied to the system to increase its volume from 20 L to 50 L?

3.2 Five moles of a monatomic ideal gas expand at constant external pressure of 1 atm. The initial gas pressure is equal to 2 atm, the initial temperature amounts to 25 °C, and the final temperature amounts to 20 °C. How much work is done by the gas? What is the change in its internal energy? How much heat is transferred from the surroundings to the system?

3.3 Two subsystems, of the volume 10 L and 2 L, respectively, contain an ideal gas in thermodynamic equilibrium. The subsystems are separated from each other with a piston. The gas temperature is constant and equal to 0 °C. The first subsystem contains 10 mol of the gas. Then we increase twice the volume of the vessel with the gas in a reversible isothermal process. Find the change in the internal energy and the work performed by each subsystem. How much heat is supplied to the whole system?

3.4 A system at constant volume V consists of three subsystems occupied by a monatomic ideal gas. The subsystems are separated from one another with pistons. The initial volume of individual subsystems amounts to $V_i^{(1)}$, $V_i^{(2)}$ and $V_i^{(3)}$, respectively. The initial temperature T_i and initial pressure p_i have the same values in each subsystem. Then we change the temperature of each subsystem to the value T_f. Find the change in the internal energy and volume, and the work performed, for each subsystem. Find also the heat supplied to the whole system.

3.5 An ideal gas in thermodynamic equilibrium occupies three subsystems, of the volume 1 L, 5 L and 10 L, respectively, separated from one another with pistons. In all subsystems, the temperature is constant and equal to 0 °C, and the initial pressure amounts to $p = 1$ atm. Then a certain amount of heat is supplied to the system in a reversible isothermal process, which causes the volume of the first subsystem to increase to the final value of 4 L. Assuming that the remaining subsystems are in equilibrium with the first subsystem, calculate the work performed and the heat supplied.

3.6 A gas satisfying the van der Waals equation of state

$$p = nRT/(V - nb) - an^2/V^2,$$

with the internal energy

$$U = 3nRT/2 - an^2/V,$$

where a and b are positive constants, expands adiabatically to the vacuum from the initial volume V_i to infinite volume ($V_f = \infty$). The number of moles n is constant. What work does the gas perform? How much heat is supplied to the system? What is the change in its internal energy and temperature?

3.7 One mole of a monatomic ideal gas expands adiabatically against constant external pressure p_{ext}. The change in the gas temperature equals ΔT. Calculate the work done by the gas and the change in its volume.

3.8 A monatomic ideal gas expands in a reversible adiabatic process. The initial temperature of the gas amounts to T_i and the mole number is equal to n. The final temperature $T_f = T_i/4$. Calculate the work done by the gas and the change in its volume.

3.9 Calculate the final pressure of a monatomic ideal gas in a reversible adiabatic process if the initial pressure $p_i = 1$ bar, initial volume $V_i = 500$ cm^3 and final volume $V_f = 2000$ cm^3.

3.10 Calculate the change in the internal energy of 1 mol of H_2O in the process of transformation of liquid water into vapour at the pressure of 1 atm. The heat of evaporation amounts to 40670 J mol^{-1} and the molar volume of water vapour in equilibrium with the liquid amounts to 30.6 L at 1 atm.

3.11 The reaction of ammonia formation from 3 mol of hydrogen and 1 mol of nitrogen has the following form:

$$N_2 + 3H_2 \rightarrow 2NH_3.$$

Knowing that during this (exothermic) reaction, which takes place at constant external pressure $p_{ext} = 1$ atm, the system gives off 92 kJ of heat to the surroundings, calculate the change in its internal energy. Calculate also the work done by the system, assuming that hydrogen, nitrogen and ammonia are ideal gases. The temperature at the beginning and at the end of the process amounts to 298 K.

3.12 Two isolated systems: (1) and (2), contain two different ideal gases whose internal energies are given by, respectively,

$$U^{(1)}(T, V, n) = \frac{3}{2}nRT,$$
$$U^{(2)}(T, V, n) = \frac{5}{2}nRT.$$

The total internal energy of the systems is equal to $U^{(1)} + U^{(2)} = 30$ kJ, and the amount of the gas in the systems is equal to $n^{(1)} = 2$ mol and $n^{(2)} = 3$ mol, respectively. Then the systems are brought into contact with each other through a stiff diathermal wall. Find the final values of the temperature and internal energy for each system.

3.13 Two isolated systems, of the volume V each, contain a photon gas whose internal energy $U = \gamma V T^4$. The initial temperature of the photon gas is equal to $T^{(1)}$ and $T^{(2)}$, respectively. Then the systems are brought into thermal contact with each other through a stiff diathermal wall. Find the final temperature of the systems.

3.14 A metal block, of the mass of 1 kg, was heated to a temperature of 400 K and then put into 0.3 kg of water. The temperature of water increased from 294 K to 300 K. Calculate the ratio of the specific heat of the metal and water, assuming that they do not depend on temperature (specific heat is the heat capacity of 1 kg of a substance). What properties of a substance determine its cooling capabilities?

3.15 The internal energy of $n = 4$ mol of a substance is given by the following expression: $U = ATV^3$, where $A = 10\ \mathrm{J\,K^{-1}\,cm^{-9}}$. Derive the formula for the internal energy of the substance for an arbitrary value of the mole number n.

3.16 $U = aVT^4$ for $n = 2$ mol, where $a = 10\ \mathrm{J\,K^{-4}\,cm^{-3}}$. Derive the formula for U for an arbitrary value of n.

3.17 An electric current of 1 A flows through a heating coil, connected to the voltage of 12 V, for 3000 s. The whole work of the electric current is transformed into heat in the heating coil and used to warm up a certain substance by 5.5 K. What is the heat capacity of the substance?

3.18 The equation of state of a certain gas has the following form:

$$\frac{pV}{nRT} = 1 + \frac{nB(T)}{V},$$

where the function $B(T)$ is positive at high temperatures and negative at low temperatures. Calculate the work performed by the gas during a reversible isothermal expansion, and compare it with the work of the ideal gas at the same temperature.

3.19 Calculate the work performed by a gas which satisfies the van der Waals equation of state

$$p = \frac{nRT}{V - nb} - \frac{an^2}{V^2},$$

during a reversible isothermal expansion. Compare it with the work performed by the ideal gas under the same conditions. How do the constants a and b affect the pressure and work? Do they increase or decrease these quantities compared to the ideal gas case? Expand $1/(V - nb)$ in a power series of nb/V and compare the resultant equation of state with the equation of state from Exercise 3.18.

3.20 The molar heat capacity at constant volume, c_v, of a substance is given by the formula

$$c_v = A + BT,$$

where A and B are some constants. Calculate the change in the internal energy of the substance in the process of isochoric heating from the temperature T_i to T_f.

3.21 The molar heat capacity at constant volume, c_v, of a substance is given by the formula

$$c_v = A + BT - CT^{-2},$$

where A, B and C are some constants. Calculate the change in the internal energy of the substance in the process of isochoric heating from the temperature T_i to T_f.

3.22 The heat capacity per unit volume, c_v, for the photon gas is given by the following formula:

$$c_v = 4\gamma T^3,$$

where γ is a constant. Calculate the change in the internal energy of the photon gas contained in the volume V, in the process of isochoric heating from the temperature T_i to T_f.

3.23 A vessel of the volume V contains a two-atomic ideal gas at the pressure p_i and temperature T_i. Calculate how much heat can be delivered to the vessel without the risk of damaging it if the resistance of the walls amounts to p_1. Substitute the following data: $V = 24$ L, $p_i = 1$ atm, $T_i = 298$ K, $p_1 = 100$ atm. If the vessel contained hydrogen its damage might have tragic consequences, since the reaction: $H_2 + \frac{1}{2}O_2 \rightarrow H_2O$ is explosive, that is, it proceeds very fast with a great amount of heat released.

3.24 A vessel of constant volume contains a monatomic ideal gas at constant temperature T. We let Δn moles of the gas out of the vessel. Calculate the change in the internal energy of the system.

3.25 A monatomic ideal gas is contained in a vessel at constant pressure p and constant temperature T. We let Δn moles of the gas out of the vessel. Calculate the change in the internal energy of the system, the work performed and the heat supplied.

Chapter 4
Entropy and Irreversibility of Thermodynamic Processes

4.1 Second Law of Thermodynamics

The second law of thermodynamics was formulated due to the observation of spontaneous processes occurring in nature. We know that water freezes in winter but never in summer. Heat flows from a warmer body to a cooler body (Fig. 2.2). A gas expands spontaneously, filling the whole vessel uniformly, when its pressure is higher than the external pressure. We never observe molecules that collect spontaneously in one part of the vessel, leaving the remaining part empty (Fig. 2.1). These examples show that some processes occur spontaneously and some other do not. We are used to this fact so much that we do not even wonder why it happens like that. When we push a block on a rough surface we observe a flow of heat to the surroundings due to the friction (Fig. 2.6). Imagine now a hypothetical situation that the heat given off to the surroundings flows back to the block and then it is transformed somehow into mechanical work, which shifts the block to its original position. Such a process would not violate the energy conservation principle, but we know very well that it does not occur in nature. Why is it impossible to transform chaotic motion of atoms of the substrate into ordered motion of the block? What is behind the fact that some processes never occur in nature, even though they do not violate the energy conservation principle? These and many other examples of observation of various macroscopic systems indicate that there exists a fundamental law responsible for irreversibility of processes occurring in nature. The search for the answers to such fundamental questions led to the formulation of the second law of thermodynamics. Implications of this law go beyond our understanding of the processes discussed above. They concern the problem of the very existence of time and the arrow of time, giving a partial answer to the question why the flow of time is unidirectional. The last problem is not discussed in this book, however. To express in the language of thermodynamics the empirical fact that some processes occur spontaneously, i.e., without an external agent, and some other do not, it is necessary to introduce a new state function, called *entropy*, for which the symbol S is usually used. Entropy is an extensive quantity, as are the internal energy U, volume V, and the amount of substance n. For a pure homogeneous substance, such as gas,

R. Hołyst, A. Poniewierski, *Thermodynamics for Chemists, Physicists and Engineers*,
DOI 10.1007/978-94-007-2999-5_4, © Springer Science+Business Media Dordrecht 2012

U, V and n uniquely define the state of the system in thermodynamic equilibrium. Therefore, we can treat S as a function of these state parameters. For the time being, we are not concerned with the determination of the entropy of a given system, but only with its general properties.

The Second Law of Thermodynamics
There exists an extensive state function, called entropy, S, whose change in an adiabatic process satisfies the inequality

$$\Delta S \geq 0,$$

where $\Delta S = 0$ only if the process is reversible.

The second law of thermodynamics states only that entropy cannot decrease in adiabatic processes. The statement that entropy never decreases is therefore incorrect. It is true only if heat cannot be transferred between the system and surroundings. We will see later that in the case of diathermal walls, an infinitesimal change in the entropy of the system in a reversible process is proportional to the heat transferred to the system, and the proportionality coefficient is the inverse of its temperature. This means that when heat is transferred from the system to the surroundings in a reversible isothermal process the entropy of the system decreases.

Entropy is a well defined quantity for equilibrium states only. Let us assume that we can determine the entropy of an adiabatically isolated system in two equilibrium states: A and B. If $S_B > S_A$, we can say, on the basis of the second law of thermodynamics, that a spontaneous process from the state A to the state B is possible. Thermodynamics does not tell us how to realize such a process but it allows to foresee whether a given process can occur spontaneously.

As an example, we consider an adiabatically isolated vessel with a movable piston. One part of the vessel is filled with a gas and the other part is empty (see Fig. 3.1). There is no external force acting on the piston. Initially the piston is blocked and the gas pressure is equal to p_i. When the blockade is removed the gas starts to expand and the expansion continues until a new equilibrium state, of the pressure $p_f < p_i$, is reached. In the new state, the whole vessel is filled with the gas. In this process, neither heat is transferred to the surroundings nor work is performed by the system, which means that the state of the surroundings does not change. The process is irreversible because in order to restore the system to its initial state an external agent is needed. We know intuitively that the gas will not return to the initial state spontaneously. If we calculate the initial and final entropy of the system we will find out that $S_f > S_i$. We will come to the same conclusion if we include a constant force acting on the piston, which exerts the pressure $p_{\text{ext}} < p_i$ on the gas. Then the gas performs the work $W^* = p_{\text{ext}}(V_f - V_i)$ on the surroundings (see Sect. 3.4.2). Let us assume that the surroundings can give back the whole energy stored in them, by performing the work $W = -W^*$ on the system. Since the final gas pressure $p_f \geq p_{\text{ext}}$, the work W is too small to restore the system to its initial state. This is because the kinetic energy of the piston, which moves with accelerated motion, is dissipated into the gas due to the internal friction (viscosity).

If an adiabatic process proceeds quasi-statically then the pressure exerted on the piston by the gas is balanced by an external force acting on the piston. The direction of the process can be reversed at any moment, therefore such a process is reversible. According the second law of thermodynamics the entropy of the system does not change during a reversible process, hence, $S_f = S_i$. We showed in Sect. 3.4 that in the ideal gas case, T and V satisfies the relation: $TV^{2/f} = const$ for a quasi-static adiabatic process (see (3.35)). Since the temperature is proportional to the internal energy of the ideal gas, we can rewrite this relation as $UV^{2/f} = const'$. If the entropy is to be constant in this process it must depend on U and V through the variable $UV^{2/f}$. We will show this in Sect. 4.3.3, where we calculate the entropy of the ideal gas.

Many conclusions concerning a given system can be drawn from the second law of thermodynamics by the following reasoning. We consider the system and its direct surroundings, with which it can interact, as one isolated system. Thus, our system and its surroundings are subsystems of that system, to which we assign indices (1) and (2), respectively. Since entropy is an extensive quantity, the total entropy of the system, S, is a sum of the entropy of its subsystems, denoted $S^{(1)}$ and $S^{(2)}$. We assume that initially the subsystems are isolated from each other and that both are in thermodynamic equilibrium. Then we allow them to interact with each other, for instance, by a transfer of heat. After some time a new equilibrium state of the whole system is established. Since an isolated system is also an adiabatically isolated one, the second law of thermodynamics applies to it. Therefore, the total change in the entropy of the whole system must satisfy the inequality $\Delta S \geq 0$, hence

$$\Delta S = \Delta S^{(1)} + \Delta S^{(2)} \geq 0. \tag{4.1}$$

In the case $\Delta S = 0$, the system proceeds from the initial state to the final state through a sequence of equilibrium states, i.e., the process is reversible. $\Delta S > 0$ means that some irreversible changes have occurred in the system or surroundings. In that case, the system and surroundings can be restored to their original states only if some additional energy is supplied.

Often an additional assumption is made that the surroundings (subsystem (2)) is much larger than the original system (subsystem (1)). This means that any interaction of the original system with the surroundings does not change practically the intensive parameters of the surroundings. For instance, the heat capacity of the surroundings is so large that any transfer of heat from the system to the surroundings does not change practically the temperature of the latter. Similarly, when the system performs work on the surroundings by changing its volume, the pressure in the surroundings does not change practically because of a great difference in the volume of the system and surroundings. A system whose heat capacity and volume can be considered infinite compared to those of any system studied is called a *heat and volume reservoir*. The atmosphere is a good example of such a reservoir.

4.1.1 Entropy Maximum Principle for Isolated Systems

Here we examine the consequences of the second law of thermodynamics in the case of an isolated system composed of m subsystems, each of which is initially also an isolated system. For simplicity, we assume that all subsystems contain the same pure substance. The equilibrium state of each subsystem is defined by its internal energy $U^{(i)}$, volume $V^{(i)}$ and mole number $n^{(i)}$, where $i = 1, \ldots, m$. Since the internal energy, volume and amount of substance are extensive parameters, we obtain the values of U, V and n for the composite system by summing up the contributions from all subsystems, i.e.,

$$U = \sum_{i=1}^{m} U^{(i)}, \tag{4.2}$$

$$V = \sum_{i=1}^{m} V^{(i)}, \tag{4.3}$$

$$n = \sum_{i=1}^{m} n^{(i)}. \tag{4.4}$$

We do the same with the entropy of the composite system, S, which is also an extensive parameter, hence

$$S = \sum_{i=1}^{m} S^{(i)}\big(U^{(i)}, V^{(i)}, n^{(i)}\big), \tag{4.5}$$

where $S^{(i)}$ is the entropy of the ith subsystem.

Then we remove all internal walls between the subsystems, leaving only the external walls which isolate the composite system from the surroundings. When the internal constraints are removed a flow of energy and matter usually occurs in the system, and the volume of the subsystems change. After some time the system reaches a new equilibrium state. In this process, U, V and n do not change because the system is isolated from the surroundings all the time. The entropy of the system in the final state, S_f, depends only on U, V and n, i.e., $S_f = S(U, V, n)$, and from the second law of thermodynamics, it follows that

$$\Delta S = S(U, V, n) - \sum_{i=1}^{m} S^{(i)}\big(U^{(i)}, V^{(i)}, n^{(i)}\big) \geq 0. \tag{4.6}$$

Note that $\Delta S = 0$ only if both the temperature and pressure had the same values in all subsystems before the removal of the internal walls, otherwise, the entropy of the composite system increases when the constraints are removed. From the above considerations the following conclusion can be drawn.

Entropy Maximum Principle
After the removal of internal constraints, present initially in an isolated system, the system reaches an equilibrium state that corresponds to the maximum of entropy on the set of all possible equilibrium states of the system with the constraints.

By the system with the constraints we understand an isolated system composed of subsystems isolated from one another with internal walls. Possible equilibrium states of the system are defined by the extensive parameters: $U^{(i)}$, $V^{(i)}$, $n^{(i)}$, for $i = 1, \ldots, m$, which satisfy conditions (4.2)–(4.4). The entropy of the composite system

$$S\big(U^{(1)}, V^{(1)}, n^{(1)}, \ldots, U^{(m)}, V^{(m)}, n^{(m)}\big) = \sum_{i=1}^{m} S^{(i)}\big(U^{(i)}, V^{(i)}, n^{(i)}\big), \tag{4.7}$$

assumes the maximum value for a certain set of the state parameters: $U^{(i)}_{\max}$, $V^{(i)}_{\max}$, $n^{(i)}_{\max}$, $i = 1, \ldots, m$. It follows from the entropy maximum principle (which is a consequence of the second law of thermodynamics) that this set of the state parameters corresponds to the equilibrium state reached by the composite system after the removal of the internal constraints.

Note that we do not have to remove all the internal constraints imposed on the system. For instance, we can replace the walls that completely isolate the subsystems from one another with diathermal walls. Then the volume and amount of substance in the subsystems do not change but the internal energy can change due to the flow of heat. The entropy maximum principle holds also in such cases. After the removal of some constraints, the system reaches a new equilibrium state that corresponds to the maximum of entropy in the presence of the remaining constraints.

The entropy maximum principle can be easily extended to an arbitrary number of components. Then condition (4.4) must be satisfied for each component.

4.2 Conditions of Thermodynamic Equilibrium

We show in this section that the conditions of thermodynamic equilibrium can be derived from the entropy maximum principle. First, we discuss the condition of thermal equilibrium, which leads to the thermodynamic definition of temperature and the relation between entropy and heat.

4.2.1 Thermal Equilibrium

We consider an isolated system composed of two subsystem: (1) and (2), which are initially isolated from each other. The equilibrium states of the subsystems are defined by the extensive parameters $U^{(i)}$, $V^{(i)}$, $n^{(i)}$, $i = 1, 2$, with the conditions:

$$U = U^{(1)} + U^{(2)}, \tag{4.8}$$
$$V = V^{(1)} + V^{(2)}, \tag{4.9}$$
$$n = n^{(1)} + n^{(2)}, \tag{4.10}$$

where U, V and n are constant. The entropy of the composite system is a sum of the entropy of its subsystems, i.e.,

$$S = S^{(1)}\big(U^{(1)}, V^{(1)}, n^{(1)}\big) + S^{(2)}\big(U^{(2)}, V^{(2)}, n^{(2)}\big). \tag{4.11}$$

Then the subsystems are brought into thermal contact through a rigid diathermal wall. The system reaches a new equilibrium state, and the internal energy of the subsystems changes, in general; due to condition (4.8) we have

$$\mathrm{d}U^{(1)} = -\mathrm{d}U^{(2)}. \tag{4.12}$$

Since

$$\mathrm{d}S^{(i)} = \left(\frac{\partial S^{(i)}}{\partial U^{(i)}}\right)_{V^{(i)}, n^{(i)}} \mathrm{d}U^{(i)}, \tag{4.13}$$

the change in the entropy of the system due to an infinitesimal transfer of energy amounts to

$$\mathrm{d}S = \mathrm{d}S^{(1)} + \mathrm{d}S^{(2)} = \big[s_u^{(1)}\big(u^{(1)}, v^{(1)}\big) - s_u^{(2)}\big(u^{(2)}, v^{(2)}\big)\big]\mathrm{d}U^{(1)}, \tag{4.14}$$

where

$$s_u^{(i)}\big(u^{(i)}, v^{(i)}\big) = \left(\frac{\partial S^{(i)}}{\partial U^{(i)}}\right)_{V^{(i)}, n^{(i)}} \tag{4.15}$$

and relation (4.12) was used. The quantity $s_u^{(i)}$ is an intensive state function, as a derivative of an extensive state function (entropy) with respect to an extensive parameter (internal energy), therefore it depends only on the molar quantities: $u^{(i)} = U^{(i)}/n^{(i)}$ and $v^{(i)} = V^{(i)}/n^{(i)}$.

From the entropy maximum principle, we know that the equilibrium state reached by the system after the removal of the internal constraints corresponds the maximum of the entropy S. In the case considered, the volume of the subsystems and the mole numbers do not change, and the only independent parameter is $U^{(1)}$. The necessary condition for the maximum (or minimum) of a function is that its first derivative vanishes, hence the condition $\mathrm{d}S = 0$ follows. Using this condition in (4.14), we find that the subsystems (1) and (2) are in thermal equilibrium with each other if

$$s_u^{(1)}\big(u^{(1)}, v^{(1)}\big) = s_u^{(2)}\big(u^{(2)}, v^{(2)}\big). \tag{4.16}$$

This means that relation (4.15) defines an intensive parameter which has the same value for the systems in thermal equilibrium. On the other hand, we already know from the zeroth law of thermodynamics that such a property has temperature (see Sect. 2.3.2), i.e.,

$$t^{(1)} = t^{(2)}, \tag{4.17}$$

where $t^{(i)}$ denotes the empirical temperature of the ith subsystem. Thus, relation (4.15) can be used as a thermodynamic definition of temperature, called the *thermodynamic temperature*, for which we use the symbol T.

This statement does not mean, however, that the thermodynamic temperature of the ith subsystem is equal to $s_u^{(i)}$, as we must take into account the convention used for ages that heat flows from the body of a higher temperature to the body of a lower temperature. To include this condition, we assume that the initial empirical temperatures of the subsystems were slightly different, and after the subsystems were brought into thermal contact, the internal energy of the subsystem (1) has increased slightly by $\Delta U^{(1)} > 0$. Because of (4.13) and (4.15) the change in the entropy of the ith subsystem amounts to

$$\Delta S^{(i)} \approx s_u^{(i)} \Delta U^{(i)}, \tag{4.18}$$

where $\Delta U^{(2)} = -\Delta U^{(1)}$. In the process considered, $\Delta U^{(1)} = Q$, where $Q > 0$ denotes the heat transferred from the subsystem (2) to the subsystem (1). According to the second law of thermodynamics, the change in the entropy of the whole system equals

$$\Delta S \approx \left(s_u^{(1)} - s_u^{(2)}\right) Q > 0, \tag{4.19}$$

since the process is irreversible. Because the heat flows from (2) to (1) we conclude that the higher temperature corresponds to the subsystem (2). Note that inequality (4.19) is satisfied if we assume $s_u^{(i)} = 1/T^{(i)}$, since $s_u^{(1)} > s_u^{(2)}$ for $T^{(1)} < T^{(2)}$. To summarize, we have shown that the following definition of the thermodynamic temperature:

$$\frac{1}{T^{(i)}} = \left(\frac{\partial S^{(i)}}{\partial U^{(i)}}\right)_{V^{(i)},n^{(i)}}, \tag{4.20}$$

satisfies both the condition of equal temperatures in the state of thermal equilibrium and the condition that the flow of heat occurs in the direction of decreasing temperature.

In Sect. 2.3.2, we introduced the absolute temperature as a positive quantity, on the basis of the ideal gas equation of state (see (2.27)). We notice that in order the thermodynamic temperature is consistent with the absolute temperature in this respect, we have to postulate that entropy is a monotonically increasing function of the internal energy of an isolated system at constant volume and at a constant mole number. If it was not so the thermodynamic temperature could assume negative values. However, only the mechanical-statistical definition of entropy given by Boltzmann:

$$S = k_B \ln W, \tag{4.21}$$

provides theoretical justification of this postulate. The quantity W denotes the number of all possible microscopic realizations (microstates) of a macroscopic isolated system in thermodynamic equilibrium, which are consistent with its parameters of state: U, V and n. If the state corresponding to the lowest internal energy is not degenerate then $W = 1$ and the entropy $S = 0$. It can be shown that W grows quickly when the internal energy of the system increases, which means that also its entropy increases. The coefficient k_B is called the *Boltzmann constant* and it is equal to the ratio of the gas constant R and the Avogadro constant N_A:

$$k_B = \frac{R}{N_A} = 1{,}380\,6504\,(24) \times 10^{-23}\ \mathrm{J\,K^{-1}}. \tag{4.22}$$

Relation Between a Reversible Flow of Heat and Entropy We have shown that if two subsystems of an isolated system are in thermal equilibrium then an infinitesimal transfer of energy between the subsystems does not change the entropy of the whole system (see (4.14) and (4.16)), which means that it proceeds in a reversible way. On the other hand, the entropy of each subsystem changes in accord with relation (4.13). In the case discussed above, the energy is transferred in the form of heat. We denote by $đQ$ the infinitesimal amount of heat transferred from the subsystem (2) to the subsystem (1), hence, $\mathrm{d}U^{(1)} = đQ$ and $\mathrm{d}U^{(2)} = -đQ$. Using relations (4.13) and (4.20) for $i = 1$, we get

$$\mathrm{d}S^{(1)} = \frac{1}{T} đQ, \tag{4.23}$$

where $T = T^{(1)} = T^{(2)}$. For the subsystem (2), we obtain a similar expression but the sign changes, i.e., $\mathrm{d}S^{(2)} = -đQ/T$.

Now we can treat the subsystem (1) as the system of our interest and the subsystem (2) as its surroundings. Omitting the index (1), we rewrite (4.23) as

$$\mathrm{d}S = \frac{1}{T} đQ. \tag{4.24}$$

This result means that if the system is in thermal equilibrium with the surroundings at the temperature T then the infinitesimal transfer of heat, $đQ$, between the system and surroundings proceeds in a reversible way, and the entropy of the system changes in accord with (4.24). As we know, a reversible process is quasi-static. A transfer of heat proceeds quasi-statically if during the process we control the temperature of the surroundings to be equal to the temperature of the system. Similarly, in the case of quasi-static work, we adjust the pressure in the surroundings to be equal to the pressure in the system. According to the first law of thermodynamics, we have (see (3.5))

$$dU = đQ - p\mathrm{d}V \tag{4.25}$$

in a quasi-static process at a constant mole number. Substituting $đQ$ from relation (4.24), we obtain

$$dU = T\mathrm{d}S - p\mathrm{d}V. \tag{4.26}$$

This relation means that the internal energy of the system is a function of its entropy and volume, at a constant mole number. Due to the postulate that entropy is a monotonically increasing function of the internal energy, we can invert the relation between the internal energy and entropy. Then S becomes a function of U and V, at constant n, and (4.26) can be expressed as follows:

$$\mathrm{d}S = \frac{1}{T}\mathrm{d}U + \frac{P}{T}\mathrm{d}V. \tag{4.27}$$

Relation (4.26) results from the combination of the first law of thermodynamics (Eq. (4.25)) with the expression for a quasi-static flow of heat, $đQ = T\mathrm{d}S$, which is a consequence of the second law. Now it is easy to show that any quasi-static process must be reversible. In a quasi-static adiabatic process, $đQ = 0$ by definition, hence, $\mathrm{d}S = đQ/T = 0$. Since the entropy does not change, the process is reversible. If a quasi-static process is not adiabatic we can include the surroundings, to form an isolated composite system together with our system. Then any quasi-static process in our system is a quasi-static adiabatic process in the composite system, therefore, it is a reversible process.

4.2.2 Mechanical Equilibrium

We show below that the condition of mechanical equilibrium can also be derived from the entropy maximum principle. As in Sect. 4.2.1, we consider an isolated system composed of two subsystems which are initially isolated from each other. Then we bring the subsystems into mechanical and thermal contact through a movable diathermal wall. The numbers of moles in the subsystems are constant but the internal energy and volume can change. From conditions (4.8) and (4.9), we have:

$$\mathrm{d}U^{(1)} = -\mathrm{d}U^{(2)}, \tag{4.28}$$

$$\mathrm{d}V^{(1)} = -\mathrm{d}V^{(2)}. \tag{4.29}$$

Applying (4.27) to the subsystems, we derive the following expression for the differential of the entropy of the composite system:

$$\mathrm{d}S = \mathrm{d}S^{(1)} + \mathrm{d}S^{(2)} = \left(\frac{1}{T^{(1)}} - \frac{1}{T^{(2)}}\right)\mathrm{d}U^{(1)} + \left(\frac{p^{(1)}}{T^{(1)}} - \frac{p^{(2)}}{T^{(2)}}\right)\mathrm{d}V^{(1)}, \tag{4.30}$$

where $T^{(i)}$ and $p^{(i)}$ denote, respectively, the temperature and pressure in the ith subsystem. After the removal of the constraints, the composite system reaches a new equilibrium state corresponding to the maximum of entropy. Since $\mathrm{d}U^{(1)}$ and $\mathrm{d}V^{(1)}$ are independent quantities, the following relations result from the condition $\mathrm{d}S = 0$:

$$T^{(1)} = T^{(2)}, \tag{4.31}$$

$$p^{(1)} = p^{(2)}. \tag{4.32}$$

They express equality of temperatures (thermal equilibrium) and pressures (mechanical equilibrium).

4.2.3 Equilibrium with Respect to the Matter Flow

As before, we consider an isolated system composed of two subsystems, isolated initially from each other. Then we bring them into contact through a rigid diathermal wall which is permeable to matter. We can realize this by making tiny holes in the wall, by which the molecules can get across. For simplicity, we consider only molecules of one kind, but the reasoning used can be easily extended to an arbitrary number of components. In such a case, we have to assume that the walls are permeable to molecules of one component but impermeable to molecules of all other components. After the removal of the internal constraints the volume of the subsystems does not change but heat and matter can flow from one subsystem to the other. From conditions (4.8) and (4.10), we get

$$dU^{(1)} = -dU^{(2)}, \tag{4.33}$$

$$dn^{(1)} = -dn^{(2)}. \tag{4.34}$$

The entropy of the ith subsystem, $S^{(i)}$, is a function of the internal energy $U^{(i)}$, volume $V^{(i)}$, and mole number $n^{(i)}$. Since the volume of the subsystems is constant, we have

$$dS^{(i)} = \left(\frac{\partial S^{(i)}}{\partial U^{(i)}}\right)_{V^{(i)},n^{(i)}} dU^{(i)} + \left(\frac{\partial S^{(i)}}{\partial n^{(i)}}\right)_{U^{(i)},V^{(i)}} dn^{(i)}. \tag{4.35}$$

We know already that the derivative of entropy withe respect to the internal energy is equal to $1/T^{(i)}$. The derivative with respect to the mole number is denoted as

$$\left(\frac{\partial S^{(i)}}{\partial n^{(i)}}\right)_{U^{(i)},V^{(i)}} = -\frac{\mu^{(i)}}{T^{(i)}}, \tag{4.36}$$

where $\mu^{(i)}$ is called the *chemical potential*. As temperature and pressure, the chemical potential is an intensive parameter of state. Now the differential of the entropy of the composite system adopts the following form:

$$dS = dS^{(1)} + dS^{(2)} = \left(\frac{1}{T^{(1)}} - \frac{1}{T^{(2)}}\right)dU^{(1)} - \left(\frac{\mu^{(1)}}{T^{(1)}} - \frac{\mu^{(2)}}{T^{(2)}}\right)dn^{(1)}. \tag{4.37}$$

After the removal of the internal constraints the system reaches a new equilibrium state that corresponds to the maximum of its entropy. Using the condition $dS = 0$ and the fact that $dU^{(1)}$ and $dn^{(1)}$ are independent quantities, we conclude that the subsystems are in equilibrium with each other if

$$T^{(1)} = T^{(2)}, \tag{4.38}$$

$$\mu^{(1)} = \mu^{(2)}. \tag{4.39}$$

The first equality expresses the condition of thermal equilibrium. If also the chemical potentials are equal then an infinitesimal amount of matter can flow from one subsystem to the other in a reversible way. This means that the subsystems are in equilibrium with respect to the matter flow.

Let us assume now that before the internal constraints were removed the temperatures of the subsystems were equal: $T^{(1)} = T^{(2)} = T$, but the chemical potentials differed slightly. Proceeding in a similar way as in the case of the heat flow, we express the change in the entropy of the ith subsystem after the removal of the constraints as follows:

$$\Delta S^{(i)} \approx -\frac{\mu^{(i)}}{T}\Delta n^{(i)}, \tag{4.40}$$

hence

$$\Delta S \approx -\frac{\mu^{(1)} - \mu^{(2)}}{T} \Delta n^{(1)}, \tag{4.41}$$

where we have used (4.10). Suppose now that initially $\mu^{(1)} > \mu^{(2)}$. Since the process is irreversible, $\Delta S > 0$ and we conclude that $\Delta n^{(1)} < 0$. This means that the matter flows from the subsystem (1), of a higher chemical potential, to the subsystem (2), of a lower chemical potential.

4.3 Entropy as a Function of State Parameters

4.3.1 Fundamental Relation of Thermodynamics

Our consideration concerning composite systems have been based on the assumption that we know the entropy of the subsystems as a function of the extensive parameters: the internal energy, volume and mole number. Application of the entropy maximum principle to an isolated system led us to the conditions of thermodynamic equilibrium and allowed to determine the intensive parameters related to each type of equilibrium: thermal, mechanical and with respect to the matter flow. In this section, we concentrate on simple systems, by which we mean uniform systems without any internal constraints. A gas in a vessel without internal walls can serve as an example of such a simple system. For simplicity, we consider only one-component systems. If a given simple system is isolated from the surroundings then its equilibrium state is completely defined by the internal energy U, volume V, and the amount of substance n. From the point of view of thermodynamics, complete information about the system is contained in the relation between its entropy S, and U, V and n, which is called the *fundamental relation*. The fundamental relation can be expressed as

$$S = S(U, V, n), \tag{4.42}$$

i.e., in the form of a functional dependence of the entropy on the internal energy, volume and amount of substance. The infinitesimal change in the entropy (its differential) has the following form (see (4.27) and (4.36)):

$$\mathrm{d}S = \frac{1}{T}\mathrm{d}U + \frac{p}{T}\mathrm{d}V - \frac{\mu}{T}\mathrm{d}n, \tag{4.43}$$

where

$$\left(\frac{\partial S}{\partial U}\right)_{V,n} = \frac{1}{T}, \tag{4.44}$$

$$\left(\frac{\partial S}{\partial V}\right)_{U,n} = \frac{p}{T}, \tag{4.45}$$

$$\left(\frac{\partial S}{\partial n}\right)_{U,V} = -\frac{\mu}{T}. \tag{4.46}$$

Due to the monotonic dependence of the entropy on the internal energy, Eq. (4.42) can be solved with respect to U for any S, at constant V and n. This leads to an alternative form of the fundamental relation:

$$U = U(S, V, n), \tag{4.47}$$

where S, V and n are the independent state parameters. The expression for $\mathrm{d}U$ results from a simple transformation of (4.43), i.e.,

$$\mathrm{d}U = T\mathrm{d}S - p\mathrm{d}V + \mu\mathrm{d}n, \tag{4.48}$$

hence

$$\left(\frac{\partial U}{\partial S}\right)_{V,n} = T, \tag{4.49}$$

$$\left(\frac{\partial U}{\partial V}\right)_{S,n} = -p, \tag{4.50}$$

$$\left(\frac{\partial U}{\partial n}\right)_{S,V} = \mu. \tag{4.51}$$

Comparing (4.48) with the expression: $\mathrm{d}U = \text{đ}Q + \text{đ}W + \text{đ}Z$ (see (3.4)), we find that

$$\text{đ}Q = T\mathrm{d}S, \qquad \text{đ}W = -p\mathrm{d}V, \qquad \text{đ}Z = \mu\mathrm{d}n. \tag{4.52}$$

We already know the first two contributions to $\mathrm{d}U$ as the quasi-static *heat* and *work*. The third contribution, related to the flow of matter, is the quasi-static *chemical work*. Equation (4.51) is the formal definition of the chemical potential μ. In practice, it is not very useful because it is rather difficult to control the entropy if a transfer of matter between the system and surroundings is possible. In a reversible adiabatic process, $\mathrm{d}S = 0$ because $\text{đ}Q = 0$. However, it is difficult to realize such a process if the walls are permeable to molecules which also carry some energy. In Sect. 4.2.3, we showed that a reversible flow of matter between the system and surroundings requires that their temperatures and chemical potentials are equal. In the next chapter, we provide a more practical definition of the chemical potential.

4.3.2 Euler Relation

Since S, U, V and n are extensive parameters, they are proportional to the mass of the system (see Sects. 2.1.2 and 2.4.4). This means that if U, V and n are multiplied by an arbitrary positive factor m, the entropy S is multiplied by the same factor, i.e.,

$$S(mU, mV, mn) = mS(U, V, n). \tag{4.53}$$

Substituting $m = 1/n$, we get

$$S(U, V, n) = nS(U/n, V/n, 1) = ns(u, v), \tag{4.54}$$

where $u = U/n$ and $v = V/n$ are the molar internal energy and molar volume, respectively, and $s(u, v) = S(u, v, 1)$ denotes the molar entropy. Substituting (4.54) into (4.44) and (4.45), we note that the dependence on n cancels out, hence

$$\left(\frac{\partial S}{\partial U}\right)_{V,n} = \left(\frac{\partial s}{\partial u}\right)_{v} = \frac{1}{T}, \tag{4.55}$$

$$\left(\frac{\partial S}{\partial V}\right)_{U,n} = \left(\frac{\partial s}{\partial v}\right)_{u} = \frac{p}{T}, \tag{4.56}$$

and

$$\mathrm{d}s = \frac{1}{T}\mathrm{d}u + \frac{p}{T}\mathrm{d}v. \tag{4.57}$$

Alternatively, we can treat u as a function of s and v: $u = u(s, v)$, and from (4.57), we get

$$\mathrm{d}u = T\mathrm{d}s - p\mathrm{d}v. \tag{4.58}$$

Note that we can also derive relations (4.57) and (4.58) directly from (4.43) and (4.48), substituting $n = 1$ and replacing S, U and V with s, u and v, respectively.

Intuitively, it is rather obvious that the properties of a uniform macroscopic system in thermodynamic equilibrium should not depend on its mass. Therefore, the fundamental relation in the form $s = s(u, v)$ or $u = u(s, v)$ should contain the whole essential information about the system, because the dependence on n can be taken into account in a simple way by (4.54). On the other hand, the reader may be concerned about the fact that the chemical potential has disappeared from our consideration. To explain this, we differentiate both sides of relation (4.53) with respect to m and substitute $m = 1$ at the end. Using relations (4.44)–(4.46), we get

$$\frac{1}{T}U + \frac{p}{T}V - \frac{\mu}{T}n = S, \tag{4.59}$$

hence

$$U = TS - pV + \mu n. \tag{4.60}$$

Identity (4.60) is called the *Euler relation*. It is simply a consequence of the fact that S, U, V and n are extensive parameters. Dividing the Euler relation by n, we obtain the following expression for the chemical potential:

$$\mu = u - Ts + pv. \tag{4.61}$$

If we choose u and v to be independent variables then

$$\mu(u, v) = u - s(u, v)T(u, v) + vp(u, v). \tag{4.62}$$

Alternatively, we can take s and v as independent variables and then $\mu = \mu(s, v)$.

In many practical problems, T and p are treated as independent parameters of state, instead of u and v or s and v. In the next chapter, we show how to exchange the roles of independent and dependent variables without any loss of information about a thermodynamic system. We also show that the chemical potential can be treated as a function of temperature and pressure. Thus, in a one-component system, the chemical potential is not an independent parameter but has a definite value for given values of temperature and pressure.

4.3.3 Entropy of the Ideal Gas

To give an example of the fundamental relation $s = s(u, v)$, we calculate the entropy of the ideal gas, using expression (4.57). We recall that the ideal gas is defined by the following equations of state (see Sects. 2.4.1 and 2.4.4):

$$pv = RT, \tag{4.63}$$

$$u = \frac{f}{2}RT, \tag{4.64}$$

where f denotes the number of degrees of freedom of a single molecule; for instance, $f = 3$ for monatomic molecules. In (4.57), u and v are the independent variables, therefore we have to express $1/T$ and p/T as their functions. From Eqs. (4.63) and (4.64), we get

$$\frac{p}{T} = \frac{R}{v}, \tag{4.65}$$

$$\frac{1}{T} = \frac{fR}{2u}, \tag{4.66}$$

and substituting these relations into (4.57), we obtain

$$\mathrm{d}s = \frac{1}{2}fR\frac{\mathrm{d}u}{u} + R\frac{\mathrm{d}v}{v}. \tag{4.67}$$

Note that $\mathrm{d}u/u = \mathrm{d}\ln(u/u_0)$, and similarly, $\mathrm{d}v/v = \mathrm{d}\ln(v/v_0)$, where u_0 and v_0 are some constants. Finally, $s(u, v)$ for the ideal gas has the following simple form:

$$s = s_0 + \frac{f}{2}R\ln\frac{u}{u_0} + R\ln\frac{v}{v_0}, \tag{4.68}$$

where $s_0 = s(u_0, v_0)$. The constants s_0, u_0, v_0 define the reference state, which can be arbitrarily chosen. The molar entropy has the same physical dimension as the gas constant, i.e., $\mathrm{J\,K^{-1}\,mol^{-1}}$. Because of the relation between entropy and the thermodynamic temperature (see (4.55)) the value of s depends on the choice of the temperature unit. For instance, if we increased the unit of temperature by a factor of 10 then the value of temperature expressed in the new units would decrease 10 times and the value of the gas constant would increase by the same factor. This results from equations of states (4.63) and (4.64), in which temperature appears in the product RT. We also notice that since s is constant in a reversible adiabatic process, the equation of the adiabat: $uv^{2/f} = const$, follows from expression (4.68), which is to be compared with (3.35).

Using equations of state (4.63) and (4.64), we can express the molar entropy of the ideal gas as a function of temperature and pressure:

$$s = s_0 + \frac{1}{2}(f+2)R\ln\frac{T}{T_0} - R\ln\frac{p}{p_0}, \tag{4.69}$$

where $T_0 = 2u_0/fR$, $p_0 = RT_0/v_0$. If we then substitute (4.69) into (4.61) and use the equations of state again we obtain the following expression for the chemical potential of the ideal gas as a function of T and p:

$$\mu = \mu_0 \frac{T}{T_0} - \frac{1}{2}(f+2)RT \ln \frac{T}{T_0} + RT \ln \frac{p}{p_0}, \tag{4.70}$$

where $\mu_0 = \mu(T_0, p_0) = [(f+2)R/2 - s_0]T_0$. It is easy to verify that the following relations hold:

$$\left(\frac{\partial \mu}{\partial T}\right)_p = -s, \tag{4.71}$$

$$\left(\frac{\partial \mu}{\partial p}\right)_T = v. \tag{4.72}$$

Although we have derived them for the ideal gas, we show in the next chapter that they are true in general.

4.3.4 Relation Between Entropy and Heat Capacity

We recall first the definition of the heat capacity given in Sect. 3.2.1:

$$C_X = \left(\frac{đQ}{\mathrm{d}T}\right)_X, \tag{4.73}$$

where X denotes the state parameter which is constant during the process. Relation (4.73) can also be expressed as $đQ = C_X \mathrm{d}T$, for a process at constant X. If heat is transferred between the system and surroundings reversibly then $đQ = T\mathrm{d}S$. Assuming that T, X and n are the state parameters of the system considered, we get

$$\mathrm{d}S = \left(\frac{\partial S}{\partial T}\right)_{X,n} \mathrm{d}T + \left(\frac{\partial S}{\partial X}\right)_{T,n} \mathrm{d}X + \left(\frac{\partial S}{\partial n}\right)_{T,X} \mathrm{d}n. \tag{4.74}$$

For a process at constant X and n (closed system), only the first term on the right-hand side remains. Comparing the two expressions for $đQ$, we obtain

$$C_X = T\left(\frac{\partial S}{\partial T}\right)_{X,n}. \tag{4.75}$$

Usually we are interested in the molar heat capacity:

$$c_X = T\left(\frac{\partial s}{\partial T}\right)_X. \tag{4.76}$$

In particular, the heat capacity at constant volume, c_v, and at constant pressure, c_p, is given, respectively, by

$$c_v = T\left(\frac{\partial s}{\partial T}\right)_v, \tag{4.77}$$

$$c_p = T\left(\frac{\partial s}{\partial T}\right)_p. \tag{4.78}$$

For the ideal gas, we substitute s in the form given by (4.68) (with $u = fRT/2$) or by (4.69), which gives $c_v = fR/2$ and $c_p = (f/2 + 1)R$ (cf. (3.13) and (3.25)).

4.4 Changes in Entropy in Reversible Processes

The second law of thermodynamics not only postulates existence of entropy but also provides practical means for calculations of its changes. In the case of reversible processes, the entropy of the system can change only if the system interacts thermally with the surroundings, which follows from the relation

$$đQ = T\mathrm{d}S. \tag{4.79}$$

In an adiabatic process, $đQ = 0$, hence also $\mathrm{d}S = 0$.

4.4.1 *Isothermal Process*

It is easy to determine the change in the entropy in a reversible isothermal process ($T = const$). An example of such a process is the isothermal expansion of a gas due to the heat supplied. When the process occurs at constant temperature we get, from relation (4.79),

$$\Delta S = \frac{Q}{T}, \tag{4.80}$$

where Q denotes the heat supplied to the system. $\Delta S > 0$ if the system absorbs heat and $\Delta S < 0$ if heat is given off by the system. In the case of the ideal gas, the heat absorbed or given off by the system in a reversible isothermal process is given by expression (3.41), hence the change in the entropy amounts to

$$\Delta S = nR\ln\frac{V_f}{V_i}, \tag{4.81}$$

where V_i and V_f denote the initial and final volume of the gas, respectively.

In general, we have (see (4.43))

$$\mathrm{d}S = \frac{1}{T}\mathrm{d}U + \frac{P}{T}\mathrm{d}V \tag{4.82}$$

for a closed system ($n = const$). If we assume the temperature and volume as the independent state parameters, then the internal energy $U = U(T, V)$ and

$$\mathrm{d}S = \frac{1}{T}\left(\frac{\partial U}{\partial T}\right)_V \mathrm{d}T + \frac{1}{T}\left[\left(\frac{\partial U}{\partial V}\right)_T + p\right]\mathrm{d}V. \tag{4.83}$$

In a isothermal process $\mathrm{d}T = 0$, hence

$$\mathrm{d}S = \frac{1}{T}\left[\left(\frac{\partial U}{\partial V}\right)_T + p\right]\mathrm{d}V. \tag{4.84}$$

To calculate the change in the entropy, we have to know the dependence of the internal energy on the volume. In the case of the ideal gas, U depends only on the temperature and a $p/T = nR/V$, hence

$$\mathrm{d}S = \frac{nR}{V}\mathrm{d}V. \tag{4.85}$$

Integrating over V from V_i to V_f, we arrive at (4.81). It is just another way of derivation of the same result, in which we use directly the form of the entropy differential. We can also obtain (4.81) directly from the expression for the molar entropy of the ideal gas derived previously (see (4.68)).

4.4.2 Isochoric and Isobaric Processes

We assume again a closed system. If we take T and V as the independent parameters of state then

$$\mathrm{d}S = \left(\frac{\partial S}{\partial T}\right)_V \mathrm{d}T + \left(\frac{\partial S}{\partial V}\right)_T \mathrm{d}V. \tag{4.86}$$

In an isochoric process, $\mathrm{d}V = 0$ and the derivative of entropy with respect to temperature can be expressed in terms of the heat capacity at constant volume (see (4.75) and (4.77)), hence

$$\mathrm{d}S = \frac{nc_v}{T}\mathrm{d}T, \tag{4.87}$$

where c_v is a function of T and the molar volume v. Integrating from the initial temperature T_i to the final temperature T_f, we arrive at the following expression:

$$\Delta S = n\int_{T_i}^{T_f} \frac{c_v(T, v)}{T}\mathrm{d}T. \tag{4.88}$$

In particular, if c_v is independent of temperature, as in the case of the ideal gas, or the dependence on temperature can be neglected in a given temperature range, then

$$\Delta S = nc_v \ln\frac{T_f}{T_i}. \tag{4.89}$$

In the case of a reversible isobaric process, we proceed in a similar way. We treat entropy as a function of temperature and pressure, hence

$$\mathrm{d}S = \left(\frac{\partial S}{\partial T}\right)_V \mathrm{d}T + \left(\frac{\partial S}{\partial p}\right)_T \mathrm{d}p. \tag{4.90}$$

In the isobaric process, $\mathrm{d}p = 0$, hence

$$\mathrm{d}S = \frac{nc_p}{T}\mathrm{d}T, \tag{4.91}$$

where we have used (4.75) and (4.78). The change in the entropy of the system in a reversible isobaric process amounts to

$$\Delta S = n \int_{T_i}^{T_f} \frac{c_p(T, p)}{T} \mathrm{d}T, \tag{4.92}$$

and if the dependence of c_p on temperature can be neglected then

$$\Delta S = n c_p \ln \frac{T_f}{T_i}. \tag{4.93}$$

4.4.3 Evaporation of Liquids

We consider evaporation of a liquid, for instance, water. If at the temperature $T = 373.15$ K (100 °C) and pressure of 1 atm we supply the heat $Q = 40.66$ kJ in a reversible process, then 1 mol of liquid water changes into water vapour. The entropy of the system consisting of liquid water and its vapour, S_{sys}, increases by $\Delta S_{\mathrm{sys}} = Q/T = 109\ \mathrm{J\,K^{-1}}$. Since the process is reversible, the total entropy of the system and surroundings (from which the heat Q is drawn) does not change ($\Delta S_{\mathrm{sys}} + \Delta S_{\mathrm{sur}} = 0$), hence $\Delta S_{\mathrm{sur}} = -109\ \mathrm{J\,K^{-1}}$. This trick is often used to determine how the entropy of a given system or its surroundings has changed. If we treat the given system together with its surroundings as one isolated system, then the total entropy of that isolated system cannot decrease in any process, and in the case of a reversible processes, it does not change.

Trouton noticed that the change in the molar entropy of a substance in the process of evaporation is approximately equal to 85 $\mathrm{J\,mol^{-1}\,K^{-1}}$ for many substances, e.g., cyclohexane, bromine, benzene, carbon tetrachloride or hydrogen sulphide. If we know the boiling point of a liquid it is usually possible to estimate the heat of evaporation for that liquid from the Trouton rule. The rule is not general, as it does not apply to water, for instance. Nevertheless, it can be used even in this case as a rough estimate. What is the main reason that the entropy of the system increases in the process of evaporation? Substantial contribution comes from the increase in the system volume due to the change of the liquid into vapour. As we know, one mole of liquid water occupies the volume of 18 $\mathrm{cm^3}$. If we approximate water vapour by the ideal gas then, at the temperature of 373.15 K and pressure of 1 atm, we obtain the molar volume of 30.6 $\mathrm{L\,mol^{-1}}$. When 1 mol of water evaporates completely the change in its entropy can be determined from expression (4.81). For the data given above, we obtain $\Delta S \approx 62\ \mathrm{J\,K^{-1}\,mol^{-1}}$. This example shows that in the process of water evaporation, the change in its molar volume is responsible for about 50 % of the total change in its entropy. It should be added, however, that water is a special substance with a high boiling point and a low melting point, at a very small molar mass. When a typical substance evaporates, the change in its entropy due to the change in the molar volume is even greater in proportion to the total change in the entropy of the system than in the case of water.

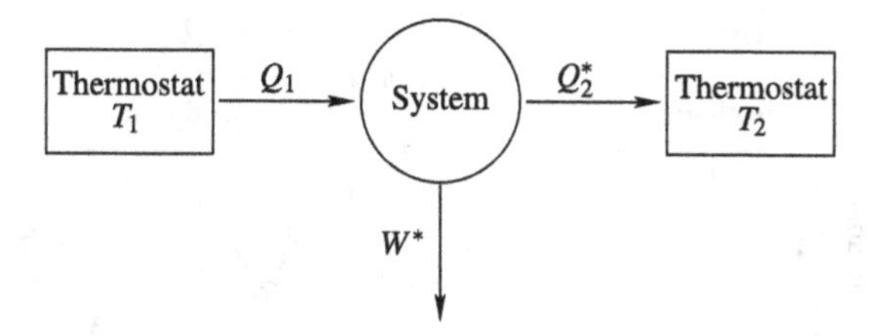

Fig. 4.1 Operation principle of the Carnot engine. The *arrows* show the direction of the energy flow in the form of heat and work, for $T_1 > T_2$. The system performs the work W^* at the expense of a part of the heat received, Q_1, and the rest of that heat, Q_2^*, is transferred to the radiator

4.5 Heat Devices

Heat devices utilize the heat drawn from the surroundings to perform useful work or utilize the work supplied to force the flow of heat in the direction of increasing temperature. Operation of heat devices is based on thermodynamic processes that proceed cyclically. The process is called a *cycle* if the final state of the system is identical to the initial state. A *thermostat* is a system in thermodynamic equilibrium with the property that a finite amount of heat drawn from it or supplied to it does not change its temperature. In other words, a thermostat is a heat reservoir (see Sect. 4.1). When a system absorbs a small amount of heat, Q, at constant volume, its temperature changes by $\Delta T \approx Q/(nc_v)$. In the case of a thermostat, $n \to \infty$ and $\Delta T \to 0$, which means that its size is considered infinite compared to the size of any system interacting with it. We recall once more that Q and W denote the heat supplied to a system and work done on the system, whereas $Q^* = -Q$ and $W^* = -W$ denote, respectively, the heat given off and the work done by the system.

4.5.1 Heat Engine and the Carnot Cycle

A heat engine consists of three basic components: a *heat reservoir* (in the car, heat comes from the combustion of petrol or another fuel), a *working substance* that performs work (e.g., a gas in a cylinder with a movable piston) and a *radiator*. Here we consider the operation of an idealized heat engine in which one thermostat, of the temperature T_1, serves as a heat reservoir, and another thermostat, of the temperature $T_2 < T_1$, serves as a radiator. The working substance in the engine forms the system which interacts with the thermostats and performs work. The principle of operation of such an engine is shown schematically in Fig. 4.1. The system receives the heat Q_1 from the thermostat at the temperature T_1 in a reversible isothermal process. A part of that heat is used by the system to perform the work $W^* = -W > 0$, and the rest of it, denoted $Q_2^* = -Q_2 > 0$, is transferred to the thermostat at the temperature T_2, again in a reversible isothermal process. The full cycle consists of the following reversible processes.

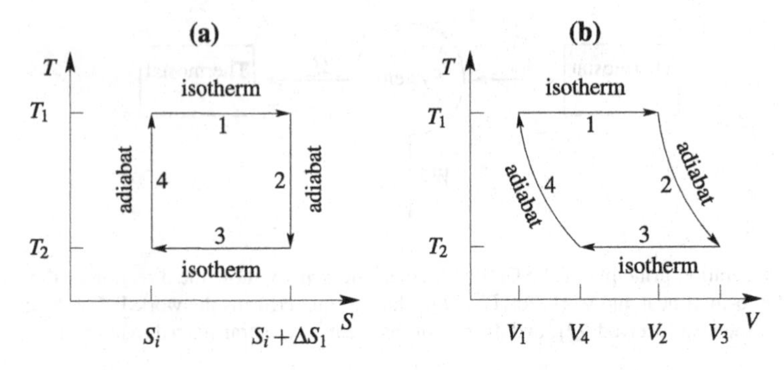

Fig. 4.2 Carnot cycle consists of four reversible processes: two isothermal processes (*1* and *3*) and two adiabatic processes (*2* and *4*), whose directions are marked with arrows. (**a**) Carnot cycle in the variables S and T. $S_i = S(T_1, V_1)$ denotes the entropy of the system in the initial state at the temperature T_1 and volume V_1. $\Delta S_1 = Q_1/T_1$ is the change in the entropy of the system in the isothermal process at $T = T_1$. Since the change in the entropy in the whole cycle equals zero, $\Delta S_2 = Q_2/T_2 = -\Delta S_1$ in the isothermal process at $T = T_2$. (**b**) Carnot cycle in the variables T and V; V_1, V_2, V_3 and V_4 denote the initial volume of the system in the consecutive processes

1. Isothermal expansion at the temperature T_1; the volume increases from V_1 to V_2, and the entropy increases by $\Delta S_1 = Q_1/T_1$ (see (4.80)).
2. Adiabatic expansion ($\Delta S = 0$); the volume increases from V_2 to V_3, and the temperature decreases from T_1 to T_2.
3. Isothermal compression at the temperature T_2; the volume decreases from V_3 to V_4, and the entropy decreases by $\Delta S_2 = Q_2/T_2$.
4. Adiabatic compression ($\Delta S = 0$); the volume decreases from V_4 to V_1 and the temperature increases from T_2 to T_1.

The engine operating as described above is called the *Carnot engine*. Figure 4.2 presents the cycle of the Carnot engine (Carnot cycle) on two diagrams. In Fig. 4.2a, the independent variables are: the entropy S and temperature T. This diagram is particularly simple because the isothermal processes are represented by the horizontal lines (isotherms), and the adiabatic processes are represented by the vertical lines (adiabats). Figure 4.2b shows the Carnot cycle in the variables T and V. In this figure, the isotherms are horizontal lines but the shape of the adiabats is different. When a full cycle is completed the system returns to its initial state, and since entropy is a state function, $\Delta S = 0$. On the other hand, ΔS is equal to the sum of the changes in the entropy in the isothermal processes, because entropy does not change in reversible adiabatic processes, hence $\Delta S = \Delta S_1 + \Delta S_2 = 0$ and we arrive at the following relation:

$$\frac{Q_1}{T_1} + \frac{Q_2}{T_2} = 0. \tag{4.94}$$

Engine Efficiency An important quantity that can be easily determined for the Carnot cycle is the *engine efficiency* η_e. It is a dimensionless quantity defined as the ratio of the work W^* performed by the engine on the surroundings to the heat Q_1 received from the heat reservoir:

$$\eta_e = \frac{W^*}{Q_1}. \tag{4.95}$$

The more work the engine performs from a given amount of the heat received, the greater its efficiency is. Since the internal energy is also a state function, it does not change in the whole cycle ($\Delta U = 0$). On the other hand, ΔU for the whole cycle is equal to the sum of the changes in U in the consecutive processes. In the isothermal process at the temperature T_1, the system (working substance) absorbs the heat Q_1, and in the isothermal process at the temperature T_2, it gives off the heat $Q_2^* = -Q_2$. The system performs work both in the isothermal and adiabatic processes because its volume changes, and the total work performed during the whole cycle is equal to $W^* = -W$. Summing up all contributions to ΔU, we get:

$$\Delta U = Q_1 + W + Q_2 = 0. \tag{4.96}$$

This identity can also be written as

$$Q_1 = W^* + Q_2^*, \tag{4.97}$$

which expresses simply the energy conservation principle for the system and surroundings. The total energy transferred from the system to the surroundings in the form of work W^* and heat Q_2^* must be equal to the heat Q_1 supplied to the system. Relation (4.94), derived from the second law of thermodynamics, can be expressed in the convenient form:

$$\frac{Q_2}{Q_1} = -\frac{T_2}{T_1}. \tag{4.98}$$

Dividing (4.97) by Q_1 and using (4.98), we get, from definition (4.95), the following expression for the efficiency of the Carnot engine:

$$\eta_e = 1 - \frac{T_2}{T_1}. \tag{4.99}$$

It is the maximum efficiency of any engine that works between two thermostats characterized by the temperatures T_1 and T_2. This is because the Carnot cycle consists of reversible processes. We will show in the next chapter that the maximum work done by the system from a given amount of heat supplied corresponds to a reversible process. We conclude from expression (4.99) that the lower the temperature of the radiator, T_2, relative to the temperature of the heat reservoir, T_1, the greater the efficiency the Carnot engine works with, and in the limit $T_2 \to 0$ the efficiency $\eta_e \to 1$.

4.5.2 Efficiency of the Carnot Cycle and Thermodynamic Temperature

Note that expression (4.99) can serve as a definition of the thermodynamic temperature because it provides a method of its measurement. Measuring the work performed by the Carnot engine and the heat supplied to it, we can determine the ratio of any two temperatures independently on the working substance used in the engine. A question may be raised whether the thermodynamic temperature used in formula (4.99) is identical to the absolute temperature that appears in the equations of state of the ideal gas. So far we have not discriminated between these two temperature scales, assuming tacitly that they correspond to the same physical quantity. However, to prove it, we should determine the efficiency of the Carnot engine which uses the ideal gas as a working substance, without invoking the second law of thermodynamics. For this purpose, we use only the equations of state of the ideal gas and the first law of thermodynamics.

To discriminate for the time being between the two temperatures, we use the symbol T' for the absolute temperature present in the equations of state of the ideal gas, and the symbol T for the thermodynamic temperature in formula (4.99). From (4.96), we determine the work W^* in the Carnot cycle:

$$W^* = Q_1 + Q_2. \tag{4.100}$$

Recall that the heat received by the ideal gas in a reversible isothermal process amounts to (see (3.41))

$$Q = -nRT' \ln \frac{V_i}{V_f}, \tag{4.101}$$

hence for the heat Q_1 and Q_2, we obtain, respectively,

$$Q_1 = -nRT_1' \ln \frac{V_1}{V_2}, \tag{4.102}$$

$$Q_2 = -nRT_2' \ln \frac{V_3}{V_4}. \tag{4.103}$$

Using (4.102) and (4.103) in (4.100) and in definition (4.95), we get

$$\eta_e = 1 - \frac{T_2' \ln(V_3/V_4)}{T_1' \ln(V_2/V_1)}. \tag{4.104}$$

We need to show now that $V_3/V_4 = V_2/V_1$. The equation of the adiabat for the ideal gas, expressed in the variables p and V (see (3.34)), leads to the following relations:

$$p_2 V_2^{1+2/f} = p_3 V_3^{1+2/f}, \tag{4.105}$$

$$p_1 V_1^{1+2/f} = p_4 V_4^{1+2/f}, \tag{4.106}$$

where V_i and p_i, for $i = 1, 2, 3, 4$, are the parameters of state of the ideal gas at the beginning of the ith process (see Fig. 4.2b). Dividing the first relation by the second one, we get

$$\frac{p_2 V_2}{p_1 V_1}\left(\frac{V_2}{V_1}\right)^{2/f} = \frac{p_3 V_3}{p_4 V_4}\left(\frac{V_3}{V_4}\right)^{2/f}. \tag{4.107}$$

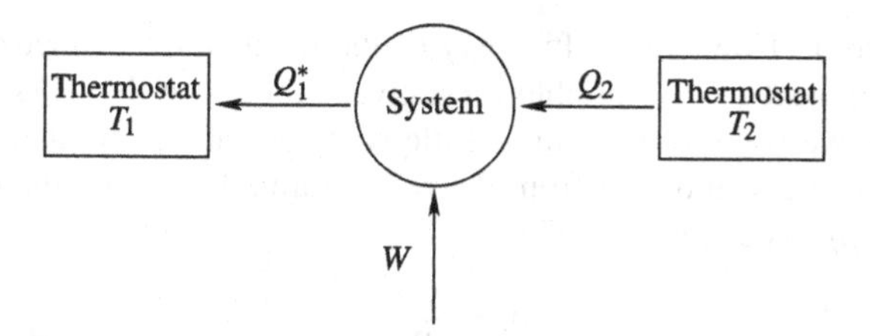

Fig. 4.3 Operation principle of the refrigerator or heat pump. The *arrows* show the direction of the energy flow in the form of heat and work, for $T_1 > T_2$. The system draws the heat Q_2 due to the work W done on it and gives off the heat Q_1^* to the thermostat at $T = T_1$. In the case of the refrigerator, its interior is represented by the thermostat at $T = T_2$. In the case of the heat pump, the thermostat at $T = T_1$ represents the interior of the room to be warmed due to the heat drawn from the cooler surroundings

Since V_1 and V_2 correspond to the isotherm $T = T_1$, and V_3 and V_4 correspond to the isotherm $T = T_2$, we have $p_1 V_1 = p_2 V_2$ and $p_3 V_3 = p_4 V_4$. Using these identities in (4.107), we find that

$$\frac{V_2}{V_1} = \frac{V_3}{V_4}, \tag{4.108}$$

hence (see (4.104))

$$\eta_e = 1 - \frac{T_2'}{T_1'}. \tag{4.109}$$

From the comparison of (4.109) with (4.99), we conclude that

$$\frac{T_2'}{T_1'} = \frac{T_2}{T_1}. \tag{4.110}$$

We have just proved that the thermodynamic temperature, whose definition is based on the fundamental relation between S, U, V and n ($T^{-1} = (\partial S/\partial U)_{V,n}$), and the absolute temperature used in the equations of state of the ideal gas, correspond to the same physical quantity. The fact that only the ratios of two temperatures appear in (4.110) should not be surprising because the choice of the temperature unit is arbitrary. However, if we use the same units then the identity $T = T'$ holds.

4.5.3 Refrigerator and the Heat Pump

What is going to happen when we reverse the run of the Carnot engine (see Fig. 4.1), that is, if we reverse the directions of the energy flow, as shown in Fig. 4.3? Now the system receives the heat Q_2 from the thermostat at the temperature T_2 and gives off the heat Q_1^* to the thermostat at the temperature $T_1 > T_2$, due to the work W done on the system. This is the operation principle of the *refrigerator*. We can reverse the natural direction of the heat flow from a warmer body to a cooler one, but it requires that some work is done by an external agent. Most often it is the work of the electric current. How is the efficiency of the refrigerator defined? The idea is to draw as much heat Q_2 as possible for a given amount of the work W received. In other words, we want to perform as little work on the system as possible for a given amount of the heat drawn from the refrigerator. Therefore, the efficiency of the refrigerator, η_r, is defined as follows:

$$\eta_r = \frac{Q_2}{W}. \tag{4.111}$$

In order to determine the efficiency of the refrigerator that operates according to the Carnot cycle, we use again both laws of thermodynamics. From relations (4.96) and (4.98), we obtain

$$\eta_r = \frac{T_2}{T_1 - T_2}. \tag{4.112}$$

For typical temperatures used in household refrigerators, $\eta_r > 1$. If T_2 decreases at constant T_1 then in order to cool a system at the temperature T_2, more and more work is required, because η_r decreases and $W = Q_2/\eta_r$. Thus, if we want to draw a finite amount of heat from the system when $T_2 \to 0$ then $\eta_r \to 0$ and $W \to \infty$. This means that it is impossible to reach the absolute zero by means of a device that operates using the Carnot cycle. This statement is related to the *third law of thermodynamics*, which is discussed in Sect. 4.7.

Another heat device whose operation is based on the Carnot cycle can be used for heating of a building or a room. It operates in a similar way as the refrigerator does but now we are interested in the heat $Q_1^* = -Q_1$ supplied by the working substance (the system) to the thermostat at $T = T_1$, due to the heat Q_2 drawn from the cooler surroundings, at $T = T_2$, and the work W done on the system. Such a device is called the *heat pump*. Everybody knows that the refrigerator heats the room it occupies. If it was inserted into a wall, instead of a window, for instance, in such a way that its back was facing the room and its door was open to the outside, then we would obtain a primitive heat pump. The refrigerator would cool the surroundings (with a miserable effect) due to the work done by the electric current, heating the room at the same time.

What is the efficiency of the heat pump? An efficient heat pump should transfer a large amount of heat, Q_1^*, to the room at the cost of as small amount of work, W, as possible. Therefore, the efficiency of the heat pump is defined as follows:

$$\eta_p = \frac{Q_1^*}{W}. \tag{4.113}$$

Using relations (4.96) and (4.98), we get

$$\eta_p = \frac{T_1}{T_1 - T_2}. \tag{4.114}$$

Note that in contrast to the engine efficiency η_e, which is always smaller than unity, we have $\eta_p = 1/\eta_e > 1$. The heat pump is a very efficient heating device, many times more efficient than a typical electric heater, which changes the whole work of the electric current into heat. It is easy to verify that the efficiency of the refrigerator and the efficiency of the heat pump are related to each other by the formula $\eta_p = 1 + \eta_r$.

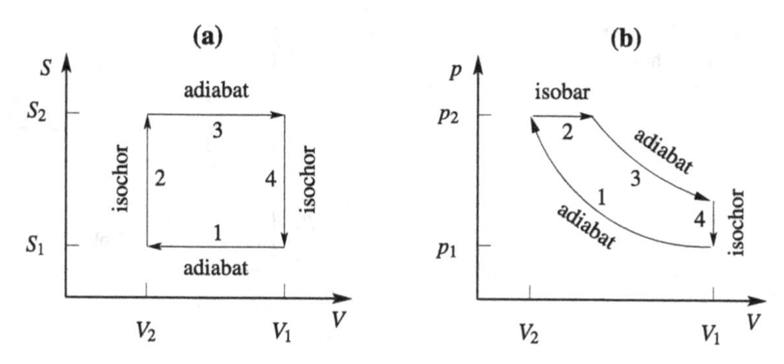

Fig. 4.4 Principle of operation of four-stroke petrol and diesel engines. The *arrows* show the direction of processes. (**a**) Otto cycle (adiabatic–isochoric) for the petrol driven engine. Combustion of the mixture (process 2) occurs at constant volume. (**b**) Diesel cycle (adiabatic–isobaric–isochoric) for the Diesel (high-pressure) engine. Combustion of the mixture (process 2) occurs at constant pressure

4.5.4 Other Thermodynamic Cycles

To design a heat engine, it is necessary to define the thermodynamic cycle on which the engine operation is based. Usually the working substance filling the engine cylinders is a mixture of air and fuel. In typical engines, the fuel is injected into cylinders in the form of tiny droplets which start to evaporate. The ignition of the fuel is then generated by a sparking-plug or high temperature obtained by very fast compression of a gas. The heat released during the fuel combustion causes the gas in the cylinder to expand and shift the piston. At the end of the cycle, the piston returns to its initial position and the whole cycle repeats. Most often petrol is used as fuel. The operation of the four-stroke petrol driven engine is described roughly by the Otto cycle, shown in Fig. 4.4a. The Otto cycle consists of the following reversible processes.

1. Adiabatic compression of a mixture of air and fuel.
2. Isochoric heating of the gas. This process corresponds to the petrol combustion, which proceeds so fast that the system does not change practically its volume.
3. Adiabatic expansion during which the system performs work.
4. Isochoric cooling.

In the last process, the products of the fuel combustion are removed from the system and fresh air is sucked in before the new compression starts. The replacement of the working substance is not represented on the diagram, however. Processes 1 and 3 can be considered adiabatic because they occur so quickly that the transfer of heat between the gas and surroundings can be neglected. The Diesel cycle describing the operation of the Diesel (oil-burning, high-pressure) engine is slightly different. It is shown in Fig. 4.4b on the Vp diagram. In this case, the combustion of the mixture (process 2) occurs at constant pressure instead of constant volume, as in the case of the Otto cycle.

The cycle which describes the operation of the steam engine is more complex. It is shown in Fig. 4.5 on the Vp diagram. It consists of five reversible processes.

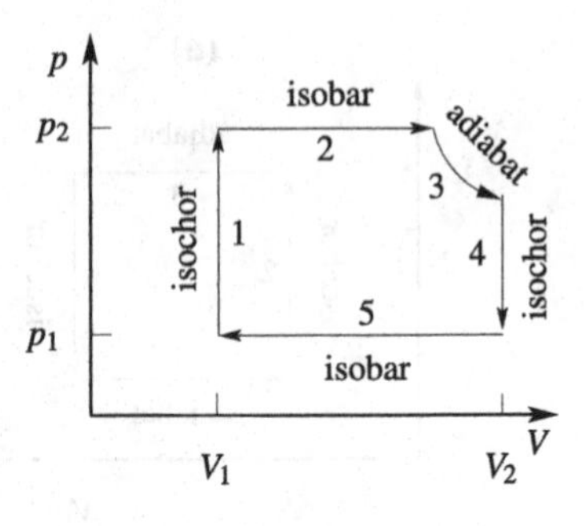

Fig. 4.5 Cycle of the steam engine. The *arrows* show the direction of processes

In the first isochoric process, the steam pressure increases. In the second process, steam expands at constant pressure. In the third process, the inflow of steam from the boiler is cut off and then steam expands adiabatically. In the fourth process, steam is let out at constant volume and the pressure decreases. Finally, the piston returns to the original position in a isobaric process.

4.6 Changes in Entropy in Irreversible Processes

4.6.1 Irreversible Flow of Heat

We consider two identical systems: (1) and (2), whose heat capacity $C_V = nc_v$ is independent of temperature. The initial temperatures of the systems are T_1 and T_2, and we assume that $T_1 > T_2$. If the systems are brought into contact through a rigid diathermal wall and isolated from the surroundings, then heat flows from the system (1) to the system (2) until thermodynamic equilibrium is reached at a temperature T_f. We know from experience that such a process is irreversible but we want to show this explicitly, calculating the change in the total entropy of the two systems. The final temperature results from the first law of thermodynamics:

$$\Delta U_1 + \Delta U_2 = C_V(T_f - T_1) + C_V(T_f - T_2) = 0, \tag{4.115}$$

hence

$$T_f = \frac{T_1 + T_2}{2}. \tag{4.116}$$

Since we can calculate changes in entropy in reversible processes, we have to construct a reversible process which takes the system (1) from the initial state to the final state, and another reversible process which does so with the system (2). We know that when the heat capacity C_V does not depend on temperature, the change in the entropy of a system in a reversible isochoric process is given by (4.89), hence

$$\Delta S_1 + \Delta S_2 = C_V\left(\ln\frac{T_f}{T_1} + \ln\frac{T_f}{T_2}\right) = C_V \ln\frac{(T_1 + T_2)^2}{4T_1T_2}. \tag{4.117}$$

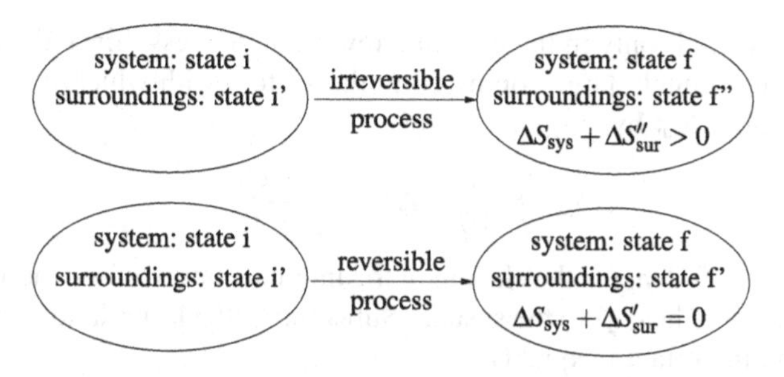

Fig. 4.6 To calculate the change in the entropy of a system in an irreversible process from the state i to the state f, we construct a reversible process from i to f. The change in the entropy of the system, ΔS_{sys}, in the reversible process is the same as in the irreversible process, since entropy is a state function. However, the change in the entropy of the surroundings is different for each process, and it amounts to $\Delta S'_{sur}$ and $\Delta S''_{sur}$, respectively, because the final state of the surroundings is different

Since $(T_1 + T_2)^2 > 4T_1T_2$, the inequality

$$\Delta S_1 + \Delta S_2 > 0 \tag{4.118}$$

holds, in accord with the second law of thermodynamics. This means that the process of the heat flow from a system at higher temperature to a system at lower temperature is irreversible.

We have already mentioned that true reversible processes do not exist in nature. For instance, to carry out the reversible process described above, we would have to transfer the heat infinitely slowly, which would take infinite time. In reality, we can carry out processes that are almost reversible, and it depends only on our technical capabilities and patience how close they approach the ideal reversible processes. As we know, entropy is a state function so its change between two given equilibrium states does not depend on the process but only on those states. In contrast to a reversible process, which is represented by a curve (or path) in a low-dimensional space of state parameters, an irreversible process does not have such a simple representation. Nevertheless, the change in entropy, or another function of state, is well defined if the initial state and final state are equilibrium states. Therefore, idealized reversible processes actually enable us to calculate changes in entropy and other state functions in the case of irreversible processes, which is illustrated schematically in Fig. 4.6.

Let us consider now a system in which an isothermal process occurs, that is, the system is in contact with a thermostat at the temperature T. During the process the system receives the heat Q from the thermostat. The system and thermostat (surroundings) form together an isolated system. According to the second law of thermodynamics, we have

$$\Delta S_{sys} + \Delta S_{sur} \geq 0, \tag{4.119}$$

and the equality holds only in the case of a reversible process. Since the thermostat is by definition infinitely large compared to the system with which it interacts, we can approximate ΔS_{sur} by

$$\Delta S_{\text{sur}} \approx \frac{\partial S_{\text{sur}}}{\partial U_{\text{sur}}} \Delta U_{\text{sur}} = \frac{\Delta U_{\text{sur}}}{T}, \tag{4.120}$$

where $\Delta U_{\text{sur}} = -Q$ denotes the change in the internal energy of the thermostat due to the transfer of the heat Q to the system. Substituting the last relation into (4.119), we arrive at an important inequality:

$$\Delta S_{\text{sys}} \geq \frac{Q}{T}. \tag{4.121}$$

It means that a change in the entropy of the system in contact with a thermostat is always greater or equal to the ratio of the heat received by the system and the temperature of the thermostat. The equality holds only for reversible processes. For $Q = 0$, we simply recover the inequality $\Delta S_{\text{sys}} \geq 0$ for adiabatic processes, i.e., the second law of thermodynamics.

4.6.2 Free Gas Expansion

We consider an ideal gas isolated adiabatically from the surroundings. Initially the gas occupies the volume V_i. Then it is allowed to expand freely ($p_{\text{ext}} = 0$) up to the volume V_f. How does the entropy of the system change? Since the expansion is free, the gas performs no work, hence, $\Delta U = 0$ and its temperature T does not change. However, since the process is irreversible, the entropy of the gas must increase.

To determine the change in the entropy of the gas in this process, we have to find a reversible process which takes the system from the state of the volume V_i and temperature T to the state of the volume V_f and the same temperature. The process in question is an isothermal process, therefore the change in the entropy amounts to (see (4.81))

$$\Delta S_{\text{rev}} = nR \ln \frac{V_f}{V_i}, \tag{4.122}$$

where the index 'rev' stands for *reversible*. The change in the entropy of the surroundings in this reversible process equals $-\Delta S_{\text{rev}}$ because the total entropy of the system and surroundings does not change.

In the irreversible process of free gas expansion, no heat flows to the system because it is isolated adiabatically from the surroundings. However, the change in the entropy of the system must be the same as in the reversible process considered, because the initial state and final state are the same in both cases, and entropy is a state function, hence

$$\Delta S_{\text{irr}} = nR \ln \frac{V_f}{V_i}. \tag{4.123}$$

The change in the entropy of the surroundings is different, however, because the total entropy of the system and surroundings must increase. This means that the final state of the surroundings is different in each process considered (see Fig. 4.6).

4.6.3 *Irreversible Chemical Reaction*

The change in entropy in a chemical reaction depends on the pressure and temperature at which the reaction occurs. According to the second law of thermodynamics, a given chemical reaction occurs spontaneously if the total entropy of the system and surroundings increases during the reaction. As an example, we consider the reaction of water formation from hydrogen and oxygen:

$$2H_2 + O_2 \rightarrow 2H_2O, \tag{4.124}$$

which proceeds at the pressure of 1 bar and temperature of 25 °C. During this reaction the entropy of the system changes by $\Delta S_{sys} = -327\ \mathrm{J\,K^{-1}\,mol^{-1}}$. The fact that the entropy decreases should not be surprising. Since a liquid is formed in the reaction of two gases, the volume of the system decreases significantly. The reaction is strongly exothermic; the heat given off to the surroundings amounts to $Q^* = 572$ kJ, which increases the entropy of the surroundings by $\Delta S_{sur} = 1918\ \mathrm{J\,K^{-1}\,mol^{-1}}$. The total increase in the entropy of the system and surroundings is positive and amounts to $\Delta S_{sys} + \Delta S_{sur} = 1591\ \mathrm{J\,K^{-1}\,mol^{-1}}$, which means that the reaction occurs spontaneously. However, the reverse reaction of water decomposition into oxygen and hydrogen cannot occur in these conditions, because it would mean that the total entropy of the system and surroundings decreases, in contradiction with the second law of thermodynamics.

4.7 Third Law of Thermodynamics

In thermodynamic processes, we can only determine the difference between the entropy of a given equilibrium state of the system and the entropy of a certain reference state. However, close to the absolute zero temperature, changes in the entropy exhibit characteristic behaviour. This observation is formulated as a separate law of thermodynamics.

The Third Law of Thermodynamics
The difference between the entropy of two states of the same temperature T, which can be linked by a reversible process, tends to zero when $T \rightarrow 0$.

The following formulation of the third law, less general then the first one, is also in use.

The entropy of the ideal crystal tends to zero when $T \to 0$.

It may happen, in general, that the entropy of a system tends to a constant (dependent on that system) when $T \to 0$. It occurs when the system is not an ideal crystal at the absolute zero, but a frozen disorder exists in it. Then its entropy at $T = 0$ differs from zero. For instance, the entropy of nitrogen oxide tends to 4.77 $\mathrm{J K^{-1} mol^{-1}}$ when its temperature tends to zero. One of the consequences of the third law of thermodynamics is that various quantities expressed in terms of the first derivatives of entropy, such as the heat capacity C_V and C_p or the thermal expansion coefficient, tend to zero when $T \to 0$.

4.8 Exercises

4.1 A cyclic process occurring in 1 mol of a monatomic ideal gas, consists of the following stages: (1) the pressure increases from p_1 to p_2 at constant volume $V = V_1$, (2) the volume increases from V_1 to V_2 at constant pressure $p = p_2$, (3) the pressure decreases from p_2 to p_1 at $V = V_2$ and (4) the volume decreases from V_2 to V_1 at $p = p_1$. Calculate the work done in the process, the change in the internal energy of the gas and the heat transferred to or from the system at each stage and in the whole process.

4.2 Calculate the efficiency of the engine whose operation is based on the cycle described in Exercise 4.1.

4.3 Try to invent your own cycle and then calculate its efficiency for the ideal gas.

4.4 Draw the Carnot cycle on the diagram: entropy versus temperature, and show that the work done in the whole cycle is equal to the area of the rectangle that represents the cycle.

4.5 An ideal gas, in the amount of 5 mol, expands reversibly at the temperature of 25 °C. The pressure decreases from 2 atm to 1 atm. Calculate the change in the entropy of the system and surroundings.

4.6 The same system as in Exercise 4.5 expands irreversibly against constant external pressure of 1 atm. The initial and final temperature is the same and amounts to 25 °C. Calculate the work done by the system, change in its internal energy and heat transferred from the surroundings to the system. How does the entropy of the system and surroundings change?

4.7 A system with the heat capacity $C_V = nc_v$, independent of temperature, is to be cooled down to the temperature T_2 by means of the Carnot engine. The initial temperature of the system amounts to $T_1 > T_2$. The temperature of the radiator, T_2, is constant. Calculate the work to be done.

4.8 Two identical systems with the heat capacity C_V, independent of temperature, are initially at the temperatures T_1 and T_2, respectively, where $T_1 > T_2$. Find the final temperature of the systems, assuming that heat is transferred from the first system to the second one in a reversible way by means of the Carnot engine. What would be the final temperature of the systems if the transfer of heat occurred in a irreversible process through a rigid diathermal wall separating the systems? Make use of the fact that in the first case the total entropy of the systems is constant, whereas in the second case the total internal energy is constant.

4.9 Calculate the change in the entropy of 1 mol of a substance whose temperature changes from T_i to T_f and its heat capacity at constant volume is given by the formula

$$c_v = A + BT - CT^{-2},$$

where A, B and C are some constants.

4.10 The heat capacity per unit volume, c_v, of the perfect blackbody changes with temperature as

$$c_v = \frac{C_v}{V} = 4\gamma T^3,$$

where γ is a constant. Calculate the change in the entropy of the perfect blackbody whose temperature changes from T_i to T_f.

4.11 We want to maintain at home a temperature of 23 °C, while the outdoor temperature amounts to 0 °C. How much can we reduce the electricity bill, using a heat pump instead of an electric heater?

4.12 The temperature of water inside an artesian well in an Australian desert amounts to 5 °C, while the temperature of air amounts to 20 °C. A heat machine of the efficiency of the Carnot engine performs the work of 500 kJ. How much heat does it transfer to the well?

4.13 A certain thermodynamic system is characterized by the following equation of state:

$$U = B\frac{S^3}{nV},$$

where B is a constant. Find the dependence of the pressure and temperature of the system on its internal energy, volume and mole number.

4.14 Using the equation of state of the photon gas: $U = \gamma V T^4$, determine its entropy S as a function of T and V, and also as a function of U and V. Then, determine the pressure of the photon gas as a function of T.

4.15 Find the entropy of a monatomic van der Waals gas, using the equations of state:

$$p = \frac{RT}{v-b} - \frac{a}{v^2}, \qquad u = \frac{3}{2}RT - \frac{a}{v},$$

where $v = V/n$ and $u = U/n$.

4.16 The entropy of the systems A and B is given by

$$S_A = \left(\frac{n_A V_A U_A}{D}\right)^{1/3} \quad \text{and} \quad S_B = \left(\frac{n_B V_B U_B}{D}\right)^{1/3},$$

respectively, where D is a constant. The temperatures and pressures in the systems are the same: $p_A = p_B$ and $T_A = T_B$. Show by direct calculation that the total entropy of the fused system is equal to $S = S_A + S_B$. Make use of the fact that the pressure and temperature are functions of the molar internal energy and molar volume, and that the relations $n = n_A + n_B$, $U = U_A + U_B$ and $V = V_A + V_B$ hold for the total mole number n, total internal energy U, and total volume V. Find the dependence of S on n, V and U.

4.17 A vessel divided into two equal parts by an internal wall is at rest on a flat surface. One half of the vessel is occupied by a gas, whereas the other half is empty. There is no friction between the vessel and the surface. What will happen to the vessel if the internal wall is removed suddenly? Will the vessel move, and if so, in what direction? If the vessel moves will it be able to stop later if there is no friction? Draw a picture showing the motion of the vessel and gas. To solve this problem, make use of the fundamental laws of mechanics and thermodynamics.

Chapter 5
Thermodynamic Potentials

5.1 Legendre Transformation of the Internal Energy and Entropy

The first and second laws of thermodynamics, discussed in the two previous chapters, show which processes are allowed by the energy conservation principle and occur spontaneously, and which do not. A given process occurs spontaneously if the change in the total entropy of the system and its surroundings is not negative. The problem is that it is rather difficult, or even impossible, to watch all changes that occur in the surroundings. On the other hand, in many physical or chemical experiments, either the temperature or pressure of the surroundings is constant or both parameters are constant. Then does it exist a simpler way to describe spontaneous processes? Is it possible to describe such processes only in terms of state functions of the system? We will see that state functions called *thermodynamic potentials* serve exactly for this purpose.

Description of the system in terms of any thermodynamic potential is equivalent to the fundamental relation for that system, i.e., $S = S(U, V, n)$ or $U = U(S, V, n)$.[1] Strictly speaking, the relation between a given thermodynamic potential and a uniquely defined set of independent state parameters, called the *natural variables* of that potential, is equivalent to the fundamental relation $U = U(S, V, n)$. We will see that at least one natural variable is an intensive parameter. The choice of the appropriate thermodynamic potential depends on the intensive parameters of the system which are fixed by the interaction with the surroundings.

The thermodynamic potentials are derived from the fundamental relation by means of a mathematical procedure called the *Legendre transformation*, which appears in many fields of physics. For example, description of a mechanical system in terms of the Lagrangian is equivalent to the description of that system in terms of the Hamiltonian. The latter is related to the former by the Legendre transformation which replaces the velocities of the bodies as independent variables with their momenta.

[1] For simplicity, we consider here only one-component systems.

R. Hołyst, A. Poniewierski, *Thermodynamics for Chemists, Physicists and Engineers*,
DOI 10.1007/978-94-007-2999-5_5,

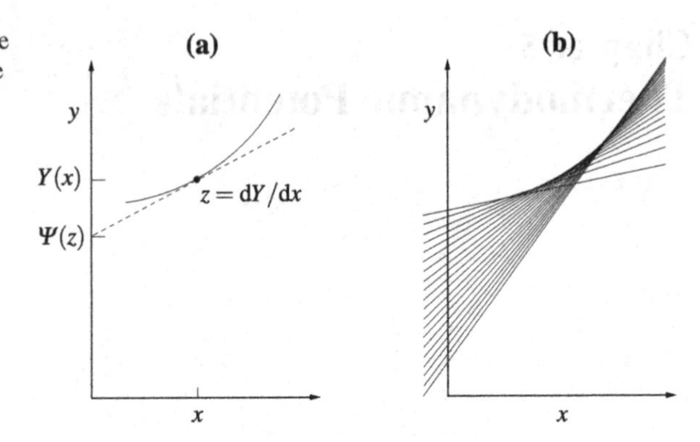

Fig. 5.1 Idea of the Legendre transformation. (**a**) The curve $y = Y(x)$ is represented as a set of the points $(x, Y(x))$. (**b**) The same curve is represented as a set of the tangents to the curve

5.1.1 Definition of the Legendre Transformation

Functions of One Variable The idea of the Legendre transformation for a function of one variable is presented in Fig. 5.1. If $Y(x)$ is a convex or concave function, i.e., its derivative $\mathrm{d}Y/\mathrm{d}x$ is a monotonic function of x, then the curve $y = Y(x)$ can be represented either as a set of the points $(x, Y(x))$ in the xy plane, or as a set of the tangents to this curve. The tangent to the curve at the given point x is defined by the slope, $z = \mathrm{d}Y/\mathrm{d}x$, and by the point of intersection of the tangent with the y axis, $\Psi(z)$. It follows from Fig. 5.1a that the slope satisfies the relation: $z = (Y - \Psi)/x$, hence

$$\Psi(z) = Y(x) - zx. \tag{5.1}$$

Ψ is a function of z because the relation $z = \mathrm{d}Y/\mathrm{d}x$ can be inverted, to give $x = x(z)$. The transformation defined by (5.1) together with the relation

$$z = \frac{\mathrm{d}Y}{\mathrm{d}x} \tag{5.2}$$

is called the *Legendre transformation* of the function $Y(x)$, and $\Psi(z)$ is the *Legendre transform* of that function. The relation $\psi = \Psi(z)$ defines a curve in the $z\psi$ plane, i.e., a set of the points $(z, \Psi(z))$, in the same way as the relation $y = Y(x)$ defines a curve in the xy plane. Both relations can be treated as equivalent representations of the same curve. Transition from one representation to the other is defined by formulae (5.1) and (5.2). Differentiating (5.1) with respect to z and using (5.2), we get

$$\frac{\mathrm{d}\Psi}{\mathrm{d}z} = -x. \tag{5.3}$$

We can also derive this result, taking the differential of Ψ. Since $\mathrm{d}Y = z\mathrm{d}x$, we have

$$\mathrm{d}\Psi = \mathrm{d}Y - z\mathrm{d}x - x\mathrm{d}z = -x\mathrm{d}z. \tag{5.4}$$

We can see now that the change of a given function into its Legendre transform replaces the independent variable with the dependent variable and vice versa. We can recover the function $Y(x)$ from $\Psi(z)$, using the formula

$$Y = \Psi + xz \tag{5.5}$$

together with relation (5.3). Note that the inverse transformation, defined by (5.3) and (5.5), has the same form as the original Legendre transformation, defined by (5.1) and (5.2), except the sign.

Functions of Several Variables For simplicity, we consider a function of two variables: $Y(x_1, x_2)$. Now there are two partial derivatives: $z_1 = (\partial Y/\partial x_1)_{x_2}$ and $z_2 = (\partial Y/\partial x_2)_{x_1}$, hence

$$dY = z_1 dx_1 + z_2 dx_2. \tag{5.6}$$

We can perform a partial Legendre transformation, i.e., with respect to one variable at a constant value of the second variable, for instance, with respect to x_1 at constant x_2:

$$\Psi(z_1, x_2) = Y(x_1, x_2) - z_1 x_1. \tag{5.7}$$

Then

$$d\Psi = -x_1 dz_1 + z_2 dx_2, \tag{5.8}$$

where $x_1 = -(\partial\Psi/\partial z_1)_{x_2}$, $z_2 = (\partial\Psi/\partial x_2)_{z_1}$. We have simply interchanged the roles of x_1 and z_1, whereas x_2 remains an independent variable. Note also that z_2 is a function of x_1 and x_2 in (5.6), whereas in (5.8), it is as a function of z_1 and x_2. In thermodynamics, we often deal with such a situation.

The second possibility is the full Legendre transformation of the function Y with respect to both variables. It is defined in an analogous way as in the case of one variable, i.e.,

$$\Psi(z_1, z_2) = Y(x_1, x_2) - z_1 x_1 - z_2 x_2. \tag{5.9}$$

We should remember about the condition that the transformation of (x_1, x_2) into (z_1, z_2) can be inverted. It follows from (5.6) and from definition (5.9) that

$$d\Psi = -x_1 dz_1 - x_2 dz_2, \tag{5.10}$$

where $x_1 = -(\partial\Psi/\partial z_1)_{z_2}$, $x_2 = -(\partial\Psi/\partial z_2)_{z_1}$. In this case, we have interchanged simultaneously the roles of x_1, x_2 and z_1, z_2 as independent and dependent variables.

Generalization to functions of more variables is straightforward. For instance, in the case of three variables, we can perform partial Legendre transformations with respect to one or two variables, or the full Legendre transformation with respect to three variables. It is easy to calculate that there are seven possibilities altogether.

Legendre Transformation of the Internal Energy All thermodynamic potentials result from partial Legendre transformations of the internal energy

$$U = U(S, V, n), \tag{5.11}$$

whose differential is given by

$$dU = T dS - p dV + \mu dn. \tag{5.12}$$

Thus, the natural variables of the internal energy are three extensive parameters: the entropy S, volume V and mole number n. The intensive parameters related to them are: the temperature T, pressure p (or rather $-p$) and chemical potential μ. They are equal to the respective partial derivatives of U (see (4.49)–(4.51)). Fundamental relation (5.11) contains complete information about the system in thermodynamic equilibrium. When we change from the internal energy to one of the thermodynamic potentials we do not lose any information about the system, however, the state parameters used as independent variables change. Not all thermodynamic potentials have practical meaning. We will also see that the full Legendre transformation with respect to all three variables leads to a function identical to zero, which follows from the fact that U, S, V and n are extensive parameters. Below we discuss the most important thermodynamic potentials: the Helmholtz free energy, enthalpy, Gibbs free energy and the grand thermodynamic potential.

5.1.2 Helmholtz Free Energy

When the system is in thermal contact with a heat reservoir its temperature is equal to the temperature of the reservoir. Therefore, it is convenient to have T, V and n as independent variables, which means that we have to interchange the entropy and temperature as the independent and dependent variable. We know from (5.12) that

$$T = \left(\frac{\partial U}{\partial S}\right)_{V,n}, \tag{5.13}$$

hence, the thermodynamic potential which results from the partial Legendre transformation of the internal energy with respect to the entropy, called the *Helmholtz free energy*, is defined as follows:

$$F = U - TS. \tag{5.14}$$

Taking the differential of both sides of (5.14) and using (5.12), we get

$$\mathrm{d}F = -S\mathrm{d}T - p\mathrm{d}V + \mu\mathrm{d}n. \tag{5.15}$$

From (5.15) we conclude that T, V and n are the independent variables, i.e., $F = F(T, V, n)$, whereas S, p and μ are functions of these variables, defined by the relations:

$$\left(\frac{\partial F}{\partial T}\right)_{V,n} = -S, \qquad \left(\frac{\partial F}{\partial V}\right)_{T,n} = -p, \qquad \left(\frac{\partial F}{\partial n}\right)_{T,V} = \mu. \tag{5.16}$$

Comparing the relations:

$$\left(\frac{\partial U}{\partial S}\right)_{V,n} = T \quad \text{and} \quad \left(\frac{\partial F}{\partial T}\right)_{V,n} = -S, \tag{5.17}$$

we notice that the entropy and temperature have been interchanged as the independent and dependent variable. Both $U = U(S, V, n)$ and $F = F(T, V, n)$ contain complete information about the system in thermodynamic equilibrium. In this sense, they are equivalent to each other. Nevertheless, the choice of one or the other depends on the actual physical situation.

A relation exists between the work W done on a closed system ($n = const$) and the change in the Helmholtz free energy in a reversible isothermal process. From the definition of F and the first law of thermodynamics, it follows that in such a process

$$\Delta F = \Delta U - T\Delta S = \Delta U - Q = W, \tag{5.18}$$

where $Q = T\Delta S$ is the heat received by the system. For that reason F is called the *free energy*. It is the part of the internal energy that can be entirely used up by a system at constant temperature to perform work.

5.1.3 Enthalpy

The next thermodynamic potential results from the partial Legendre transformation of the internal energy with respect to volume. It is called the *enthalpy*, H:

$$H = U + pV. \tag{5.19}$$

It was already introduced in Sect. 3.3. Note that the plus sign in front of pV results from the minus sign in the relation

$$\left(\frac{\partial U}{\partial V}\right)_{S,n} = -p. \tag{5.20}$$

Using (5.12), it is easy to show that

$$\mathrm{d}H = T\mathrm{d}S + V\mathrm{d}p + \mu\mathrm{d}n. \tag{5.21}$$

We infer from the form of $\mathrm{d}H$ that S, p and n are the independent variables, i.e., $H = H(S, p, n)$, whereas T, V and μ are given by the relations:

$$\left(\frac{\partial H}{\partial S}\right)_{p,n} = T, \qquad \left(\frac{\partial H}{\partial p}\right)_{S,n} = V, \qquad \left(\frac{\partial H}{\partial n}\right)_{S,p} = \mu. \tag{5.22}$$

Transforming the internal energy into enthalpy, we interchange V and p as the independent and dependent variable. In a closed system at constant pressure, the change in the enthalpy is equal to the heat received by the system, because

$$\Delta H = \Delta U + p\Delta V = \Delta U - W = Q, \tag{5.23}$$

where we have used definition (5.19) and the first law of thermodynamics.

5.1.4 Gibbs Free Energy

The thermodynamic potential obtained from the partial Legendre transformation of the internal energy with respect to the entropy S and volume V is called the *Gibbs free energy*, G, or simply the *Gibbs function*. It is defined by the formula

$$G = U - TS + pV. \tag{5.24}$$

The Gibbs free energy can also be expressed as $G = F + pV$ or $G = H - TS$, therefore, sometimes it is referred to as the *free enthalpy*. Taking the differential of both sides of (5.24) and using (5.12), we get

$$dG = -SdT + Vdp + \mu dn. \tag{5.25}$$

From (5.25), we infer that T, p and n are the independent parameters, i.e., $G = G(T, p, n)$, whereas S, V and μ are given by the relations:

$$\left(\frac{\partial G}{\partial T}\right)_{p,n} = -S, \qquad \left(\frac{\partial G}{\partial p}\right)_{T,n} = V, \qquad \left(\frac{\partial G}{\partial n}\right)_{T,p} = \mu. \tag{5.26}$$

Unlike the Helmholtz free energy and enthalpy, which are functions of one intensive and two extensive parameters, G depends on two intensive parameters: T and p, and on the extensive parameter n. Now the chemical potential has a simple interpretation in terms of G. It is related to the change in the Gibbs free energy caused by addition of an infinitesimal amount of matter to the system at constant temperature and pressure.

To derive the relation between the chemical potential, temperature and pressure, we use the Euler relation (see (4.60))

$$U = TS - pV + \mu n. \tag{5.27}$$

From the Euler relation, which simply reflects the fact that U, S, V and n are extensive parameters, another form of the Gibbs free energy follows, i.e.,

$$G = \mu n. \tag{5.28}$$

Thus, in the case of one-component systems, the chemical potential is identical with the molar Gibbs free energy $g = G/n$. Taking the differential of the two sides of (5.28), we get $dG = \mu dn + nd\mu$, which must agree with (5.25), hence

$$nd\mu = -SdT + Vdp. \tag{5.29}$$

Dividing both sides by n, we obtain an important relation called the *Gibbs–Duhem equation*:

$$d\mu = -sdT + vdp, \tag{5.30}$$

where $s = S/n$ and $v = V/n$ denote the molar entropy and molar volume, respectively. The Gibbs–Duhem equation can also be obtained from (5.25) when $n = 1$ is substituted. We conclude from (5.30) that in one-component systems the chemical potential is a function of temperature and pressure, and

$$\left(\frac{\partial \mu}{\partial T}\right)_{p} = -s, \qquad \left(\frac{\partial \mu}{\partial p}\right)_{T} = v. \tag{5.31}$$

In Sect. 4.3.3, we derived relations (5.31) for the ideal gas, and here we have shown that they hold in general.

Then we calculate the change in the Gibbs free energy in a reversible process at constant temperature and pressure. From the definition of G (see (5.24)) and the first law of thermodynamics, we get

$$\Delta G = \Delta U - T\Delta S + p\Delta V = \Delta U - Q + p\Delta V = W + p\Delta V, \tag{5.32}$$

where $Q = T\Delta S$, and W denote the total work performed on the system. Until now we have considered only mechanical work of the form $-p\Delta V$. However, also other kinds of work can be performed on the system, for instance, the work of the electric current. Denoting by $W' = W + p\Delta V$ all kinds of work different from $-p\Delta V$, we can express (5.32) as follows:

$$\Delta G = W'. \tag{5.33}$$

Formula (5.33) is very useful in electrochemistry because measuring the voltage of an electrochemical cell, we can determine the change in the Gibbs free energy in a chemical reaction (see Chap. 11).

5.1.5 Grand Thermodynamic Potential

The last thermodynamic potential to be discussed here results from the partial Legendre transformation of the internal energy with respect to the entropy and mole number:

$$\Omega = U - TS - \mu n. \tag{5.34}$$

Ω is called the *grand thermodynamic potential*. It follows from its definition that $d\Omega = dU - d(TS) - d(\mu n)$, hence

$$d\Omega = -SdT - pdV - nd\mu, \tag{5.35}$$

where we have used (5.12). The state parameters T, V and μ are the independent variables, i.e., $\Omega = \Omega(T, V, \mu)$, whereas S, p and n are given by the relations:

$$\left(\frac{\partial \Omega}{\partial T}\right)_{V,\mu} = -S, \quad \left(\frac{\partial \Omega}{\partial V}\right)_{T,\mu} = -p, \quad \left(\frac{\partial \Omega}{\partial \mu}\right)_{T,V} = -n. \tag{5.36}$$

Using the Euler relation, we can also express Ω as follows:

$$\Omega = -pV, \tag{5.37}$$

hence, $\Omega/V = -p$. The potential Ω is rarely used in phenomenological thermodynamics. However, it is very useful in statistical thermodynamics if we consider a system at constant temperature and volume, which in addition can exchange matter with the surroundings. Then we say that the system is contact with a heat and matter reservoir, which fixes the value of the temperature and chemical potential of the system.

Both the Gibbs free energy and the grand thermodynamic potential depend on two intensive parameters and one extensive parameter. Then is it possible to define a thermodynamic potential which depends only on intensive parameters? Such a potential must be a function of T, p and μ because it is given by the full Legendre transformation of the internal energy with respect to the three extensive parameters: S, V and n, which is $U - TS + pV - \mu n$. Note, however, that because of the Euler relation, it is identical to zero, which simply reflects the fact that T, p and μ are not independent of one another in one-component systems.

5.1.6 Massieu Functions

We can also consider the Legendre transformation of entropy. It is defined in a similar way as in the case of the internal energy. The state functions formed in this way are called the *Massieu functions*. The Massieu functions, as the thermodynamic potentials, also provide full information about a system in thermodynamic equilibrium. Here we give one example of a Massieu function, which will be used in Sect. 5.2.2, but then we will not use them any more. However, we recommend the reader to derive the remaining five Massieu functions for a one-component system.

To obtain the Massieu functions, we start from the fundamental relation $S = S(U, V, n)$ and use the expression for the entropy differential (cf. (5.12))

$$\mathrm{d}S = \frac{1}{T}\mathrm{d}U + \frac{p}{T}\mathrm{d}V - \frac{\mu}{T}\mathrm{d}n. \tag{5.38}$$

The intensive parameter coupled with the internal energy is $1/T$, hence, the Legendre transformation of the entropy with respect to the internal energy leads to the function

$$\Sigma = S - \frac{1}{T}U. \tag{5.39}$$

It is easy to see that $\Sigma = -F/T$. From relations (5.38) and (5.39), we get

$$\mathrm{d}\Sigma = -U\mathrm{d}\left(\frac{1}{T}\right) + \frac{p}{T}\mathrm{d}V - \frac{\mu}{T}\mathrm{d}n = \frac{U}{T^2}\mathrm{d}T + \frac{p}{T}\mathrm{d}V - \frac{\mu}{T}\mathrm{d}n. \tag{5.40}$$

Substituting $n = 1$, we obtain the following relation:

$$\mathrm{d}\sigma = \frac{u}{T^2}\mathrm{d}T + \frac{p}{T}\mathrm{d}v, \tag{5.41}$$

where $\sigma = \Sigma/n$, $u = U/n$ and $v = V/n$, from which we infer that σ is a function of T and v. If we know u and p as functions of T and v, then we can determine the function $\sigma(T, p)$. We use relation (5.41) in Sect. 5.2.2, to determine the thermodynamic potentials for the ideal gas.

5.2 Natural Variables

It often happens in thermodynamics that the same physical quantity is considered as a function of different variables. It is one of the reasons that the formalism used in thermodynamics may be considered difficult. In this section, we give more attention to this problem.

For instance, the dependence of entropy on U, V and n contains complete information about a given system in thermodynamic equilibrium, and the relation $S = S(U, V, n)$ is called the fundamental relation. Let us assume now that we have managed to determine the molar entropy of the system as a function of temperature and pressure: $s = s(T, p)$. Since entropy is an extensive parameter, we have $S(T, p, n) = ns(T, p)$. Is this information about the system sufficient to recover the fundamental relation? The answer is negative. To find out why, we first look for the thermodynamic potential which depends on T, p and n. We know that it is the Gibbs free energy G. As we also know, the relation $G = G(T, p, n)$ is equivalent to the fundamental relation, i.e., it also contains complete information about the system. If we knew the chemical potential as a function of T and p we would also know $G(T, p, n)$, since $G = n\mu(T, p)$ (see (5.28)), and we could recover the fundamental relation $S = S(U, V, n)$ by means of the inverse Legendre transformation. However, it follows from the Gibbs–Duhem equation:

$$\mathrm{d}\mu = -s(T, p)\mathrm{d}T + v(T, p)\mathrm{d}p, \tag{5.42}$$

that to determine $\mu(T, p)$, both the molar entropy and molar volume as functions of T and p are needed. Therefore, the relation $s = s(T, p)$ or $S = S(T, p, n)$ contains only partial information about the system.

By *natural variables* we understand the parameters of state whose relation with one of the state functions: the entropy, the internal energy, a thermodynamic potential or a Massieu function, contains complete information about the system in thermodynamic equilibrium.

5.2.1 Equivalent Representations of the Fundamental Relation

The fundamental relation $U = U(S, V, n)$ and the relations between thermodynamic potentials and their natural variables represent equivalent forms of description of a system in thermodynamic equilibrium, i.e., the relations: $F = F(T, V, n)$, $H = H(S, p, n)$, $G = G(T, p, n)$ and $\Omega = \Omega(T, V, \mu)$ are equivalent to the fundamental relation. In other words, they are equivalent representations of the fundamental relation. As we have already mentioned, in the case of one-component systems, there exist six thermodynamic potentials. For the remaining two potentials, which are included only for completeness, there are no commonly accepted names, therefore, we assume the symbols $\Psi = \Psi(S, V, \mu)$ and $\Theta = \Theta(S, p, \mu)$. Below we present a short summary of all equivalent representations of the fundamental relation. The potentials F, H and Ψ are partial Legendre transforms of the function $U(S, V, n)$ with respect to one extensive parameter, and the potentials G, Ω and Θ, with respect to two extensive parameters.

Fundamental relation: $U = U(S, V, n)$

$$\mathrm{d}U = T\mathrm{d}S - p\mathrm{d}V + \mu\mathrm{d}n,$$

$$\left(\frac{\partial U}{\partial S}\right)_{V,n} = T, \qquad \left(\frac{\partial U}{\partial V}\right)_{S,n} = -p, \qquad \left(\frac{\partial U}{\partial n}\right)_{S,V} = \mu.$$

Helmholtz free energy: $F = F(T, V, n)$

Legendre transform of $U(S, V, n)$ with respect to S;

$$F = U - TS, \qquad \mathrm{d}F = -S\mathrm{d}T - p\mathrm{d}V + \mu\mathrm{d}n,$$

$$\left(\frac{\partial F}{\partial T}\right)_{V,n} = -S, \qquad \left(\frac{\partial F}{\partial V}\right)_{T,n} = -p, \qquad \left(\frac{\partial F}{\partial n}\right)_{T,V} = \mu.$$

Enthalpy: $H = H(S, p, n)$
Legendre transform of $U(S, V, n)$ with respect to V;

$$H = U + pV, \qquad dH = TdS + Vdp + \mu dn,$$
$$\left(\frac{\partial H}{\partial S}\right)_{p,n} = T, \qquad \left(\frac{\partial H}{\partial p}\right)_{S,n} = V, \qquad \left(\frac{\partial H}{\partial n}\right)_{S,p} = \mu.$$

Potential Ψ: $\Psi = \Psi(S, V, \mu)$
Legendre transform of $U(S, V, n)$ with respect to n;

$$\Psi = U - \mu n, \qquad d\Psi = TdS - pdV - nd\mu,$$
$$\left(\frac{\partial \Psi}{\partial S}\right)_{V,\mu} = T, \qquad \left(\frac{\partial \Psi}{\partial V}\right)_{S,\mu} = -p, \qquad \left(\frac{\partial \Psi}{\partial \mu}\right)_{S,V} = -n.$$

Gibbs free energy: $G = G(T, p, n)$
Legendre transform of $U(S, V, n)$ with respect to S and V;

$$G = U - TS + pV, \qquad dG = -SdT + Vdp + \mu dn,$$
$$\left(\frac{\partial G}{\partial T}\right)_{p,n} = -S, \qquad \left(\frac{\partial G}{\partial p}\right)_{T,n} = V, \qquad \left(\frac{\partial G}{\partial n}\right)_{T,p} = \mu.$$

Grand thermodynamic potential: $\Omega = \Omega(T, V, \mu)$
Legendre transform of $U(S, V, n)$ with respect to S and n;

$$\Omega = U - TS - \mu n, \qquad d\Omega = -SdT - pdV - nd\mu,$$
$$\left(\frac{\partial \Omega}{\partial T}\right)_{V,\mu} = -S, \qquad \left(\frac{\partial \Omega}{\partial V}\right)_{T,\mu} = -p, \qquad \left(\frac{\partial \Omega}{\partial \mu}\right)_{T,V} = -n.$$

Potential Θ: $\Theta = \Theta(S, p, \mu)$
Legendre transform of $U(S, V, n)$ with respect to V and n;

$$\Theta = U + pV - \mu n, \qquad d\Theta = TdS + Vdp - nd\mu,$$
$$\left(\frac{\partial \Theta}{\partial S}\right)_{p,\mu} = T, \qquad \left(\frac{\partial \Theta}{\partial p}\right)_{S,\mu} = V, \qquad \left(\frac{\partial \Theta}{\partial \mu}\right)_{S,p} = -n.$$

5.2.2 Thermodynamic Potentials for the Ideal Gas

The molar entropy of the ideal gas, $s(u, v)$, was already calculated in Sect. 4.3.3. Now we want to determine the thermodynamic potentials. To do it, it is convenient to use relation (5.41). Substituting the equations of state: $u = fRT/2$ and $pv = RT$, where f denotes the number of degrees of freedom of one molecule, we get

$$d\sigma = \frac{fR}{2T}dT + \frac{R}{v}dv. \tag{5.43}$$

It is easy to verify that relation (5.43) is satisfied by the function:

$$\sigma(T, v) = \sigma_0 + \frac{1}{2} f R \ln \frac{T}{T_0} + R \ln \frac{v}{v_0}, \tag{5.44}$$

where the constants σ_0, T_0 and v_0 define the reference state. From relation (5.39), we determine the molar entropy: $s = \sigma + u/T$. Expressing T by u, we recover (4.68):

$$s(u, v) = s_0 + \frac{1}{2} f R \ln \frac{u}{u_0} + R \ln \frac{v}{v_0}, \tag{5.45}$$

where $s_0 = \sigma_0 + f R/2$. Then, by inverting the relation $s = s(u, v)$, we obtain:

$$u(s, v) = u_0 \left(\frac{v_0}{v} \right)^{2/f} \exp\left[\frac{2(s - s_0)}{f R} \right]. \tag{5.46}$$

Helmholtz Free Energy The molar Helmholtz free energy, $\phi = F/n$, follows from the relation $\phi/T = -\sigma$:

$$\frac{\phi(T, v)}{T} = \frac{\phi_0}{T_0} - \frac{1}{2} f R \ln \frac{T}{T_0} - R \ln \frac{v}{v_0}, \tag{5.47}$$

where $\phi_0 = \phi(T_0, v_0) = -\sigma_0 T_0$. Since F is an extensive quantity, we have

$$F(T, V, n) = n F(T, V/n, 1) = n \phi(T, V/n). \tag{5.48}$$

Enthalpy The molar enthalpy is equal to $h = H/n = u + pv = (1 + 2/f)u$. The natural variables for h are the molar entropy and pressure, therefore, we have to express u as a function of s and p. The substitution of $v/v_0 = (u/u_0)(p_0/p)$ into (5.45) gives

$$s(u, p) = s_0 + \frac{1}{2}(f + 2) R \ln \frac{u}{u_0} - R \ln \frac{p}{p_0}, \tag{5.49}$$

where $u_0 = f R T_0/2$, $p_0 = R T_0/v_0$, hence,

$$h(s, p) = h_0 \left(\frac{p}{p_0} \right)^{2/(f+2)} \exp\left[\frac{2(s - s_0)}{(f + 2)R} \right], \tag{5.50}$$

where $h_0 = (1 + 2/f)u_0$. Since $H(S, p, n)$ is an extensive quantity, we have

$$H(S, p, n) = n H(S/n, p, 1) = n h(S/n, p). \tag{5.51}$$

Gibbs Free Energy For one-component systems, the molar Gibbs free energy is equal to the chemical potential, and since $G = F + pV$, we have $\mu = \phi + pv$. Using (5.47) and the equation of state: $pv = RT$, we obtain (cf. (4.70)):

$$\frac{\mu(T, p)}{T} = \frac{\mu_0}{T_0} - \frac{1}{2}(f + 2) R \ln \frac{T}{T_0} + R \ln \frac{p}{p_0}, \tag{5.52}$$

where $\mu_0 = \phi_0 + R T_0$, and

$$G(T, p, n) = n G(T, p, 1) = n \mu(T, p). \tag{5.53}$$

Grand Thermodynamic Potential The natural variables for Ω are T, V and μ. Since $\Omega/V = -p$ we have

$$\Omega(T, V, \mu) = -Vp(T, \mu). \tag{5.54}$$

To obtain the pressure as a function of T and μ, we invert relation (5.52) at constant temperature, hence

$$p(T, \mu) = p_0 \left(\frac{T}{T_0}\right)^{(f+2)/2} \exp\left(\frac{\mu}{RT} - \frac{\mu_0}{RT_0}\right). \tag{5.55}$$

Finally, we notice that all constants that appear in Eqs. (5.45)–(5.55) can be expressed by three independent constants, which we have chosen to be σ_0, T_0 and v_0. We recommend the reader to verify that using an arbitrary thermodynamic potential, one can derive the equations of state of the ideal gas.

5.3 Free-Energy Minimum Principle

Using the second law of thermodynamics, we derived in Sect. 4.1.1 the entropy maximum principle for an isolated system composed of a certain number of subsystems. When the internal constraints imposed initially on the subsystems are removed, the system tends to a new equilibrium state. In other words, a spontaneous process occurs in the system. The entropy maximum principle enables us to determine the state parameters of the subsystems that correspond to the equilibrium state of the system without internal constraints. The following question can be raised. Since the relation between any thermodynamic potential and its natural variables is equivalent to the fundamental relation $U = U(S, V, n)$ or $S = S(U, V, n)$, is it possible to formulate, for a given potential, a principle analogous to the entropy maximum principle for isolated systems? We will show that in the case of the Helmholtz and Gibbs free energy.

5.3.1 Systems at Constant Temperature

We want to find the condition which any spontaneous process occurring at constant temperature must satisfy. To do this, we consider a closed system at constant volume and in thermal contact with a heat reservoir whose temperature is denoted by T^r. The system considered and the reservoir form together an isolated system, which means that the condition

$$\Delta\big(U + U^r\big) = 0, \tag{5.56}$$

holds, where U and U^r denote the internal energy of the system and reservoir, respectively. According to the second law of thermodynamics, the change in the total entropy of the system and reservoir in any process satisfies the inequality

$$\Delta S + \Delta S^r \geq 0. \tag{5.57}$$

We treat the size of the reservoir as infinite compared to the size of the system. Therefore, the transfer of the energy $\Delta U^r = -\Delta U$ between the system and reservoir does not change the temperature of the latter. From the point of view of the reservoir, ΔU^r can be treated as an infinitesimal amount of energy, hence

$$\Delta S^r = \frac{\Delta U^r}{T^r}. \tag{5.58}$$

Substituting (5.58) into (5.57) and replacing ΔU^r with $-\Delta U$, we get the following inequality:

$$\Delta\big(U - T^r S\big) \le 0. \tag{5.59}$$

Since $U - T^r S$ is equal to the Helmholtz free energy of the system considered ($F = U - TS$) provided that $T = T^r$, we can formulate the following conclusion, which is a consequence of the second law of thermodynamics.

Corollary 5.1 *When a spontaneous process occurs in a closed system at constant volume and in contact with a heat reservoir, the change in the Helmholtz free energy of the system satisfies the inequality*

$$\Delta F \le 0,$$

where $\Delta F = 0$ *only in the case of a reversible process.*

A given process can occur spontaneously in a closed system at constant volume and at constant temperature if the initial state of the system is an equilibrium state in the presence of some internal constraints. Therefore, we consider a system composed of m subsystems which are initially separated from one another with rigid diathermal walls, and the whole system is in thermal contact with a heat reservoir at the temperature T^r. The temperature of each subsystem is equal to the temperature of the reservoir, i.e., $T^{(i)} = T^r$, for $i = 1, \ldots, m$. The total volume of the system, V, and the total mole number, n, are constant:

$$\sum_{i=1}^{m} V^{(i)} = V, \qquad \sum_{i=1}^{m} n^{(i)} = n. \tag{5.60}$$

The Helmholtz free energy is an extensive quantity, hence

$$F\big(T^r, V^{(1)}, n^{(1)}, \ldots, V^{(m)}, n^{(m)}\big) = \sum_{i=1}^{m} F^{(i)}\big(T^r, V^{(i)}, n^{(i)}\big), \tag{5.61}$$

where $F^{(i)}$ and F correspond to the ith subsystem and the whole system, respectively. After the internal constraints are removed the system tends to a new equilibrium state with a lower value of F. The *Helmholtz free-energy minimum principle* follows from Corollary 5.1.

Corollary 5.2 *A closed system at constant volume and in thermal contact with a heat reservoir, after the removal of internal constraints, reaches a new equilibrium state that corresponds to the minimum of the Helmholtz free energy on the set of all equilibrium states of the system in the presence of the constraints.*

By a system with constraints we understand a system composed of subsystems separated by internal walls. Equilibrium states of such a system are defined by the parameters $V^{(i)}$ and $n^{(i)}$, $i = 1, \dots, m$, which satisfy conditions (5.60), and by the condition of thermal equilibrium between the subsystems and the heat reservoir, i.e., $T^{(i)} = T^r$. To prove statement 5.2, we take the state parameters that minimize F. If they did not correspond to the equilibrium state of the system without constraints then, after the removal of the constraints, the system would have to tend spontaneously to a certain equilibrium state of a higher value of F, which would contradict the condition $\Delta F \leq 0$. The Helmholtz free-energy minimum principle is merely a conclusion from the second law of thermodynamics. However, it is closer to the actual experimental situation and easier to apply than the entropy maximum principle.

Using the necessary condition for a minimum of F ($\mathrm{d}F = 0$) at constant temperature, we get

$$\mathrm{d}F = \sum_{i=2}^{m} \left[-\left(p^{(i)} - p^{(1)}\right)\mathrm{d}V^{(i)} + \left(\mu^{(i)} - \mu^{(1)}\right)\mathrm{d}n^{(i)} \right] = 0, \tag{5.62}$$

where we have used (5.15) and (5.60). From condition (5.62), it follows that pressures and chemical potentials are the same in all subsystems. In this way, we have recovered the conditions of mechanical equilibrium and equilibrium with respect to the flow of matter, derived in Sect. 4.2 from the entropy maximum principle. The equality of temperatures is ensured by thermal contact of the system with the heat reservoir.

Work in the Isothermal Process Now we consider a system in thermal contact with a heat reservoir. The system performs work W^* on the surroundings, from which it is thermally insulated. The system and reservoir form together an adiabatically isolated system, hence, the inequality (5.57) must be satisfied. The work performed by the system amounts to

$$W^* = -\Delta\left(U + U^r\right) = -\Delta U - T^r \Delta S^r, \tag{5.63}$$

where we have used (5.58). Using (5.63) and (5.57), we get

$$W^* \leq -\Delta F, \tag{5.64}$$

where F is the Helmholtz free energy of the system at the temperature $T = T^r$. In the case of a reversible process, $W^* = -\Delta F$ (cf. (5.18)), hence

$$W^*_{\mathrm{irr}} < W^*_{\mathrm{rev}} = -\Delta F, \tag{5.65}$$

where W^*_{rev} and W^*_{irr} denote, respectively, the work in a reversible and irreversible process between the same two states.

5.3.2 Systems at Constant Temperature and Pressure

In practice, we often deal with processes occurring both at constant temperature and pressure. We can imagine a system in thermal contact with the surroundings and closed with a movable piston, to ensure the equality of temperatures and pressures in the system and surroundings. Such a system is said to be in contact with a heat and volume reservoir. Since the size of the reservoir is treated as infinite compared to the size of the system, a transfer of heat or a change in the volume of the system does not influence the temperature and pressure of the reservoir.

We assume that the system in thermal and mechanical contact with the heat and volume reservoir, at the temperature T^r and pressure p^r, is closed. The system and reservoir form together an isolated system, hence

$$\Delta(U+U^r)=0 \quad \text{and} \quad \Delta(V+V^r)=0, \tag{5.66}$$

where U and V correspond to the system, and U^r and V^r correspond to the reservoir. From the second law of thermodynamics, it follows that

$$\Delta S+\Delta S^r \geq 0. \tag{5.67}$$

Because of infinite size of the reservoir, changes in its internal energy, $\Delta U^r = -\Delta U$, or volume, $\Delta V^r = -\Delta V$, can be treated as infinitesimal quantities, hence

$$\Delta S^r = \frac{\Delta U^r}{T^r} + \frac{p^r \Delta V^r}{T^r} = -\frac{\Delta U}{T^r} - \frac{p^r \Delta V}{T^r}. \tag{5.68}$$

Substituting (5.68) into inequality (5.67), we get

$$\Delta(U - T^r S + p^r V) \leq 0. \tag{5.69}$$

The function $U - T^r S + p^r V$ is equal to the Gibbs free energy of the system ($G = U - TS + pV$) provided that $T = T^r$ and $p = p^r$. Therefore, the following conclusion can be formulated.

Corollary 5.3 *When a spontaneous process occurs in a closed system in contact with a heat and volume reservoir, the change in the Gibbs free energy of the system satisfies the inequality*

$$\Delta G \leq 0,$$

where $\Delta G = 0$ only in the case of a reversible process.

The minimum principle for the potential G is obtained in an analogous way as in the case of F. We consider a closed system composed of subsystems, separated from one another with internal walls which allow a flow of heat and changes in the volume of the subsystems. The whole system is in contact with the heat and volume reservoir, i.e., $T^{(i)} = T^r$ and $p^{(i)} = p^r$ for all subsystems. The condition

$$\sum_{i=1}^{m} n_k^{(i)} = n_k, \tag{5.70}$$

must be satisfied for each component, where the total mole numbers, n_k, are constant. Here we consider the general case of a multi-component system. Recall that the chemical potential of a one-component system is a function of temperature and pressure. The total Gibbs free energy, which is an extensive quantity, is given by the following expression:

$$G\big(T^r, p^r, \{n_k^{(1)}\}, \ldots, \{n_k^{(m)}\}\big) = \sum_{i=1}^{m} G^{(i)}\big(T^r, p^r, \{n_k^{(i)}\}\big), \tag{5.71}$$

where $\{n_k^{(i)}\}$ denotes the set of mole numbers of all components in the ith subsystem. From Corollary 5.3, the *Gibbs free-energy minimum principle* follows.

Corollary 5.4 *A closed system in thermal contact with a heat and volume reservoir, after the removal of internal constraints, reaches a new equilibrium state that corresponds to the minimum of the Gibbs free energy on the set of all equilibrium states of the system in the presence of the constraints.*

An equilibrium state of the system with the constraints is defined by the mole numbers $\{n_k^{(i)}\}$, which satisfy conditions (5.70), and by the conditions of thermal and mechanical equilibrium: $T^{(i)} = T^r$ and $p^{(i)} = p^r$, for $i = 1, \ldots, m$. After the removal of the constraints, a flow of matter between the subsystems usually occurs.

Using the necessary condition for a minimum of G ($\mathrm{d}G = 0$) at constant temperature and pressure, we get

$$\mathrm{d}G = \sum_{i=1}^{m} \mathrm{d}G^{(i)} = 0, \tag{5.72}$$

where

$$\mathrm{d}G^{(i)} = \sum_{k} \mu_k^{(i)} \mathrm{d}n_k^{(i)}, \tag{5.73}$$

and we have generalized expression (5.25) to an arbitrary number of components; $\mu_k^{(i)}$ denotes the chemical potential of the kth component in the ith subsystem. Because of condition (5.70), for each component k, we have

$$\sum_{i=1}^{m} \mathrm{d}n_k^{(i)} = 0. \tag{5.74}$$

Finally, we transform condition (5.72) into the following form:

$$\mathrm{d}G = \sum_{k} \sum_{i=2}^{m} \big(\mu_k^{(i)} - \mu_k^{(1)}\big) \mathrm{d}n_k^{(i)} = 0. \tag{5.75}$$

From (5.75), we conclude that in thermodynamic equilibrium, the chemical potential of each component must have the same value in all subsystems, i.e., $\mu_k^{(i)} = \mu_k^{(1)}$ for all indices i and k. In this way, using the Gibbs free-energy minimum principle, we have recovered the condition of equilibrium with respect to the flow of matter, derived earlier form the entropy maximum principle (see Sect. 4.2). The equality of temperatures and pressures in the subsystems is ensured by the contact with the heat and volume reservoir.

Work in the Isothermal–Isobaric Process We consider a system in thermal and mechanical contact with the heat and volume reservoir at the temperature T^r and pressure p^r. The system performs the work W^* on the surroundings but no heat is transferred between them. The system and reservoir form together an adiabatically isolated system, hence, inequality (5.67) must be satisfied for any spontaneous process. Moreover, we assume that the total volume of the system and reservoir does not change. This means that the work performed by the system on the surroundings is not the work due to the change of its volume. We have

$$W^* = -\Delta\left(U + U^r\right) = -\Delta U - T^r \Delta S^r + p^r \Delta V^r = -\Delta U - T^r \Delta S^r - p^r \Delta V, \tag{5.76}$$

where we have used (5.68) and the condition $\Delta(V + V^r) = 0$. From (5.76) and inequality (5.67), it follows that

$$W^* \leq -\Delta G, \tag{5.77}$$

where G is the Gibbs free energy of the system at the temperature T^r and pressure p^r. If the process is reversible then $W^* = -\Delta G$ (cf. (5.33)), hence

$$W^*_{\mathrm{irr}} < W^*_{\mathrm{rev}} = -\Delta G, \tag{5.78}$$

where W^*_{rev} and W^*_{irr} denote, respectively, the non-volume work in a reversible and irreversible process between the same two states.

5.4 Examples of Application of Thermodynamic Potentials

The problems we have discussed so far show that to describe a one-component uniform system in thermodynamic equilibrium, we need six state parameters, i.e., three extensive parameters: S, V and n, and three coupled to them intensive parameters: T, p and μ. In each couple of the parameters: (S, T), (V, p) and (n, μ), we can interchange the independent and dependent variable, using an appropriate thermodynamic potential. If we assume that the mole number is constant we are left with two couples of the state parameters: (S, T) and (V, p). Therefore, we have four possible sets of independent variables: S and V, T and V, S and p, T and p, which correspond to U, F, H and G, respectively. Derivation of various thermodynamic relations often consists in differentiation of one parameter with respect to another parameter at a constant value of a third parameter. To do this, a few simple rules are used, which are presented and derived below.

5.4.1 Rules of Calculation of Some Partial Derivatives

We assume that there are four state parameters: X, Y, Z and W, and only two of them are independent parameters. Then it can be shown that the following relations between the partial derivatives hold:

$$\left(\frac{\partial X}{\partial Y}\right)_Z = \left(\frac{\partial Y}{\partial X}\right)_Z^{-1}, \tag{5.79}$$

$$\left(\frac{\partial X}{\partial Y}\right)_Z = \left(\frac{\partial X}{\partial W}\right)_Z \left(\frac{\partial W}{\partial Y}\right)_Z, \tag{5.80}$$

$$\left(\frac{\partial X}{\partial Y}\right)_Z = -\left(\frac{\partial Z}{\partial Y}\right)_X \left(\frac{\partial Z}{\partial X}\right)_Y^{-1}, \tag{5.81}$$

$$\left(\frac{\partial Z}{\partial Y}\right)_W = \left(\frac{\partial Z}{\partial Y}\right)_X + \left(\frac{\partial Z}{\partial X}\right)_Y \left(\frac{\partial X}{\partial Y}\right)_W. \tag{5.82}$$

Derivation Relation (5.79) is simply the derivative of an inverse function. Inversion of the relation $Y = Y(X, Z)$ at constant Z gives $X = X(Y, Z)$. Then, differentiating both sides of the relation $Y = Y(X(Y, Z), Z)$ with respect to Y, we get

$$1 = \left(\frac{\partial Y}{\partial X}\right)_Z \left(\frac{\partial X}{\partial Y}\right)_Z, \tag{5.83}$$

from which (5.79) follows.

Relation (5.80) is also a known rule of differentiation of a composed function. If the variable W in the function $X(W, Z)$ is itself a function of Y and Z then differentiating the function $X(W(Y, Z), Z)$ with respect to Y at constant Z, we get (5.80).

It is easy to verify that relation (5.81) follows from (5.82) when $W = Z$ is substituted, so it remains to prove relation (5.82). Note that if in the function $Z(X, Y)$ the variable X is a function of Y and W then differentiating the composed function $Z(X(Y, W), Y)$ with respect to Y at constant W, we recover (5.82).

5.4.2 Maxwell Relations

The internal energy, $U(S, V, n)$, and three thermodynamic potentials: the Helmholtz free energy, $F(T, V, n)$, the enthalpy, $H(S, p, n)$, and the Gibbs free energy, $G(T, p, n)$, are the most often used functions of state. As we know, their first partial derivatives express relations between independent and dependent parameters of state, which follows from the form of the differentials of these functions (see Sect. 5.2.1):

$$\begin{aligned}
\mathrm{d}U &= T\mathrm{d}S - p\mathrm{d}V + \mu\mathrm{d}n, \\
\mathrm{d}F &= -S\mathrm{d}T - p\mathrm{d}V + \mu\mathrm{d}n, \\
\mathrm{d}H &= T\mathrm{d}S + V\mathrm{d}p + \mu\mathrm{d}n, \\
\mathrm{d}G &= -S\mathrm{d}T + V\mathrm{d}p + \mu\mathrm{d}n,
\end{aligned}$$

where we have assumed a one-component system. Since the mixed second order partial derivatives of a function do not depend on the order of differentiation, there must exist some relations between the first derivatives of the state parameters corresponding to these mixed derivatives. They are called the *Maxwell relations*. For a function of three variables, there are three equalities of mixed derivatives, hence, three Maxwell relations follow. In practice, the most useful Maxwell relations are obtained at a constant mole number, therefore, we restrict ourselves only to this case. From the form of the differentials $\mathrm{d}U$, $\mathrm{d}F$, $\mathrm{d}H$ and $\mathrm{d}G$, the following Maxwell relations are derived:

$$\left(\frac{\partial^2 U}{\partial S \partial V}\right)_n = \left(\frac{\partial^2 U}{\partial V \partial S}\right)_n \Rightarrow \left(\frac{\partial T}{\partial V}\right)_{S,n} = -\left(\frac{\partial p}{\partial S}\right)_{V,n}, \tag{5.84}$$

$$\left(\frac{\partial^2 F}{\partial T \partial V}\right)_n = \left(\frac{\partial^2 F}{\partial V \partial T}\right)_n \Rightarrow \left(\frac{\partial S}{\partial V}\right)_{T,n} = \left(\frac{\partial p}{\partial T}\right)_{V,n}, \tag{5.85}$$

$$\left(\frac{\partial^2 H}{\partial S \partial p}\right)_n = \left(\frac{\partial^2 H}{\partial p \partial S}\right)_n \Rightarrow \left(\frac{\partial T}{\partial p}\right)_{S,n} = \left(\frac{\partial V}{\partial S}\right)_{p,n}, \tag{5.86}$$

$$\left(\frac{\partial^2 G}{\partial T \partial p}\right)_n = \left(\frac{\partial^2 G}{\partial p \partial T}\right)_n \Rightarrow \left(\frac{\partial S}{\partial p}\right)_{T,n} = -\left(\frac{\partial V}{\partial T}\right)_{p,n}. \tag{5.87}$$

Maxwell relations are useful because they allow to express quantities that cannot be easily determined in experiment in terms of other directly measurable quantities. For instance, due to relation (5.85), we can determine the dependence of entropy on volume at constant temperature by the measurement of pressure as a function of temperature at constant volume.

5.4.3 Second Partial Derivatives of the Internal Energy and Thermodynamic Potentials

We show below that the second derivatives of thermodynamic potentials are related to the quantities that can be measured by standard methods. These quantities characterize the behaviour of a system influenced by some changes in the surroundings, for instance, a change in the volume of the system caused by an increase in the temperature of the surroundings at constant pressure. We have already met some of these quantities in Chap. 3. We assume again that the system contains a pure substance and that the mole number does not change.

Gibbs Free Energy We begin with the calculation of the second derivatives of G:

$$\left(\frac{\partial^2 G}{\partial T^2}\right)_p = -\left(\frac{\partial S}{\partial T}\right)_p, \tag{5.88}$$

$$\left(\frac{\partial^2 G}{\partial p^2}\right)_T = \left(\frac{\partial V}{\partial p}\right)_T, \tag{5.89}$$

$$\frac{\partial^2 G}{\partial p \partial T} = \left(\frac{\partial V}{\partial T}\right)_p, \tag{5.90}$$

where we have suppressed the constant parameter n, to simplify the notation. We recall that

$$c_p = \frac{T}{n}\left(\frac{\partial S}{\partial T}\right)_p, \tag{5.91}$$

denotes the molar heat capacity at constant pressure (see Sect. 4.3.4). We define two more quantities: the *isothermal compressibility*,

$$\kappa_T = -\frac{1}{V}\left(\frac{\partial V}{\partial p}\right)_T, \tag{5.92}$$

and the *thermal expansion coefficient*,

$$\alpha = \frac{1}{V}\left(\frac{\partial V}{\partial T}\right)_p. \tag{5.93}$$

The first one characterizes the relative change in the volume caused by the pressure exerted on the system at constant temperature. The higher κ_T is, the easier to compress the given substance. Gases are much more compressible than liquids and solids. The second quantity characterizes the relative change in the volume caused by an increase in the temperature of the system at constant pressure. The higher α is, the more the substance can expand. The thermal expansion coefficient appeared already in Sect. 2.3.2, where we discussed thermometers (see (2.19)). Using relations (5.88)–(5.90), we can express the second derivatives of G by c_p, κ_T and α:

$$\left(\frac{\partial^2 G}{\partial T^2}\right)_p = -\frac{n c_p}{T}, \qquad \left(\frac{\partial^2 G}{\partial p^2}\right)_T = -n v \kappa_T, \qquad \frac{\partial^2 G}{\partial p \partial T} = n v \alpha, \tag{5.94}$$

where $v = V/n$. We show below that also the second derivatives of the Helmholtz free energy, enthalpy and internal energy can be expressed in terms of c_p, κ_T and α.

Helmholtz Free Energy For the function F, we have:

$$\left(\frac{\partial^2 F}{\partial T^2}\right)_V = -\left(\frac{\partial S}{\partial T}\right)_V, \tag{5.95}$$

$$\left(\frac{\partial^2 F}{\partial V^2}\right)_T = -\left(\frac{\partial p}{\partial V}\right)_T, \tag{5.96}$$

$$\frac{\partial^2 F}{\partial V \partial T} = -\left(\frac{\partial p}{\partial T}\right)_V. \tag{5.97}$$

We know that the molar heat capacity at constant volume (see Sect. 4.3.4) is defined as

$$c_v = \frac{T}{n}\left(\frac{\partial S}{\partial T}\right)_V. \tag{5.98}$$

Then we have

$$\left(\frac{\partial p}{\partial V}\right)_T = \left(\frac{\partial V}{\partial p}\right)_T^{-1} = -\frac{1}{V\kappa_T}, \tag{5.99}$$

where we have used (5.79) and (5.92). The derivative $(\partial p/\partial T)_V$ is transformed according to (5.81):

$$\left(\frac{\partial p}{\partial T}\right)_V = -\left(\frac{\partial V}{\partial T}\right)_p \left(\frac{\partial V}{\partial p}\right)_T^{-1} = \frac{\alpha}{\kappa_T}. \tag{5.100}$$

Thus, we can express the second derivatives of F in the following form:

$$\left(\frac{\partial^2 F}{\partial T^2}\right)_V = -\frac{nc_v}{T}, \qquad \left(\frac{\partial^2 F}{\partial V^2}\right)_T = \frac{1}{nv\kappa_T}, \qquad \frac{\partial^2 F}{\partial V \partial T} = -\frac{\alpha}{\kappa_T}. \tag{5.101}$$

It can be shown that c_v can also be expressed by c_p, α and κ_T. Applying (5.82) to the parameters S, T, p and V, we get

$$\left(\frac{\partial S}{\partial T}\right)_p = \left(\frac{\partial S}{\partial T}\right)_V + \left(\frac{\partial S}{\partial V}\right)_T \left(\frac{\partial V}{\partial T}\right)_p. \tag{5.102}$$

Then using the Maxwell relation $(\partial S/\partial V)_T = (\partial p/\partial T)_V$ and (5.100), and the definitions of c_p, c_v and α, we arrive at the following relation:

$$c_p - c_v = \frac{Tv\alpha^2}{\kappa_T}. \tag{5.103}$$

The inequality $c_p > c_v$ results from the fact that when a system absorbs heat at constant pressure it also performs some work by increasing its volume, so the heat absorbed is only partially used to warm the system. On the other hand, the whole heat absorbed by a system at constant volume is used to increase its temperature. Therefore, the same change in the temperature requires more heat in the former case than in the latter. It is easy to show that in the case of the ideal gas, $\alpha = 1/T$, $\kappa_T = 1/p$, and $c_p - c_v = R$.

Enthalpy For the function H, we have

$$\left(\frac{\partial^2 H}{\partial S^2}\right)_p = \left(\frac{\partial T}{\partial S}\right)_p, \tag{5.104}$$

$$\left(\frac{\partial^2 H}{\partial p^2}\right)_S = \left(\frac{\partial V}{\partial p}\right)_S, \tag{5.105}$$

$$\frac{\partial^2 H}{\partial S \partial p} = \left(\frac{\partial T}{\partial p}\right)_S. \tag{5.106}$$

The derivative $(\partial T/\partial S)_p$ can be expressed by c_p:

$$\left(\frac{\partial T}{\partial S}\right)_p = \left(\frac{\partial S}{\partial T}\right)_p^{-1} = \frac{T}{nc_p}. \tag{5.107}$$

Analogously with the isothermal compressibility, the *adiabatic compressibility* is defined:

$$\kappa_S = -\frac{1}{V}\left(\frac{\partial V}{\partial p}\right)_S. \tag{5.108}$$

Here the entropy is constant, instead of temperature, because the compression occurs without flow of heat, and in a reversible adiabatic process the entropy does not change. Then, using (5.81), we transform $(\partial T/\partial p)_S$ as follows:

$$\left(\frac{\partial T}{\partial p}\right)_S = -\left(\frac{\partial S}{\partial p}\right)_T\left(\frac{\partial S}{\partial T}\right)_p^{-1} = \frac{Tv\alpha}{c_p}, \tag{5.109}$$

where we have used Maxwell relation (5.87) and the definitions of α and c_p. Thus, we have

$$\left(\frac{\partial^2 H}{\partial S^2}\right)_p = \frac{T}{nc_p}, \qquad \left(\frac{\partial^2 H}{\partial p^2}\right)_S = -nv\kappa_S, \qquad \frac{\partial^2 H}{\partial S\partial p} = \frac{Tv\alpha}{c_p}. \tag{5.110}$$

Finally, we show that κ_S can also be expressed by κ_T, c_p and α. To do this, we apply (5.82) to the parameters V, p, S and T, which gives

$$\left(\frac{\partial V}{\partial p}\right)_S = \left(\frac{\partial V}{\partial p}\right)_T + \left(\frac{\partial V}{\partial T}\right)_p\left(\frac{\partial T}{\partial p}\right)_S. \tag{5.111}$$

Then, using the definitions of κ_T and α, and relation (5.109), we get

$$\kappa_T - \kappa_S = \frac{Tv\alpha^2}{c_p}. \tag{5.112}$$

It is more difficult to compress a system in an adiabatic process than in the isothermal one ($\kappa_S < \kappa_T$) because its temperature increases during the adiabatic compression. Note also that the following relation results from (5.103) and (5.112):

$$\frac{\kappa_S}{\kappa_T} = \frac{c_v}{c_p}. \tag{5.113}$$

Internal Energy For the second derivatives of U, we get

$$\left(\frac{\partial^2 U}{\partial S^2}\right)_V = \left(\frac{\partial T}{\partial S}\right)_V, \tag{5.114}$$

$$\left(\frac{\partial^2 U}{\partial V^2}\right)_S = -\left(\frac{\partial p}{\partial V}\right)_S, \tag{5.115}$$

$$\frac{\partial^2 U}{\partial S\partial V} = \left(\frac{\partial T}{\partial V}\right)_S. \tag{5.116}$$

The derivatives $(\partial T/\partial S)_V$ and $(\partial p/\partial V)_S$ can be easily expressed by c_v and κ_S, respectively, whereas

$$\left(\frac{\partial T}{\partial V}\right)_S = -\left(\frac{\partial S}{\partial V}\right)_T\left(\frac{\partial S}{\partial T}\right)_V^{-1} = -\frac{T\alpha}{nc_v\kappa_T}, \tag{5.117}$$

where we have used (5.81), Maxwell relation (5.85), relation (5.100) and the definition of c_v, hence

$$\left(\frac{\partial^2 U}{\partial S^2}\right)_V = \frac{T}{nc_v}, \qquad \left(\frac{\partial^2 U}{\partial V^2}\right)_S = \frac{1}{nv\kappa_S}, \qquad \frac{\partial^2 U}{\partial S \partial V} = -\frac{T\alpha}{nc_v\kappa_T}. \tag{5.118}$$

Finally, using relations (5.103) and (5.112), we can express c_v and κ_S by c_p, κ_T and α.

To summarize, we have shown that all second derivatives of the thermodynamic potentials and the internal energy, calculated at constant mole number, can be expressed in terms of three independent quantities related to the second derivatives of the potential G, that is: c_p, κ_T and α, and by T, v and n.

5.4.4 Reversible Adiabatic Process

We want to derive the equation of the adiabat, i.e., the relation between T and p, V and p, or T and V in a reversible adiabatic process. As we know, in adiabatic processes there is no flow of heat between the system and surroundings, and since $đQ = T\mathrm{d}S$ in a reversible process, the condition $đQ = 0$ means that $S = const.$

First, we determine the dependence of pressure on temperature at constant S:

$$\mathrm{d}p = \left(\frac{\partial p}{\partial T}\right)_S \mathrm{d}T = \frac{c_p}{Tv\alpha}\mathrm{d}T, \tag{5.119}$$

where we have used (5.109). Equation (5.119) is the equation of the adiabat expressed in terms of the variables p and T. In general, we cannot integrate this equation because the coefficient at $\mathrm{d}T$ is a function of temperature and pressure. However, in the case of the ideal gas, we have $pv = RT$, $\alpha = 1/T$ and $c_p = (f+2)R/2$ (see (3.25)), hence

$$\frac{\mathrm{d}p}{p} = \frac{f+2}{2}\frac{\mathrm{d}T}{T}, \tag{5.120}$$

which leads to the following equation of the adiabat:

$$pT^{-(f+2)/2} = const. \tag{5.121}$$

If we want to express the equation of the adiabat in terms of V and p we proceed analogously, i.e.,

$$\mathrm{d}p = \left(\frac{\partial p}{\partial V}\right)_S \mathrm{d}V = -\frac{1}{V\kappa_S}\mathrm{d}V, \tag{5.122}$$

where we have used (5.108). Applying relation (5.113) to the ideal gas, we get

$$\kappa_S = \frac{f}{(f+2)p}, \tag{5.123}$$

hence

$$\frac{\mathrm{d}p}{p} = -\frac{f+2}{f}\frac{\mathrm{d}V}{V}. \tag{5.124}$$

Integrating this equation, we find the following equation of the adiabat:

$$pV^{(f+2)/f} = const, \tag{5.125}$$

which we derived already in Sect. 3.4. We can also derive equation (5.125), substituting T from the equation of state of the ideal gas into (5.121).

Finally, the equation of the adiabat in the variables V and T results from the combination of Eqs. (5.119) and (5.122):

$$\mathrm{d}T = -\frac{T\alpha}{nc_p\kappa_s}\mathrm{d}V, \tag{5.126}$$

and for the ideal gas

$$\frac{\mathrm{d}T}{T} = -\frac{2}{f}\frac{\mathrm{d}V}{V}, \tag{5.127}$$

which after integration gives

$$TV^{2/f} = const. \tag{5.128}$$

5.4.5 Free Gas Expansion

We consider a vessel divided into two parts by a rigid wall and isolated from the surroundings. One part of the vessel contains a gas in the volume V_i, whereas the second part is empty. When the internal wall is removed the gas starts to expand freely and eventually fills up the whole volume of the vessel, V_f. The initial state and final state are equilibrium states, whereas the intermediate states are not because we do not control the speed of the process. Therefore, the free gas expansion is not a quasi-static process, hence, it is not reversible. There is no energy transfer because the vessel forms an isolated system, which means that the internal energy of the gas does not change. However, its temperature can change. We treat the temperature as a function of the internal energy, volume and mole number (the latter is constant), and its change in this process amounts to

$$T_f - T_i = T(U, V_f) - T(U, V_i), \tag{5.129}$$

where T_i and T_f denote the temperature of the initial and final states, respectively. If the change in the gas volume, $\Delta V = V_f - V_i$, is small then we can expand $T(U, V_f)$ in a Taylor series around V_i, hence

$$\Delta T = T_f - T_i \approx \left(\frac{\partial T}{\partial V}\right)_U \Delta V, \tag{5.130}$$

where the derivative is calculated at $V = V_i$. To calculate this derivative, we use identity (5.81):

$$\left(\frac{\partial T}{\partial V}\right)_U = -\frac{(\partial U/\partial V)_T}{(\partial U/\partial T)_V}. \tag{5.131}$$

We know that $(\partial U/\partial T)_V = nc_v$ (see (3.12)), whereas

$$\left(\frac{\partial U}{\partial V}\right)_T = \left(\frac{\partial U}{\partial V}\right)_S + \left(\frac{\partial U}{\partial S}\right)_V \left(\frac{\partial S}{\partial V}\right)_T = -p + T\left(\frac{\partial p}{\partial T}\right)_V, \tag{5.132}$$

where we have used (5.82) and Maxwell relation (5.85). Finally, using (5.100), we get

$$\left(\frac{\partial U}{\partial V}\right)_T = -p + \frac{T\alpha}{\kappa_T}. \tag{5.133}$$

In the ideal gas case, $\alpha = 1/T$, $\kappa_T = 1/p$ and $(\partial U/\partial V)_T = 0$, hence $\Delta T = 0$. For real gases, we have

$$\Delta T \approx \frac{1}{nc_v}\left(p - \frac{T\alpha}{\kappa_T}\right)\Delta V. \tag{5.134}$$

In a similar way, we determine the change in the gas entropy in the process, i.e.

$$\Delta S = S(U, V_k) - S(U, V_p). \tag{5.135}$$

Since $(\partial S/\partial V)_U = p/T$, for a small change in the volume, we have

$$\Delta S \approx \frac{p}{T}\Delta V > 0. \tag{5.136}$$

The process is irreversible because the entropy of the gas increases in an adiabatically isolated system (see Sect. 4.6.2).

5.4.6 Joule–Thomson Process

The Joule–Thomson process is shown schematically in Fig. 5.2. The piston on the left-hand side exerts a constant pressure p_i on the gas in the left part of the vessel and pushes it through a porous plug. The gas pressure on the right-hand side of the plug, $p_f < p_i$, is also maintained constant due to the second piston. The whole system is adiabatically isolated. Due to the flow through the porous plug, the gas warms up or cools down, or its temperature does not change, depending on the initial temperature and the initial and final values of the pressure.

First, we show that the enthalpy of the system at the beginning and at the end of the process has the same value. We assume that the amount of the gas is constant and that its volume and internal energy at the beginning and at the end of the process amount to V_i, U_i and V_f, U_f, respectively. The work done on the gas in the left part of the vessel amounts to $p_i V_i$ because its volume changes from V_i to zero at the constant pressure p_i. Analogously, the work done by the gas in the right part of the vessel amounts to $p_f V_f$. The internal energy of the system at the end of the process is equal to

$$U_f = U_i + p_i V_i - p_f V_f. \tag{5.137}$$

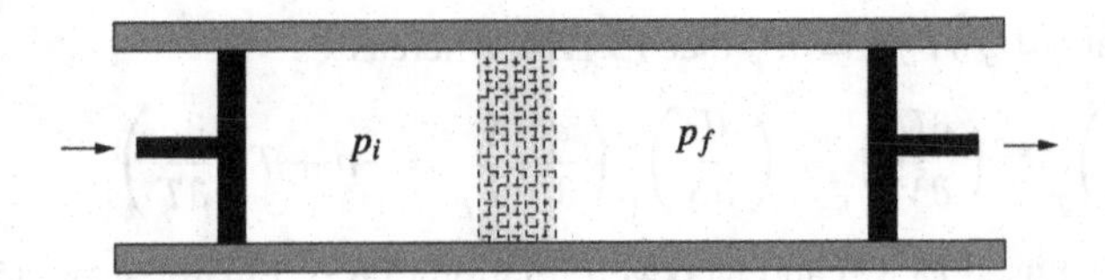

Fig. 5.2 Joule–Thomson process. A gas is pushed through a porous material separating the two parts of the vessel. The gas pressure in each part of the vessel is constant and amounts to p_i and $p_f < p_i$, respectively

As we know, the enthalpy $H = U + pV$, hence, we get

$$H_f = H_i, \tag{5.138}$$

which means that $\Delta H = 0$ in the Joule–Thomson process. We are interested how the temperature changes in this process. If the change in the gas pressure caused by the flow through the porous material is small we can calculate ΔT from the following relation:

$$\mathrm{d}T = \left(\frac{\partial T}{\partial p}\right)_H \mathrm{d}p = -\frac{(\partial H/\partial p)_T}{(\partial H/\partial T)_p}\mathrm{d}p, \tag{5.139}$$

where we have used (5.81). The denominator is equal to the heat capacity at constant pressure, C_p (see (3.23)). Then we transform the nominator, using relation (5.82):

$$\left(\frac{\partial H}{\partial p}\right)_T = \left(\frac{\partial H}{\partial p}\right)_S + \left(\frac{\partial H}{\partial S}\right)_p \left(\frac{\partial S}{\partial p}\right)_T = V + T\left(\frac{\partial S}{\partial p}\right)_T. \tag{5.140}$$

From Maxwell relation (5.87) and from the definition of α (see (5.93)), it follows that $(\partial S/\partial p)_T = -V\alpha$, and

$$\left(\frac{\partial H}{\partial p}\right)_T = V(1 - T\alpha). \tag{5.141}$$

Dividing the nominator and denominator in (5.139) by the mole number, we arrive at the following expression:

$$\mathrm{d}T = \frac{v(T\alpha - 1)}{c_p}\mathrm{d}p. \tag{5.142}$$

For the ideal gas, $\alpha = 1/T$, hence, $\mathrm{d}T = 0$. In the case of real gases, $\mathrm{d}T > 0$ if $T\alpha > 1$ and $\mathrm{d}T < 0$ if $T\alpha < 1$ because the pressure difference is positive. The temperature at which $T\alpha = 1$ is called the *inversion temperature*. In suitable conditions, the Joule–Thomson process can be used to cool the gas down to the temperature of condensation. First, however, the gas has to be cooled below its inversion temperature.

Finally, we calculate the change in the gas entropy in the Joule–Thomson process. Since the enthalpy does not change in this process, we have

$$H(S_f, p_f, n) = H(S_i, p_i, n). \tag{5.143}$$

If the pressure difference is small then

$$H(S_f, p_f, n) \approx H(S_i, p_i, n) + T_i \Delta S + V_i \Delta p, \tag{5.144}$$

where $\Delta S = S_f - S_i$, $\Delta p = p_f - p_i$, and we have used the relation: $\mathrm{d}H = T\mathrm{d}S + V\mathrm{d}p$, for $n = const$, hence

$$\Delta S \approx -\frac{V_i}{T_i}\Delta p > 0, \tag{5.145}$$

for $\Delta p < 0$. The process is irreversible because the entropy of the gas increases in an adiabatically isolated system.

5.5 Intrinsic Stability of a System

So far we have considered only the conditions of thermodynamic equilibrium for subsystems of a composite system, which follow from the entropy maximum principle or the free-energy (of Helmholtz or Gibbs) minimum principle. In other words, we have studied only the necessary condition for a maximum or minimum of a given state function, that is, $\mathrm{d}S = 0$, $\mathrm{d}F = 0$ or $\mathrm{d}G = 0$. Now we want to explore the consequences of the fact that a given equilibrium state corresponds to a maximum of entropy of a composite system. This means that we have to find the conditions for the second partial derivatives of entropy.

We consider a one-component system, from which we separate a small subsystem. We assume that the subsystem chosen can exchange heat with the complementary subsystem and can also change its volume but the mole numbers in the subsystem, n, and in the complementary subsystem, n', where $n \ll n'$, are constant. The parameters corresponding to the subsystem are denoted by s (molar entropy), u (molar internal energy) and v (molar volume), and the parameters corresponding to the complementary subsystem are denoted by s', u' and v', respectively. The molar entropy of both subsystems is the same state function, i.e., $s = s(u, v)$, $s' = s(u', v')$, because the subsystems are parts of the same system. The whole system is isolated from the surroundings, therefore, the equilibrium state of the system corresponds to a maximum of its entropy,

$$S_{\mathrm{tot}} = ns(u, v) + n's(u', v'), \tag{5.146}$$

at constant internal energy and volume, hence

$$n\mathrm{d}u + n'\mathrm{d}u' = 0, \tag{5.147}$$

$$n\mathrm{d}v + n'\mathrm{d}v' = 0. \tag{5.148}$$

As we know, the necessary condition for the maximum of the function S_{tot}, i.e.,

$$\mathrm{d}S_{\mathrm{tot}} = n\mathrm{d}s + n'\mathrm{d}s' = \left(\frac{1}{T} - \frac{1}{T'}\right)n\mathrm{d}u + \left(\frac{p}{T} - \frac{p'}{T'}\right)n\mathrm{d}v = 0, \tag{5.149}$$

where we have used (5.147) and (5.148), leads to the equality of temperatures, $T = T'$, and pressures, $p = p'$, in the subsystem chosen and in the complementary subsystem. Since equilibrium between the two subsystems corresponds to the maximum of entropy of the whole system, the second differential of S_{tot} must satisfy the following condition:

$$\mathrm{d}^2 S_{tot} = n\mathrm{d}^2 s + n'\mathrm{d}^2 s' < 0, \tag{5.150}$$

where

$$\mathrm{d}^2 s = \frac{1}{2}\left[s_{uu}(\mathrm{d}u)^2 + 2s_{uv}\mathrm{d}u\mathrm{d}v + s_{vv}(\mathrm{d}v)^2\right], \tag{5.151}$$

and an analogous expression can be written for $\mathrm{d}^2 s'$. To simplify the notation, the first derivatives of entropy with respect to u and v are denoted by s_u and s_v, respectively, and for the second derivatives we use the symbols s_{uu}, s_{uv} and s_{vv}, Note that because of conditions (5.147) and (5.148), we have $\mathrm{d}u' = -(n/n')\mathrm{d}u$ and $\mathrm{d}v' = -(n/n')\mathrm{d}v$, hence, the ratio of $n'\mathrm{d}^2 s'$ to $n\mathrm{d}^2 s$ is proportional to $n/n' \ll 1$. Therefore, we can neglect the term $n'\mathrm{d}^2 s'$ in expression (5.150) and consider the condition

$$\mathrm{d}^2 s < 0. \tag{5.152}$$

Condition (5.152) means that any small subsystem of a homogeneous system must be in thermodynamic equilibrium with the complementary subsystem.

Expression (5.151) is a quadratic form of $\mathrm{d}u$ and $\mathrm{d}v$. To determine its sign, it is convenient to get rid of the mixed term by a suitable change of variables. Because $\mathrm{d}s = (1/T)\mathrm{d}u + (p/T)\mathrm{d}v$, we have $s_u = 1/T$ and $s_v = p/T$, hence

$$\mathrm{d}T^{-1} = s_{uu}\mathrm{d}u + s_{uv}\mathrm{d}v, \tag{5.153}$$

where we have simply used the fact that $T = T(u, v)$. Substituting $\mathrm{d}u$ determined from (5.153) into (5.151), we get

$$\mathrm{d}^2 s = \frac{1}{2}\left[\frac{1}{s_{uu}}(\mathrm{d}T^{-1})^2 + \left(s_{vv} - \frac{s_{uv}^2}{s_{uu}}\right)(\mathrm{d}v)^2\right]. \tag{5.154}$$

Since $\mathrm{d}T^{-1}$ and $\mathrm{d}v$ are arbitrary, condition (5.152) is satisfied only if the coefficients at $(\mathrm{d}T^{-1})^2$ and $(\mathrm{d}v)^2$ are negative. Using (3.12), we find that

$$s_{uu} = \left(\frac{\partial T^{-1}}{\partial u}\right)_v = -T^{-2}\left(\frac{\partial u}{\partial T}\right)_v^{-1} = -\frac{1}{T^2 c_v}. \tag{5.155}$$

To associate the second coefficient with a known physical quantity, we note that putting $T = const$ in (5.153), we get

$$\left(\frac{\partial u}{\partial v}\right)_T = -\frac{s_{uv}}{s_{uu}}, \tag{5.156}$$

hence

$$s_{vv} - \frac{s_{uv}^2}{s_{uu}} = s_{vv} + s_{uv}\left(\frac{\partial u}{\partial v}\right)_T = \left(\frac{\partial s_v}{\partial v}\right)_T = \frac{1}{T}\left(\frac{\partial p}{\partial v}\right)_T. \tag{5.157}$$

Using the definition of the isothermal compressibility, we express $\mathrm{d}^2 s$ as follows:

$$\mathrm{d}^2 s = -\frac{1}{2}\left[\frac{c_v}{T^2}(\mathrm{d}T)^2 + \frac{1}{T v \kappa_T}(\mathrm{d}v)^2\right]. \tag{5.158}$$

The inequality $\mathrm{d}^2 s < 0$ holds if the conditions

$$c_v > 0 \quad \text{and} \quad \kappa_T > 0 \tag{5.159}$$

are satisfied. The first of them is called the *condition of thermal stability*. It states that a flow of heat into the system must increase its temperature.The second condition is the *condition of mechanical stability*, which means that compression of a mechanically stable system must increase its pressure. If at least one of the conditions (5.159) is not satisfied then the system cannot exist in a homogeneous form. For instance, if the temperature and pressure are such that liquid water and water vapour can exist simultaneously in the system then a change in the volume of the system at constant temperature does not change its pressure; it changes only the proportion of the liquid to the vapour. For such a system, $(\partial p/\partial v)_T = 0$, which means that the isothermal compressibility κ_T becomes infinite. Violation of an intrinsic stability condition of a system is associated with a *phase transition* (see Chap. 6).

We note finally that the condition of mechanical stability can also be derived from the Helmholtz free-energy minimum principle. Using a similar reasoning as presented above to a system in thermal contact with a heat reservoir, we arrive at the conclusion that for the molar Helmholtz free energy of a small subsystem, the following inequality must hold:

$$\mathrm{d}^2\phi > 0. \tag{5.160}$$

It results from the fact that the state of thermodynamic equilibrium between the small subsystem and the complementary subsystem corresponds to a minimum of the Helmholtz free energy of the whole system at constant temperature. Since $\phi = \phi(T, v)$ and $T = const$, the condition

$$\mathrm{d}^2\phi = \frac{1}{2}\left(\frac{\partial^2\phi}{\partial v^2}\right)_T (\mathrm{d}v)^2 = \frac{1}{2v\kappa_T}(\mathrm{d}v)^2 > 0, \tag{5.161}$$

is equivalent to the condition $\kappa_T > 0$.

5.6 Exercises

5.1 Find the Legendre transform, $\Psi(z)$, of the functions: $Y(x) = Ax^2 + Bx + C$ and $Y(x) = x + a\mathrm{e}^x$. Verify the relation $\mathrm{d}\Psi/\mathrm{d}z = -x$.

5.2 In an isobaric process at the pressure p, the change in the internal energy and volume of a system amounts to ΔU and ΔV, respectively. Find the change in the enthalpy.

5.3 In an isothermal process at the temperature T, the change in the internal energy and entropy of a system amounts to ΔU and ΔS, respectively. Find the change in the Helmholtz free energy.

5.4 Prove the following relations:

$$\left(\frac{\partial F/T}{\partial T}\right)_{V,n} = -\frac{U}{T^2}, \qquad \left(\frac{\partial G/T}{\partial T}\right)_{p,n} = -\frac{H}{T^2}.$$

$$\left(\frac{\partial H/p}{\partial p}\right)_{S,n} = -\frac{U}{p^2}, \qquad \left(\frac{\partial G/p}{\partial p}\right)_{T,n} = -\frac{F}{p^2}.$$

$$\left(\frac{\partial U/V}{\partial V}\right)_{S,n} = -\frac{H}{V^2}, \qquad \left(\frac{\partial F/V}{\partial V}\right)_{T,n} = -\frac{G}{V^2}.$$

5.5 The pressure and mole number in a system are constant. We assume that the enthalpy of the system, H, is known in the range of temperature $T_i \leq T \leq T_f$, and that the dependence of H on temperature can be neglected. Calculate the value of the Gibbs free energy at the temperature T_f, provided that its value at the temperature T_i amounts to G_i.

5.6 A certain thermodynamic system is described by the following fundamental equation:

$$U = B\frac{S^3}{nV},$$

where B is a constant. Find the dependence of the pressure and temperature of the system on its entropy, volume and mole number. What is the dependence of the Gibbs free energy of the system on its natural variables?

5.7 Find the enthalpy and the Helmholtz and Gibbs free energy of a photon gas whose internal energy $U = \gamma V T^4$, and γ is a constant.

5.8 Find the dependence of the Helmholtz free energy, $F(T, V, n)$, on volume for the van der Waals gas, using the equation of state

$$p = \frac{RT}{v - b} - \frac{a}{v^2},$$

where $v = V/n$.

5.9 What is the form of the internal energy, $U(T, V, n)$, for the van der Waals gas if $c_v = 3R/2$ and

$$p = \frac{RT}{v - b} - \frac{a}{v^2}?$$

5.10 We have determined experimentally the molar heat capacity and constant pressure as a function of temperature: $c_p = a + bT$, where a and b are some constants.

Calculate the change in the enthalpy and entropy of the system heated at constant pressure from the temperature T_i to the temperature T_f.

5.11 Prove the identity:

$$\left(\frac{\partial C_p}{\partial p}\right)_{T,n} = -T\left(\frac{\partial^2 V}{\partial T^2}\right)_{p,n}.$$

5.12 For many liquids and solids, the coefficient of thermal expansion α and the isothermal compressibility κ_T are practically constant in the range of 50 atm and a few tens of degrees centigrade around the room temperature. Using this fact, derive an approximate equation of state, $V = V(T, p)$, for liquids and solids.

5.13 Derive the Maxwell relations for the thermodynamic potentials $\Psi(S, V, \mu)$ and $\Theta(S, p, \mu)$.

5.14 What are the natural variables of the potential $\psi = \Psi/V$ (potential Ψ per unit volume)? Express the differential $d\psi$ in the natural variables.

5.15 Show that the potential $\Theta(S, p, \mu)$ can be expressed as $\Theta = ST(p, \mu)$. What is the form of dT in these variables?

5.16 Express the derivatives $(\partial v/\partial \mu)_T$ and $(\partial s/\partial \mu)_T$, where $v = V/n$ and $s = S/n$, by κ_T and α.

5.17 The equation of state of a dilute gas can be expressed as follows:

$$\frac{pv}{RT} = 1 + b(T)p.$$

Using this equation of state, derive the equation for the inversion temperature in the Joule–Thomson process.

Part II
Phase Transitions

Chapter 6
Phase Transitions in Pure Substances

6.1 Concept of Phase

The phenomena of evaporation and freezing of liquids or melting and sublimation of solids are well known to everybody from everyday experience. We can easily observe some changes of the state of matter for such substances as water or carbon dioxide. At atmospheric pressure, liquid water freezes, i.e., changes into solid (ice), at the temperature of 0 °C, and at the temperature of 100 °C it changes into gaseous state (water vapour or steam). The same substance which in some conditions exists as a liquid in different conditions can exist as a solid or gas. We observe transformations of substances which change their physical properties even though the chemical composition remains the same. In everyday life, we usually observe situations when one state of matter (gaseous, liquid or solid) changes into another state of matter, for instance, the liquid state changes into the solid or gaseous state. It is not a rule, however. Some substances can exist in different forms even in the same state of matter. For example, solid carbon can exist in the form of *graphite* or *diamond*. The difference between these two forms results from different arrangement of atoms in a periodic crystal structure, which leads to different physical properties of graphite an diamond. Another example is helium at a temperature of a few kelvins. Even at such low temperatures helium exists in the liquid state. Below the temperature of 2.17 K the liquid isotope of helium, ^{4}He, can exist in two different forms called He I and He II, depending on the pressure exerted. He I is a normal viscous liquid, whereas He II exhibits an unusual property, called *super-fluidity*, which enables the liquid to flow without friction even through very narrow capillaries. These examples show clearly that the division of matter into three states only, i.e., the liquid, solid and gaseous state, known from the basic course of physics, is insufficient to describe a huge variety of forms of matter existence. Therefore, a more general concept of *phase* is introduced.

Definition 6.1 *Phase* means any equilibrium state of a macroscopic system that is homogeneous in respect of the physical properties and chemical composition.

R. Hołyst, A. Poniewierski, *Thermodynamics for Chemists, Physicists and Engineers*,
DOI 10.1007/978-94-007-2999-5_6, © Springer Science+Business Media Dordrecht 2012

By homogeneity it is understood that properties of the system do not vary in a scale much larger than the size of atoms or molecules. For instance, liquid water, water vapour and ice are different phases of H_2O. Similarly, graphite and diamond are different phases of carbon, and He I and He II are different phases of liquid helium at low temperatures. At higher temperatures, helium exists in the gaseous state.

In what follows, we assume that the temperature and pressure are the state parameters that can be controlled, therefore, the Gibbs free energy is the suitable thermodynamic potential to describe a given system (see Chap. 5). Any homogeneous and intrinsically stable state of the system corresponds to a certain phase. It may happen, however, that for some values of temperature and pressure, the homogeneous state ceases to be intrinsically stable. Then the system must split into homogeneous subsystems corresponding to different phases. The process of transformation of one phase to another phase is called a *phase transition.*

Definition 6.2 *Phase transition* is a qualitative change occurring in a system due to variations of external conditions, such as temperature, pressure and electric or magnetic fields.

Freezing and evaporation of water, change of diamond into graphite or He I into He II are examples of phase transitions. For instance, when water is cooled at atmospheric pressure down to the temperature of 0 °C, we observe that small pieces of ice appear. Instead of a homogeneous liquid both the liquid and solid are present in the system. The system is no longer homogeneous as a whole but both liquid water and ice form homogeneous subsystems which are single phases.

6.2 Classification of Phase Transitions

Phase transitions are either of *first order* or *continuous*; the latter are also referred to as *second-order phase transitions*. Most of the phase transitions that we know from everyday life are of first order, for instance, the transition from liquid water to ice or to vapour.

6.2.1 First-Order Phase Transitions

First-order phase transitions are characterized by discontinuous changes of physical properties of a system. For instance, during a first-order phase transition the molar volume and molar entropy change discontinuously. We know very well that the density of ice is smaller than the density of liquid water because ice floats on the surface of the liquid. For most substances, however, it is the other way round, that is, the liquid phase has smaller density than the solid phase. In this respect (and not only in this one) water is an untypical substance. The discontinuity in the molar entropy results in the *heat of transition*. For instance, to change ice into liquid water, about 6 kJ of heat per mole has to be supplied. This phenomenon is commonly known as *melting*, and the corresponding heat of transition, as the *heat of melting*. Similarly, a transformation of one mole of a liquid into gas, i.e., evaporation of the liquid, requires a definite amount of heat called the *heat of evaporation*. A direct transition from the solid phase to the gas phase, called *sublimation*, is also a first-order phase transition, and the heat associated with this transition is called the *heat of sublimation*.

First-order phase transitions are associated with *phase coexistence*. For example, at atmospheric pressure, ice and water coexist at the temperature of 0 °C. Complete transformation of ice into water requires that a definite amount of heat is absorbed by the system, but during the transition the temperature of the system does not change. Liquid water gives off the same amount of heat to the surroundings when it freezes. A similar phenomenon of phase coexistence occurs in the case of a liquid and its vapour. If the liquid does not fill the whole volume of a closed vessel, from which air has been pumped out, then the space above the liquid surface is filled with its vapour which coexists with the liquid at a definite pressure. Varying the volume of the vessel at constant temperature with a piston, for instance, we change the amounts of the liquid and vapour, but the pressure in the system does not change. A decrease in the vessel volume causes condensation of a certain amount of vapour, which is associated with a flow of heat from the system to surroundings, whereas an increase in the vessel volume results in evaporation of a certain amount of liquid at the cost of heat drawn from the surroundings.

6.2.2 Continuous Phase Transitions

Phase transitions in which all properties of a system vary continuously are called *continuous phase transitions*. In contrast to first-order phase transitions, in which two phases can coexist, during a continuous transition one phase is immediately replaced by another phase. This is caused by the lack of heat of transition, because in continuous transitions, the molar entropy is a continuous function of the state parameters

A typical example of a continuous phase transition is the transition from the *paramagnetic phase* to the *ferromagnetic phase*, which occurs in iron at the temperature $T_C = 1043$ K, called the *Curie temperature*. Paramagnetic material is not magnetized in the absence of the magnetic field $\mathbf{B}$, and in the presence of $\mathbf{B}$ its magnetization $\mathbf{M}$ is proportional to $\mathbf{B}$. In the case of a ferromagnetic material, the magnetization does not vanish even when the magnetic field is switched off. The temperature dependence of the magnitude of the magnetization for $\mathbf{B} = 0$, denoted $|\mathbf{M}_0|$, is shown schematically in Fig. 6.1. For $T > T_C$, $\mathbf{M}_0 = 0$, and for $T < T_C$, $\mathbf{M}_0 \neq 0$. The magnetization $\mathbf{M}_0$ is a continuous function also at the transition temperature T_C. The molar entropy is also continuous. However, the derivative of $|\mathbf{M}_0|$

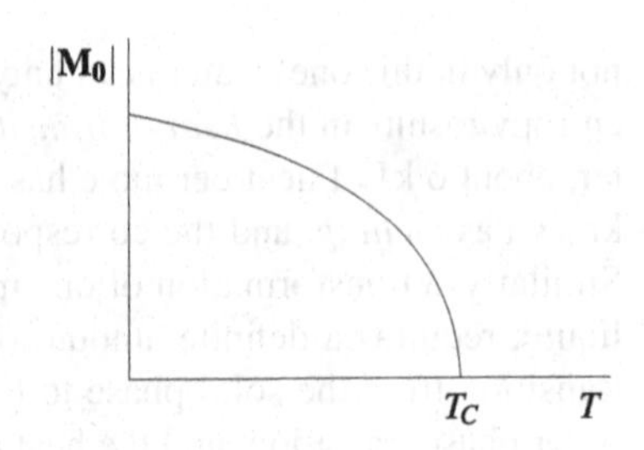

Fig. 6.1 Magnitude of the magnetization as a function of temperature at zero magnetic field, for a ferromagnetic material such as iron. T_C denotes the temperature of a continuous transition from the paramagnetic phase to the ferromagnetic phase

with respect to temperature is discontinuous at T_C. As it is shown in Fig. 6.1, for $T < T_C$, $\mathrm{d}|\mathbf{M}_0|/\mathrm{d}T \to \infty$ when $T \to T_C$, and for $T > T_C$, $\mathrm{d}|\mathbf{M}_0|/\mathrm{d}T = 0$. In continuous phase transitions, also other quantities can diverge to infinity. These quantities are related to the rates of change in physical properties of a system. For instance, heat capacity is proportional to the rate of change in entropy with temperature. To summarize, in continuous phase transitions, the properties of a system change continuously at the transition, but the rates of change of these properties are discontinuous or divergent.

In particular conditions, the transition from a liquid to gas also becomes continuous. The line of liquid–gas coexistence terminates at a *critical point* in which the difference between the two phases disappears, including the heat of transition (see Sect. 6.4.1). However, the continuous transition from the paramagnetic phase to the ferromagnetic phase has a different character, i.e., it is not related to any critical point at the end of a first-order transition line. In this case, the continuous phase transition is related to a physical quantity called the *order parameter*. In a material such as iron, microscopic magnetic dipole moments exist which can order along a common direction not only in an external magnetic field, but also spontaneously due to interactions between them. This spontaneous ordering is possible only below the Curie temperature. The magnetization of the whole sample is a resultant magnetic dipole moment; $\mathbf{M} = 0$ means lack of ordering, and $\mathbf{M} \neq 0$ means that microscopic magnetic dipole moments are ordered to a certain degree, and $|\mathbf{M}|$ is a measure of the degree of that order. Therefore, the magnetization $\mathbf{M}$ can be treated as an order parameter in ferromagnetic systems such as iron. Continuous phase transitions belong to the *order-disorder* transition type,[1] which means that they occur in systems for which an order parameter can be defined. Obviously, the choice of the order parameter depends on a given physical system. A whole branch of condensed matter physics, called *critical phenomena*, is devoted to the studies of continuous phase transitions and non-analytic behaviour of various physical quantities associated with these transitions. These interesting problems are beyond the scope of this book, however. In what follows, we discuss only first order phase transitions, which are more often met in everyday life than continuous transitions.

[1]Phase transitions of this type can also be of first-order, e.g., the transition from a liquid to a crystalline solid.

6.2.3 Ehrenfest Classification

The classification of phase transitions proposed by Ehrenfest is based on the continuity of derivatives of the chemical potential μ with respect to temperature and pressure. The chemical potential, as well as temperature and pressure, is continuous at a phase transition, which is required by the conditions of thermal, mechanical and with respect to matter flow equilibrium. However, the derivatives of μ do not have to be continuous. If the first derivatives: $s = -(\partial\mu/\partial T)_p$ and $v = (\partial\mu/\partial p)_T$ (see (5.31)), i.e., the molar entropy and molar volume, are discontinuous then the transition is of first order, according to the Ehrenfest classification. By analogy, Ehrenfest considered the transition to be of nth order if at least one of the nth order partial derivatives of μ is discontinuous at the transition, and all lower-order derivatives of μ are continuous. For instance, in a second-order transition, s and v are continuous and at least one of the second-order derivatives of μ is discontinuous at the transition.

The Ehrenfest classification is mainly of historical importance. Nowadays the transition is said to be of first order if the heat of the transition does not vanish, i.e., the molar entropy is discontinuous. All transitions in which first-order partial derivatives of μ are continuous are called continuous phase transitions (of second order according to the Ehrenfest classification). Modern theories of phase transitions study non-analytic behaviour of thermodynamic potentials in the neighbourhood of continuous phase transitions. The non-analytic behaviour shows up in divergences of various quantities, such as heat capacity, compressibility or magnetic susceptibility (in ferromagnetic systems), which are related to second-order partial derivatives of thermodynamic potentials. The Ehrenfest classification does not include this type of behaviour, however.

6.3 Conditions of Phase Coexistence

6.3.1 Two-Phase Coexistence

We consider a closed one-component system composed of two subsystems: the phases α and β. The temperature T and pressure p in the system are constant. We recall that the Gibbs free energy of a pure substance is given by (see Sect. 5.1.4):

$$G = U - TS + pV = \mu n, \tag{6.1}$$

where U, S, V, μ and n denote, respectively, the internal energy, entropy, volume, chemical potential and mole number, and

$$\mathrm{d}G = -S\mathrm{d}T + V\mathrm{d}p + \mu\mathrm{d}n. \tag{6.2}$$

Substituting $n = 1$ mol in (6.2), we obtain the Gibbs–Duhem equation (see (5.30)):

$$\mathrm{d}\mu = -s\mathrm{d}T + v\mathrm{d}p, \tag{6.3}$$

where $s = S/n$ and $v = V/n$ denote the molar entropy and molar volume. Equation (6.3) simply expresses the fact that the chemical potential is a function of temperature and pressure.

Our purpose is to derive the condition of coexistence of two phases. The Gibbs free energy of the whole system, G, amounts to:

$$G = n^\alpha \mu^\alpha(T, p) + n^\beta \mu^\beta(T, p), \tag{6.4}$$

where μ^α, μ^β and n^α, n^β denote the chemical potentials and mole numbers for the phases α and β, respectively. For each phase, the Gibbs–Duhem equation holds, i.e.,

$$\mathrm{d}\mu^\alpha = -s^\alpha \mathrm{d}T + v^\alpha \mathrm{d}p, \tag{6.5}$$

$$\mathrm{d}\mu^\beta = -s^\beta \mathrm{d}T + v^\beta \mathrm{d}p. \tag{6.6}$$

The system is in contact with a heat and volume reservoir at the temperature T and pressure p, and we have

$$n^\alpha + n^\beta = n, \tag{6.7}$$

where the total mole number in the system, n, is constant (closed system). However, the matter can flow between the two phases because there is no physical barrier between them. The process of transformation of one phase into the other phase is reversible, which means that $\Delta G = 0$ in this process (see Corollary 5.3). Using (6.4) and the condition $\Delta n^\alpha + \Delta n^\beta = 0$, we get:

$$\Delta G = \left(\mu^\alpha - \mu^\beta\right)\Delta n^\alpha = 0. \tag{6.8}$$

From the fact that Δn^α can be arbitrary, we draw the following conclusion.

Corollary 6.1 *Phases α and β coexist if their chemical potentials are equal, i.e.,*

$$\mu^\alpha = \mu^\beta.$$

The chemical potentials μ^α and μ^β are different functions of temperature and pressure. Therefore, the condition $\mu^\alpha(T, p) = \mu^\beta(T, p)$ defines a certain line in the Tp plane, which is called the *two-phase coexistence line*. The dependence of the chemical potential on temperature at constant pressure, in the neighbourhood of a phase transition, is shown schematically in Fig. 6.2. The broken line marks the transition temperature. Note that in Fig. 6.2 the lines corresponding to μ^α and μ^β do not terminate at the transition temperature. This means that the phase α can exist, i.e., can be intrinsically stable, also in a certain range of temperature above the transition temperature, even though the phase β corresponds to a lower value of the Gibbs free energy. Similarly, the phase β can exist in a certain range of temperature below the transition temperature. Such states are called *metastable states*. For instance, if very clean water is cooled slowly at atmospheric pressure it can remain in the liquid state even down to −42 °C. Such a state is called a *super-cooled liquid*. However, it is not thermodynamically stable and even a small perturbation can cause its crystallization, i.e., transition to a state of the minimal value of the Gibbs free energy. Similarly, it is possible to observe a *super-heated liquid* or *super-saturated vapour*.

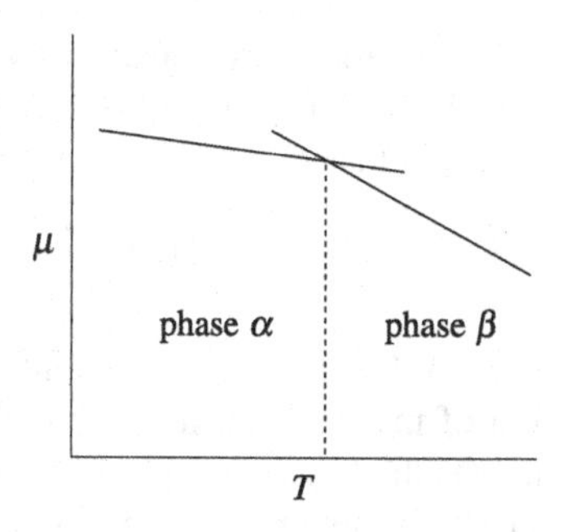

Fig. 6.2 Chemical potential as a function of temperature at constant pressure (*solid line*) in the neighbourhood of a first-order phase transition. The intersection of $\mu^\alpha(T)$ and $\mu^\beta(T)$ corresponds to the transition temperature (*broken line*)

The Gibbs free energy of a system in which two phases coexist has the same form as in the case of a homogeneous system, i.e.,

$$G = n^\alpha \mu^\alpha + n^\beta \mu^\beta = n\mu_{\mathrm{coex}}; \tag{6.9}$$

where $\mu_{\mathrm{coex}} = \mu^\alpha = \mu^\beta$ denotes the value of the chemical potential at the coexistence. Only one intensive parameter, T or p, is independent, as the second one follows from the condition $\mu^\alpha(T,p) = \mu^\beta(T,p)$. Since the total mole number is constant, the remaining state parameter of the two-phase system can be n^α, for instance. Its value can vary from $n^\alpha = 0$ to $n^\alpha = n$, but the value of G does not depend on n^α. Note that the condition of two-phase coexistence reduces the number of independent intensive parameters by one, therefore, an additional extensive parameter is needed to completely define the state of the two-phase system. Instead of n^α or n^β, we can also choose the total volume, V, or the total entropy, S, of the two-phase system, where

$$V = n^\alpha v^\alpha + n^\beta v^\beta, \tag{6.10}$$

$$S = n^\alpha s^\alpha + n^\beta s^\beta, \tag{6.11}$$

and v^α, v^β and s^α, s^β denote, respectively, the molar volume and molar entropy of the coexisting phases.

Helmholtz Free Energy of Two-Phase Systems We assume here that T, V and n are the independent parameters of state. Varying T or V at constant n, we can change n^α and n^β. If we denote by $v = V/n$ the mean molar volume of the two-phase system then, from conditions (6.7) and (6.10), we get

$$n^\alpha = n\frac{v - v^\beta}{v^\alpha - v^\beta}, \qquad n^\beta = n\frac{v^\alpha - v}{v^\alpha - v^\beta}, \tag{6.12}$$

hence, the proportion of the coexisting phases is determined by the following *lever rule*:

$$n^\alpha(v^\alpha - v) = n^\beta(v - v^\beta). \tag{6.13}$$

For instance, if a liquid and vapour coexist in a closed vessel then a decrease in the volume V results in condensation of a certain amount of vapour. The pressure in the system does not change until the gas phase disappears completely from the system.

In the process described above, G does not change because T, p and n are constant, but the Helmholtz free energy F changes. From the Euler relation (see (4.60)), it follows that $F = U - TS = -pV + \mu n$, hence, for the two-phase system we have

$$F = F^\alpha + F^\beta = -p_{\text{coex}}\left(V^\alpha + V^\beta\right) - \mu_{\text{coex}}\left(n^\alpha + n^\beta\right) = -p_{\text{coex}} V + \mu_{\text{coex}} n, \tag{6.14}$$

where $V^\alpha = n^\alpha v^\alpha$ and $V^\beta = n^\beta v^\beta$, and p_{coex} is the pressure at coexistence, i.e., the root of the equation $\mu^\alpha(T, p) = \mu^\beta(T, p)$ at a given temperature. Relation (6.14) defines the function $F(T, V, n)$ because μ_{coex} and p_{coex} depend on temperature. It is easy to show that the following relations are satisfied:

$$\left(\frac{\partial F}{\partial T}\right)_{n,V} = -S, \qquad \left(\frac{\partial F}{\partial V}\right)_{T,n} = -p_{\text{coex}}, \qquad \left(\frac{\partial F}{\partial n}\right)_{T,V} = \mu_{\text{coex}}. \tag{6.15}$$

The last two relations follow from the linear dependence of F on n and V. To derive the first relation, we notice that $F^\alpha = F^\alpha(T, V^\alpha, n^\alpha)$, where n^α and V^α depend on T, and F^β has an analogous form, hence

$$\left(\frac{\partial F^\alpha}{\partial T}\right)_{V,n} + \left(\frac{\partial F^\beta}{\partial T}\right)_{V,n} = -S^\alpha - S^\beta - p_{\text{coex}}\frac{\mathrm{d}(V^\alpha + V^\beta)}{\mathrm{d}T} + \mu_{\text{coex}}\frac{\mathrm{d}(n^\alpha + n^\beta)}{\mathrm{d}T}, \tag{6.16}$$

where $S^\alpha = n^\alpha s^\alpha$ and $S^\beta = n^\beta s^\beta$. Because of the relations $V^\alpha + V^\beta = V$ and $n^\alpha + n^\beta = n$, only the term $-S^\alpha - S^\beta = -S$ remains on the right-hand side. A change in F at constant T and n is equal to the work done on the system, i.e., $\Delta F = -p_{\text{coex}}\Delta V$.

Enthalpy of Two-Phase Systems We assume here that p, S and n are the independent parameters of state. For the given pressure p, the temperature of a first-order phase transition, that is, the temperature of the two-phase coexistence, is a root of the equation $\mu^\alpha(T, p) = \mu^\beta(T, p)$. Varying p or S at constant n, we change n^α and n^β. They are determined from relations (6.7) and (6.11):

$$n^\alpha = n\frac{s - s^\beta}{s^\alpha - s^\beta}, \qquad n^\beta = n\frac{s^\alpha - s}{s^\alpha - s^\beta}, \tag{6.17}$$

hence an alternative form of the lever rule follows:

$$n^\alpha\left(s^\alpha - s\right) = n^\beta\left(s - s^\beta\right). \tag{6.18}$$

For instance, if a liquid and vapor coexist in a closed vessel then the heat supplied to the system and constant pressure is used up to evaporate a part of the liquid and to perform work. The temperature does not change until the whole liquid phase disappears.

The thermodynamic potential whose natural variables are S, p and n is the enthalpy H. From the Euler relation, it follows that $H = U + pV = TS + \mu n$, hence, for the two-phase system we have

$$H = H^\alpha + H^\beta = T_{\text{coex}}\left(S^\alpha + S^\beta\right) + \mu_{\text{coex}}\left(n^\alpha + n^\beta\right) = T_{\text{coex}} S + \mu_{\text{coex}} n. \tag{6.19}$$

Relation (6.19) defines the function $H(S, p, n)$, since T_{coex} and μ_{coex} are functions of p. Analogously to the case of the Helmholtz free energy, we derive the following relations:

$$\left(\frac{\partial H}{\partial S}\right)_{p,n} = T_{\text{coex}}, \qquad \left(\frac{\partial H}{\partial p}\right)_{S,n} = V, \qquad \left(\frac{\partial H}{\partial n}\right)_{S,p} = \mu_{\text{coex}}. \tag{6.20}$$

Heat supplied to the system at constant pressure in a reversible process is equal to the change in its enthalpy. From relation (6.19), we have $\Delta H = T_{\text{coex}}\Delta S$, since p and n are constant, and T_{coex} and μ_{coex} depend only of p. From the last relation, we can determine the change in the entropy of the two-phase system and then the change in n^α and n^β (see (6.11)).

6.3.2 Three-Phase Coexistence

The condition of coexistence of three phases: α, β and γ, is equivalent to two conditions of two-phase coexistence, for instance, $\mu^\alpha(T, p) = \mu^\beta(T, p)$ and $\mu^\alpha(T, p) = \mu^\gamma(T, p)$, from which the equality $\mu^\beta(T, p) = \mu^\gamma(T, p)$ also follows. Each of these conditions defines a line in the Tp plane, and two independent conditions define the intersection of two lines, which is the point of three-phase coexistence called the *triple point*. For instance, the triple point of water, at which liquid water, ice and water vapour coexist, is defined by the temperature of 273.16 K (0.01 °C) and pressure of 611.73 Pa (4.59 torr). The temperature of the triple point of water is thus slightly higher than the temperature of freezing at the pressure of 1 atm. In a pure substance, at most three phases can coexist. We shall see that in the case of mixtures, the maximum number of coexisting phases depends on the number of components.

At the triple point, the intensive parameters T, p and μ, have definite values, denoted by T_{trp}, p_{trp} and μ_{trp}, respectively. The Gibbs potential of the three-phase system amounts to

$$G = n^\alpha\mu^\alpha + n^\beta\mu^\beta + n^\gamma\mu^\gamma = n\mu_{\text{trp}}, \tag{6.21}$$

which means that it does not depend on the amounts of individual phases. An equilibrium state of the system is defined by three extensive parameters, which we assume to be: S, V and n, where

$$S = n^\alpha s^\alpha + n^\beta s^\beta + n^\gamma s^\gamma, \tag{6.22}$$

$$V = n^\alpha v^\alpha + n^\beta v^\beta + n^\gamma v^\gamma, \tag{6.23}$$

$$n = n^\alpha + n^\beta + n^\gamma. \tag{6.24}$$

The inversion of these relations gives

$$n^\alpha = \frac{n}{D}\left[\left(s^\beta - s\right)\left(v^\gamma - v\right) - \left(s^\gamma - s\right)\left(v^\beta - v\right)\right], \tag{6.25}$$

$$n^\beta = \frac{n}{D}\left[\left(s^\gamma - s\right)\left(v^\alpha - v\right) - \left(s^\alpha - s\right)\left(v^\gamma - v\right)\right], \tag{6.26}$$

$$n^\gamma = \frac{n}{D}\left[\left(s^\alpha - s\right)\left(v^\beta - v\right) - \left(s^\beta - s\right)\left(v^\alpha - v\right)\right], \tag{6.27}$$

where $D = s^\alpha v^\beta - s^\beta v^\alpha + s^\beta v^\gamma - s^\gamma v^\beta + s^\gamma v^\alpha - s^\alpha v^\gamma$. For instance, if $s = s^\alpha$ and $v = v^\alpha$, then $n^\alpha = n$ and $n^\beta = n^\gamma = 0$, and analogously for the other phases. S, V and n are the natural variables of the internal energy U. Using the Euler relation, we find the following form of U for the three-phase system:

$$U = U^\alpha + U^\beta + U^\gamma = T_{\text{trp}} S - p_{\text{trp}} V + \mu_{\text{trp}} n, \tag{6.28}$$

hence

$$\left(\frac{\partial U}{\partial S}\right)_{V,n} = T_{\text{trp}}, \qquad \left(\frac{\partial U}{\partial V}\right)_{S,n} = -p_{\text{trp}}, \qquad \left(\frac{\partial U}{\partial n}\right)_{S,V} = \mu_{\text{trp}}. \tag{6.29}$$

Let us assume that we have prepared a system in which three phases coexist, for instance, ice, liquid water and water vapour. We can change the proportion of the phases, changing the volume of the system in a reversible adiabatic process ($S = const$). Alternatively, we can supply heat to the system at $V = const$, i.e., in a reversible isochoric process. In both processes, the temperature and pressure remain constant until one of the coexisting phases disappears.

6.4 Phase Diagrams

The regions in which individual phases exist are presented in the form of *phase diagrams*, usually in the plane spanned by T and p. The border lines between different regions correspond to phase transitions. In this section, we present a few examples of phase diagrams.

6.4.1 Phase Diagram of a Typical Substance

The phase diagram of a typical substance in the Tp plane is shown in Fig. 6.3. The two-dimensional regions correspond to the gaseous, liquid and solid phases. The solid lines correspond to the gas–liquid, liquid–solid and gas–solid coexistence. All two-phase coexistence lines meet at the triple point. The liquid–solid coexistence line, called the *melting line*, has a positive slope for typical substances. We will show that it occurs when the molar volume of the liquid is larger than the molar volume of the solid. In the case of water, it is the other way round, i.e., the slope of the liquid–solid coexistence line is negative (see Fig. 6.5).

Note the difference between the liquid–gas and liquid–solid coexistence lines. The first terminates in the *critical point* defined by the *critical temperature* T_{cr} and *critical pressure* p_{cr}. At the critical point, the difference between the liquid and vapour disappears. Approaching the critical point along the liquid–gas coexistence line, we observe that the density of the vapour increases and the density of the liquid decreases, and at the critical point they become equal to each other. Then the interface between the two phases disappears. The heat of evaporation also vanishes

Fig. 6.3 Phase diagram of a typical pure substance

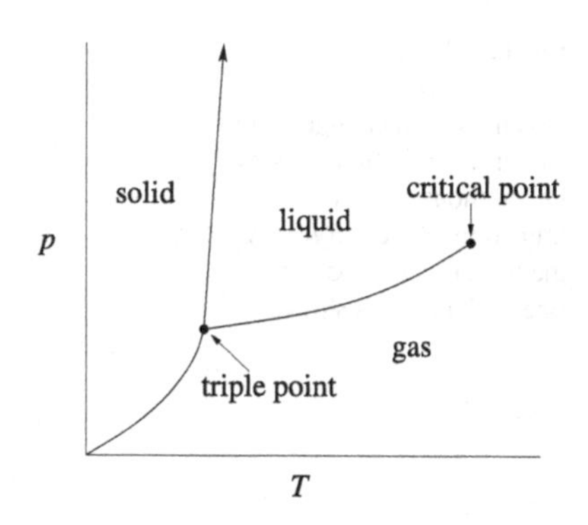

at the critical point. When $T > T_{cr}$ it is not possible to liquefy gas, irrespective of the pressure applied. Above the critical temperature, there exists one phase called the *super-critical* phase. It is a gaseous phase although its density is rather liquid like close to the critical point.

The arrow in Fig. 6.3 marks the continuation of the liquid–solid coexistence line. Why does not it terminate in a critical point, as the liquid–gas coexistence line does? In a liquid, as in a gas, the motion of molecules is chaotic. A single molecule is not related to any particular space point. In a crystalline solid, molecules can only oscillate around specific points that form a well defined three-dimensional periodic structure. Such a structure cannot appear gradually in a continuous fashion. Therefore, the line of the transition from a liquid to a crystalline solid cannot terminate in a critical point, at which the difference between two phases disappears by definition.

What does it mean in practice that the liquid–gas coexistence line terminates in a critical point? Looking at Fig. 6.3, we can see that it is possible to carry out a process which transforms a liquid into a gas or vice versa without crossing of the transition line. An example of such a process is shown in Fig. 6.4 with the broken line. For instance, the process begins in the liquid phase, close to the liquid–gas coexistence line, and ends also close to the coexistence line but on the other side in the gaseous phase. At any moment during the process, we do not observe the interface between the liquid and vapour, because the broken line does not cross the coexistence line. This shows that it is possible to transform a liquid into a vapour in such way that during the process we do not observe any qualitative changes in the system. Therefore, according to Definition 6.2, the process considered is not a phase transition.

For a temperature T between the triple and critical points, the pressure of *saturated vapour*, meaning the vapour which coexists with the liquid, is equal to $p_{coex}(T)$. It is the pressure of the vapour that fills the space above the liquid in a closed vessel from which air has been pumped out. On the other hand, a liquid in an open vessel feels atmospheric pressure. When it is heated, the process of evaporation proceeds from the surface as long as the saturated vapour pressure is smaller than atmospheric pressure. If they are equal the process of evaporation becomes

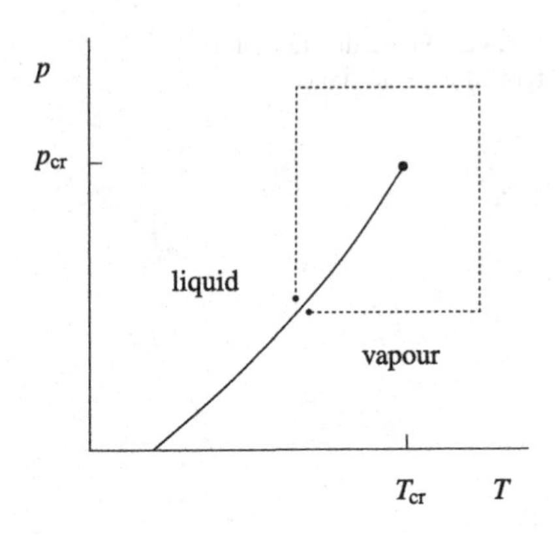

Fig. 6.4 Example of a process (*broken line*) in which a liquid changes into a vapour in a continuous way, i.e., without a phase transition. For clarity, only the liquid–gas coexistence line (*solid line*) is shown

violent because vapour bubbles form in the whole volume of the liquid. This very well known phenomenon is called *boiling*, and the temperature at which the saturated vapour pressure is equal to the external pressure is called the *boiling point*. The *normal boiling point* corresponds to the pressure of 1 atm.

A substance whose phase diagram looks like that in Fig. 6.3 is carbon dioxide, for instance. The coordinates of the triple point of CO_2 are: 217 K and 5.11 bar, and the coordinates of the critical point: 304 K and 72.8 bar. As we can see, the pressure of the triple point is much higher than atmospheric pressure. This means that CO_2 does not exist in the liquid state at atmospheric pressure but it undergoes a direct transition from the solid to gas (at 195 K at the pressure of 1 bar). For this reason, solid carbon dioxide is called the *dry ice*.

6.4.2 Phase Diagram of Water

The phase diagram of water is shown in Fig. 6.5. The black circles mark the triple and critical points. In the range of pressure up to 2000 bar, liquid water coexists with the solid phase called ice Ih. The melting line has a negative slope and is almost vertical, which means that even large variations of pressure cause only slight changes in the *melting point*. The negative slope of the melting line is related to hydrogen bonds between water molecules. In the solid phase, water molecules are more loosely packed than in the liquid phase, therefore the density of ice is smaller than the density of liquid water. In the high pressure region, liquid water can coexist with other crystalline structures of ice, marked with the numbers: III, V, VI.[2] For pressures above 2000 bar, the slope of the melting line becomes positive. For instance, the melting point of ice VII, which is not shown in Fig. 6.5, can exceed 100 °C, but this phase exists for pressures above 22 000 bar.

---

[2] It turned out that the form denoted previously as ice IV does not exist.

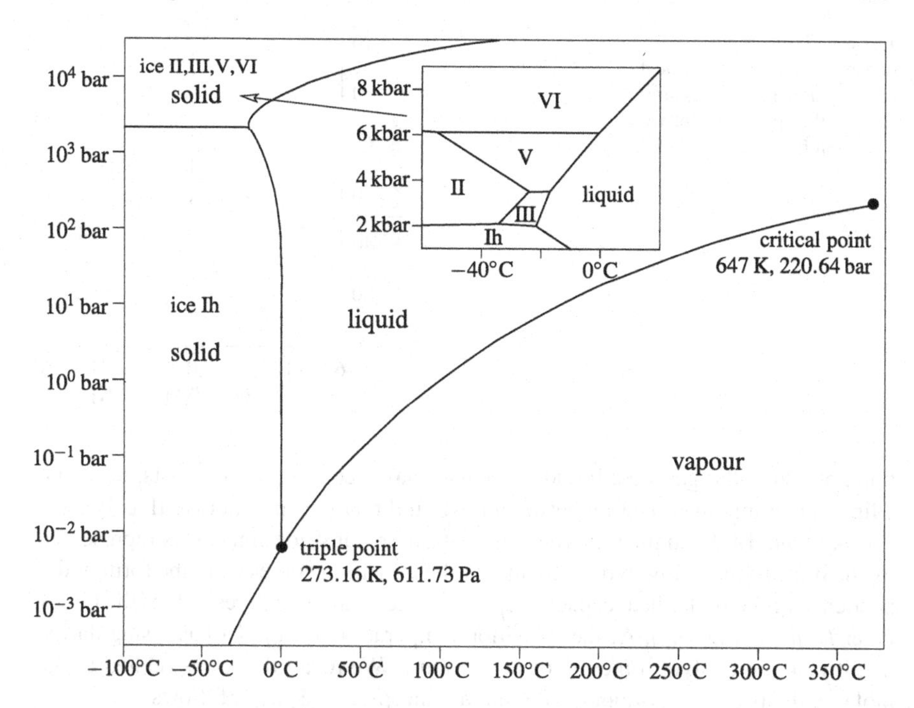

Fig. 6.5 Phase diagram of water; note the logarithmic scale of pressure. In the inset, the region marked as ice II, III, V, VI is shown in more detail. These crystalline structures of ice, not singled out in the main plot, differ from the solid phase denoted ice Ih. The phase diagram is not complete, as it does not show all known structures of ice that exist at higher pressures or lower temperatures

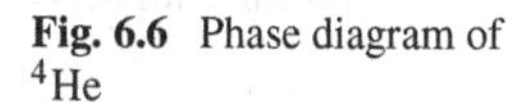
Fig. 6.6 Phase diagram of ^{4}He

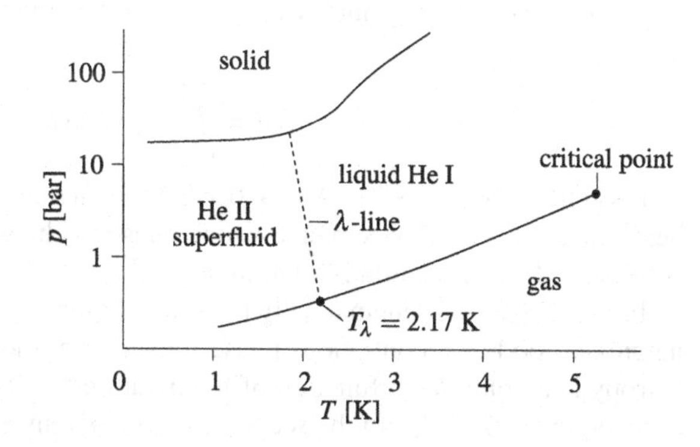

6.4.3 Phase Diagram of ^{4}He

Figure 6.6 shows schematically the phase diagram of ^{4}He. As we have already mentioned there exist two liquid phases: He I and He II, where He II denotes the superfluid phase. The He I–gas coexistence line ends in a critical point but it does not intersect the solid–liquid coexistence line, which means that there is no triple

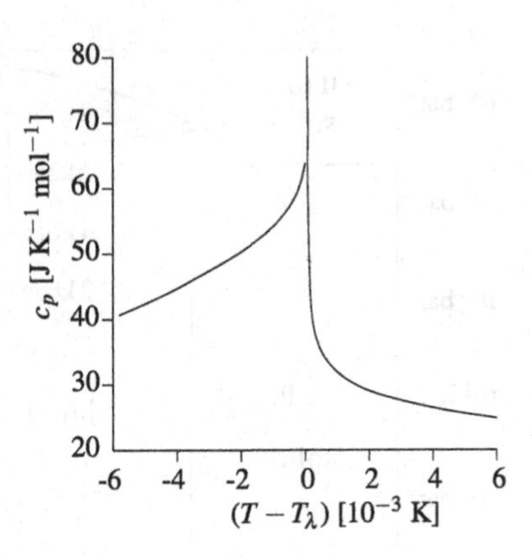

Fig. 6.7 Temperature dependence of the molar heat capacity at constant pressure close to the continuous phase transition from He I to He II

point at which the gaseous, liquid and solid phases coexist. There exists, however, a line of continuous transition between the liquid phases: He I and He II, called the λ-*transition*. To distinguish it from the first-order transition lines, it is represented by the broken line in Fig. 6.6. The name of the transition comes from the form of dependence of the molar heat capacity, c_p, on temperature, which resembles the Greek letter *lambda* (Fig. 6.7). At the transition temperature T_λ, $c_p(T)$ has a singularity, i.e., it is a non-analytic function of temperature. We recall that c_p is related to the molar enthalpy, $h = H/n$, and to the molar entropy, $s = S/n$, as follows:

$$c_p = \left(\frac{\partial h}{\partial T}\right)_p = T\left(\frac{\partial s}{\partial T}\right)_p . \tag{6.30}$$

Obviously the amount of heat needed to change the temperature of the system by $\Delta T = T_2 - T_1$ must be finite. If $T_1 < T_\lambda < T_2$ then the change in the molar enthalpy amounts to

$$\Delta h = \int_{T_1}^{T_2} c_p(T)\mathrm{d}T \tag{6.31}$$

and is finite. For $T_1 \to T_2$, $\Delta h \to 0$ which means that an infinitesimal amount of heat absorbed by the system causes one phase to change into the other phase. This is because the λ-transition is continuous.

Figure 6.8 shows schematically the shape of the functions $s(T)$ and $c_p(T)$ in the neighbourhood of a continuous phase transition at the temperature T_{tr}. The molar entropy is a continuous function of temperature but its derivative is discontinuous or divergent at $T = T_{\mathrm{tr}}$. In the second case, $c_p(T)$ diverges at the transition temperature. The type of this divergence depends on a physical system but often it has the following form near T_{tr}:

$$c_p(T) = A|T - T_{\mathrm{tr}}|^{-\alpha}, \tag{6.32}$$

where the exponent α is positive and the coefficient A does not depend on T. If $\alpha < 0$ then $c_p(T)$ has a sharp but finite maximum at the transition temperature. For instance, in the case of the λ-transition in ^{4}He (see Fig. 6.7) α has a small negative value of the order -0.01.

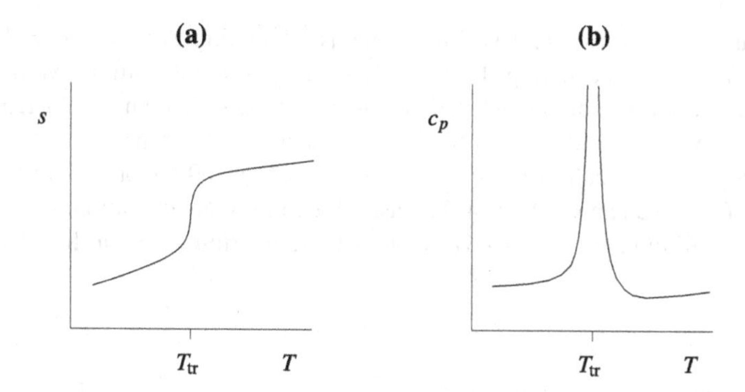

Fig. 6.8 Temperature dependence of: (**a**) the molar entropy, (**b**) the molar heat capacity at constant pressure, in the neighbourhood of a continuous phase transition

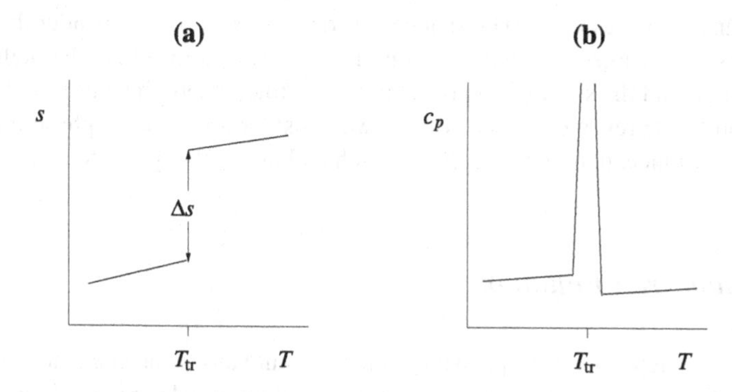

Fig. 6.9 Temperature dependence of: (**a**) the molar entropy, (**b**) the molar heat capacity at constant pressure, in the neighbourhood of a first-order phase transition; Δs denotes the jump of the molar entropy at the transition temperature T_{tr}

It is instructive to compare the behaviour of $s(T)$ and $c_p(T)$ close to the transition temperature for continuous and first-order transitions. As we know, in the latter case, $s(T)$ is discontinuous at the transition temperature, which is shown in Fig. 6.9a. Using expression (6.30), we present Δh for $T_1 < T_{tr} < T_2$ as follows:

$$\Delta h = \int_{T_1}^{T_2} T \frac{\partial s}{\partial T} \mathrm{d}T = T_2 s(T_2) - T_1 s(T_1) - \int_{T_1}^{T_2} s(T) \mathrm{d}T, \tag{6.33}$$

where we have integrated by parts. In the limit $T_1 \to T_2$, we get

$$\Delta h = T_{tr} \Delta s. \tag{6.34}$$

As we know, Δh is equal to the heat received by a system at constant pressure, and in this case it is equal to the heat of transition per mole. The dependence of c_p on temperature is shown in Fig. 6.9b. The derivative $\partial s/\partial T$ at $T = T_{tr}$ does not exist in ordinary sense, since $s(T)$ is discontinuous.[3] Dividing Δh by $\Delta T = T_2 - T_1$, for $T_1 < T_{tr} < T_2$, and taking the limit $T_1 \to T_2$, we obtain infinite value of c_p. However, if both T_1 and T_2 are smaller or larger than T_{tr} then $c_p(T)$ tends to a finite value when T_1 and T_2 tend to T_{tr}. This mean that the function $c_p(T)$ can be represented schematically by the shape shown in Fig. 6.9b. Comparing Fig. 6.9b with Fig. 6.8b, we can see that in the case of a continuous transition, $c_p(T)$ tends gradually to infinity when $T \to T_{tr}$, or to a finite maximum, as in the case of the λ-transition.

6.5 Two-Phase Coexistence Lines

Our present aim is to derive a general formula for a two-phase coexistence line. We recall that two phases can coexist only in the case of a first-order phase transition. For substances such as water (H_2O) or carbon dioxide (CO_2), we consider three two-phase coexistence lines, i.e., the evaporation (or condensation) line, the melting (or freezing) line and the sublimation (or deposition) line, where the names in brackets correspond to the reverse process. Below we consider a first-order phase transition in a pure substance, denoted $\alpha \to \beta$, at which a phase α changes into a phase β.

6.5.1 *Clapeyron Equation*

We denote the dependence of pressure on temperature along the coexistence line by $p_{\mathrm{coex}}(T)$. Two phases coexist when their chemical potentials are equal (see Corollary 6.1), hence

$$\mu^\alpha\big(T, p_{\mathrm{coex}}(T)\big) = \mu^\beta\big(T, p_{\mathrm{coex}}(T)\big) \tag{6.35}$$

Differentiating both sides of (6.35) with respect to temperature, we get:

$$\left(\frac{\partial \mu^\alpha}{\partial T}\right)_{p_{\mathrm{coex}}} + \left(\frac{\partial \mu^\alpha}{\partial p_{\mathrm{coex}}}\right)_T \frac{\mathrm{d}p_{\mathrm{coex}}}{\mathrm{d}T} = \left(\frac{\partial \mu^\beta}{\partial T}\right)_{p_{\mathrm{coex}}} + \left(\frac{\partial \mu^\beta}{\partial p_{\mathrm{coex}}}\right)_T \frac{\mathrm{d}p_{\mathrm{coex}}}{\mathrm{d}T}. \tag{6.36}$$

Then we use the relations: $(\partial\mu/\partial T)_p = -s$ and $(\partial\mu/\partial p)_T = v$, to transform (6.36) into the form called the *Clapeyron equation*:

$$\frac{\mathrm{d}p_{\mathrm{coex}}}{\mathrm{d}T} = \frac{\Delta s}{\Delta v}, \tag{6.37}$$

where $\Delta s = s^\beta - s^\alpha$ and $\Delta v = v^\beta - v^\alpha$ denote, respectively, the changes in the molar entropy and volume at the transition $\alpha \to \beta$. Since the molar enthalpy $h = \mu + Ts$, and $\Delta\mu = \mu^\beta - \mu^\alpha = 0$, hence, $\Delta h = T\Delta s$, where $\Delta h = h^\beta - h^\alpha$. This

[3] The derivative of a discontinuous function close to the point of discontinuity is represented by a distribution called the Dirac δ-function.

shows that the Clapeyron equation relates the slope of the coexistence line with the enthalpy of the transition Δh and the change in the molar volume Δv as follows:

$$\frac{\mathrm{d}p_{\mathrm{coex}}}{\mathrm{d}T} = \frac{\Delta h}{T \Delta v}. \tag{6.38}$$

The quantities Δh and Δv depend only on the temperature because they are defined on the coexistence line. If we assume that $\Delta h > 0$ for the transition $\alpha \to \beta$ then the sign of the derivative $\mathrm{d}p_{\mathrm{coex}}/\mathrm{d}T$ is determined by the sign of Δv.

6.5.2 Solid–Liquid Coexistence

We denote the solid and liquid phases with the indices s and l, respectively. To melt a solid, heat has to be supplied to the system, therefore, the *enthalpy of melting* is positive, i.e., $\Delta h = h^l - h^s > 0$. In the case of typical substances, e.g. CO_2, also $\Delta v = v^l - v^s > 0$, hence, $\mathrm{d}p_{\mathrm{coex}}/\mathrm{d}T > 0$ (Fig. 6.3). For anomalous substances, e.g. H_2O, $\Delta v < 0$, hence, also $\mathrm{d}p_{\mathrm{coex}}/\mathrm{d}T < 0$ (Fig. 6.5), which means that an increase in pressure lowers the melting point.

If Δh and Δv are approximately constant in a certain range of temperature then Eq. (6.38) can be integrated to give

$$p_{\mathrm{coex}}(T) = p_{\mathrm{coex}}(T_0) + \frac{\Delta h}{\Delta v} \ln \frac{T}{T_0}, \tag{6.39}$$

where T_0 denotes a reference temperature. When the relative change in temperature, $(T - T_0)/T_0$, is small we have $\ln(T/T_0) \approx (T - T_0)/T_0$ and Eq. (6.39) adopts a simpler form:

$$p_{\mathrm{coex}}(T) \approx p_{\mathrm{coex}}(T_0) + \frac{\Delta h}{T_0 \Delta v}(T - T_0). \tag{6.40}$$

Example 6.1 The heat of melting of ice at the temperature $T_0 = 273.15$ K is equal to $\Delta h = 6.01$ kJ/mol, and $\Delta v = -1.7\ \mathrm{cm}^3\,\mathrm{mol}^{-1}$, hence, $\Delta h/(T_0 \Delta v) = -127.7\ \mathrm{atm\,K}^{-1}$, and $p_{\mathrm{coex}}(T_0) = 1$ atm. To lower the melting point of ice by only 1 °C below zero, the pressure must exceed atmospheric pressure by almost 128 atm. We can estimate the drop in the melting point of ice caused by a skater weighting 70 kg. Assuming 30 cm for the length of a skate and 2 mm for the width of its blade, we obtain the pressure $p \approx (70 \times 9.81/6)\ \mathrm{N\,cm}^{-2}$, which is about 11 atm, provided that only one skate touches the ice surface. Thus, the drop in the melting point caused by the skater amounts to about 0.1 °C, which can explain the effect of sliding only at a temperature close to 0 °C. Then what causes sliding at lower temperatures? It is known that friction between ice and the blade of a skate is an important factor. It provides a sufficient amount of heat to form a thin layer of water. However, this effect does not explain a well known fact that it is also difficult to stand on skates. Recent studies show the existence of a phenomenon called *surface melting*. It causes the surface of ice to be covered with a microscopic liquid layer, of the thickness of a few molecules only, which exists at temperatures well below 0 °C. It is not a separate phase, however, but a surface layer of a different structure than the crystalline bulk phase.

6.5.3 *Liquid–Gas Coexistence*

We use the index g for the gas phase. In the process of evaporation, heat is supplied to the liquid, hence, the *enthalpy of evaporation*, $\Delta h = h^g - h^l$, is positive. Also $\Delta v = v^g - v^l > 0$ and usually the molar volume of the gaseous phase is much larger than the molar volume of the liquid phase. Only close to the critical point do they become comparable. Far from the critical point, we have $\Delta v \approx v^g$. Moreover, we assume that the molar volume of the vapour can be determined from the equation of state of the ideal gas: $v^g = RT/p_{\text{coex}}$. Substituting Δv into Eq. (6.38), we obtain the *Clausius–Clapeyron* equation:

$$\frac{\mathrm{d}\ln p_{\text{coex}}}{\mathrm{d}T} = \frac{\Delta h}{RT^2}. \tag{6.41}$$

If the dependence of Δh on temperature can be neglected then Eq. (6.41) can be integrated. Assuming that T_0 is a reference temperature, we obtain the following expression:

$$p_{\text{coex}}(T) = p_{\text{coex}}(T_0)\exp\left[\frac{\Delta h}{RT_0}\left(1 - \frac{T_0}{T}\right)\right]. \tag{6.42}$$

6.5.4 *Solid–Gas Coexistence*

The change in the enthalpy at the solid–gas transition, $\Delta h = h^g - h^s$, is called the *enthalpy of sublimation* and is positive. Also $\Delta v = v^g - v^s > 0$, and the molar volume of the gas is much larger than the molar volume of the solid. Therefore, we can use similar reasoning as for the liquid–gas transition and assume that $\Delta v \approx v^g$ and $v^g = RT/p_{\text{coex}}$. It leads again to Eq. (6.41) in which the enthalpy of melting is replaced by the enthalpy of sublimation. If the latter does not depend on temperature then the pressure of sublimation as a function of temperature is given by expression (6.42).

6.6 Liquid–Vapour Two-Phase Region

So far we have presented phase diagrams in the Tp plane. Then a single phase is represented by a two-dimensional region, and the two-phase coexistence is represented by a line. As we know, the equation of state of the form $f(p, v, T) = 0$ holds for a

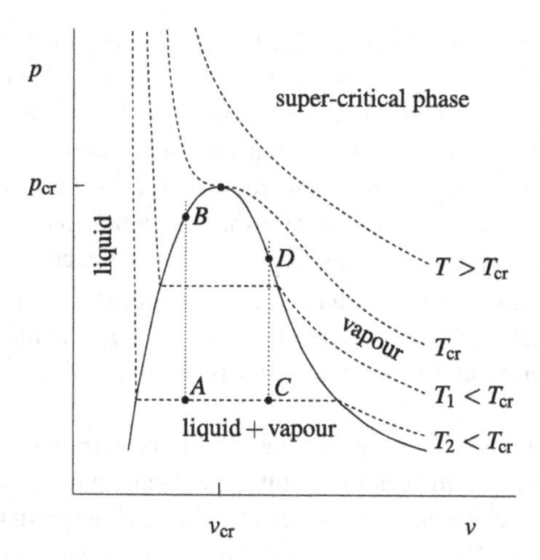

Fig. 6.10 Isotherms in the liquid and vapour one-phase regions, and in the liquid + vapour two-phase region are shown (*broken lines*). The *solid line* represents the molar volume of the liquid (*left branch*) and of the vapour (*right branch*) at the liquid–vapour coexistence, and v_{cr} and p_{cr} correspond to the critical point. *Above* the critical isotherm, $T = T_{cr}$, only the super-critical phase exists. The *vertical segments*, denoted AB and CD, correspond to the isochors $v = v_1 < v_{cr}$ and $v = v_2 > v_{cr}$, respectively

pure substance, which means that phase diagrams can also be plotted in the vp or vT planes. As an example, we consider the liquid–vapour coexistence (see Fig. 6.10). Above the critical temperature T_{cr} only the super-critical phase exists for any pressure. Below T_{cr}, a substance can exist as a liquid or as a gas (vapour). The solid line shows the molar volume of the coexisting phases as a function of pressure; the left branch, $v^l(p)$, corresponds to the liquid and the right branch, $v^g(p)$, corresponds to the vapour. The two branches merge at the critical point, i.e., $v^l = v^g = v_{cr}$. For a given pressure $p < p_{cr}$, the temperature T of the liquid-vapour coexistence satisfies the equation $p = p_{coex}(T)$, hence, v^l and v^g can also be treated as functions of temperature.

On the diagram, several isotherms are shown. The horizontal segments for $T < T_{cr}$ correspond to the *saturated vapour* pressure, $p_{coex}(T)$, and pressures $p < p_{coex}(T)$ correspond to the *unsaturated vapour*. The isotherms on the liquid side are very steep. Liquids and solids are not very compressible, and even a small change in the volume results in a huge change in the pressure. In the two-phase region, the pressure remains constant when the volume of the system changes. This is because a change in the volume at constant temperature changes only the proportion of the coexisting phases until one of them disappears. This proportion follows from the lever rule (see (6.13)):

$$n^l\left(v - v^l\right) = n^g\left(v^g - v\right), \tag{6.43}$$

where $v^l \le v \le v^g$. When the temperature approaches T_{cr} the horizontal segment of an isotherm shortens, and at T_{cr} its length reduces to a point. On the horizontal segment, $(\partial p/\partial v)_T = 0$, and at $T = T_{cr}$, the derivative $(\partial p/\partial v)_T$ vanishes only at the critical point, $v = v_{cr}$, and is negative for $v \neq v_{cr}$. Therefore, the critical point is an inflection point on the critical isotherm. For $T > T_{cr}$, the inequality $(\partial p/\partial v)_T < 0$ always holds. We recall that the isothermal compressibility is defined as $\kappa_T = -v^{-1}(\partial v/\partial p)_T$. From the shape of the isotherms, it follows that in the one-phase regions $\kappa_T > 0$, which means that the condition of mechanical stability is satisfied. Obviously, if $\kappa_T > 0$ also $\kappa_T^{-1} > 0$. Thus, the condition of mechanical stability is not satisfied in the two-phase region, where $\kappa_T^{-1} = 0$. When the coexistence line is approached from the one-phase region κ_T remains constant as long as $T < T_{cr}$. However, when the critical point is approached $\kappa_T^{-1} \to 0$ from above, which means that $\kappa_T \to +\infty$.

The divergence of the isothermal compressibility is an important characteristic of the critical point. Close to the critical point the substance becomes very compressible, and even small variations of pressure cause large changes in the molar volume. This effect leads to a phenomenon called the *critical opalescence*. It consists in strong scattering of light of different wavelength by a substance close to the critical point. Since the substance becomes very compressible an instantaneous density in any small fragment of the fluid can be very different from its average value. Close to the critical point large *fluctuations*[4] of the density exist, which spread out over distances comparable to the wavelength of light. The presence of such fluctuations causes scattering of light and the effect of fluid opalescence.

What is going to happen when a closed vessel filled partially with a liquid is heated at constant volume? We assume that the rest of the vessel is occupied only by the vapour coexisting with the liquid. Since the vessel is closed, the average molar volume, $v = V/n$, does not change. During the heating the molar volume of the liquid increases and the molar volume of the vapour decreases. Suppose first that the liquid–vapour interface is in the upper part of the vessel ($n^l \gg n^g$), hence $v = v_1 < v_{cr}$. Due to the heating the interface moves up (the segment AB in Fig. 6.10). At the temperature T_B, which corresponds to the isotherm crossing the point B, the whole vessel is filled with the liquid and further heating proceeds in the liquid region. Analogously, if the liquid-vapour interface is initially in the lower part of the vessel ($n^l \ll n^g$), i.e., $v = v_2 > v_{cr}$, then during the heating the interface moves down (the segment CD in Fig. 6.10). At the temperature T_D, which corresponds to the isotherm crossing the point D, the whole vessel is filled with the vapour and further heating proceeds in the vapour region.

The situations described above are also presented on the diagram in the Tp plane (Fig. 6.11). The solid line represents the liquid–vapour coexistence. The initial points A and C in Fig. 6.10 correspond here to one point because $T_A = T_C$. The isochors $v = v_1$ and $v = v_2$ are shown with a broken line in the one-phase regions. They branch off from the coexistence line at the points B and D. The critical isochor, $v = v_{cr}$, is also shown. In this particular case, the heating does not cause large

[4]Fluctuations are instantaneous and uncontrollable deviations of a given physical quantity from its average value.

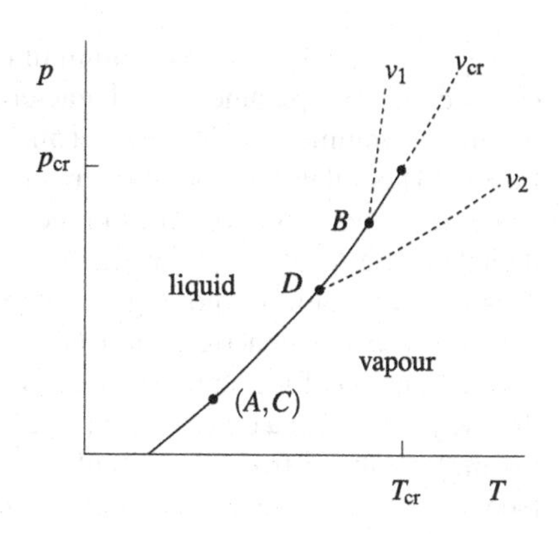

Fig. 6.11 Isochors: $v = v_1 < v_{cr}$, $v = v_2 > v_{cr}$ and $v = v_{cr}$. The points A, B, C and D are defined in Fig. 6.10

changes in the position of the interface which persists to the critical temperature. Slightly below T_{cr} the interface begins to smear out, and at T_{cr} it disappears completely because the liquid and gaseous phases become indistinguishable. The critical isochor crosses the critical point and enters directly the region of the super-critical phase.

At the critical point, both Δv and the heat of transition Δh disappear, which means that the liquid–vapour transition becomes continuous at this single point. It should be stressed, however, that many continuous transitions are not related to any critical point on a first-order transition line (see Sect. 6.4.1), for instance, the continuous transitions between the paramagnetic and ferromagnetic phases or between He I and He II.

6.7 Van der Waals Equation of State

So far our discussion of phase transitions has been of qualitative nature only. In this section, we present and study an equation of state which provides a simple mathematical model for the liquid–vapour coexistence. It is quite obvious that the ideal gas equation of state, $pv = RT$, is not suitable for this purpose because the isotherms that exhibit only monotonic behaviour cannot describe the two-phase region. Therefore, van der Waals proposed the following equation of state (see Sect. 2.4.2):

$$p = \frac{RT}{v - b} - \frac{a}{v^2}, \tag{6.44}$$

where the constants a and b are to be determined experimentally for a given substance. The van der Waals equation of state is of empirical nature but it can also be derived from a microscopic theory. Note that Eq. (6.44) reduces to the ideal gas equation of state for $a = 0$ and $b = 0$. Since the ideal gas model is based on the assumption that molecules do not interact with one another, the constants a and b must be related to intermolecular interactions.

The interaction of two molecules is strongly repulsive at a very short distance compared with the size of a molecule, and attractive at a longer distance. Due to the strong repulsion there exists a minimum volume which a given number of molecules can occupy. The parameter b characterizes this minimum volume per mole. Thus, the molar volume $v < b$ does not make any physical sense. The parameter a in Eq. (6.44) is related to the attractive interaction between molecules at a large distance. If there are N molecules in the volume V and the molecules interact in pairs then there are $N(N-1)/2$ interactions, which can be approximated by $N^2/2$ when N is large. Assuming a homogeneous system and taking into account that the internal energy is an extensive parameter, we calculate the internal energy due to attractive interactions for a unit volume, and then multiply the result by the volume. In this way we obtain a quantity proportional to $(N/V)^2V$, which is easy to convert to the molar volume from the formula $N = nN_A$, where N_A is the Avogadro constant (see Sect. 2.2.2). The reasoning presented above leads to the following approximate expression for the internal energy of the system (see (2.40)):

$$U = \frac{f}{2}nRT - \frac{an^2}{V}, \tag{6.45}$$

where the parameter $a > 0$ characterizes, in an average way, the energy of the attractive interactions. The first term comes from the kinetic energy of molecules and is the same as for the ideal gas.

Note that there is no contribution from the repulsive interactions in expression (6.45). It can be understood by analogy with the billiard balls. The balls interact strongly when they collide with one another but their potential energy of deformation changes quickly into the kinetic energy. Thus, the total energy averaged over the time is practically equal to their kinetic energy. If the balls interacted also indirectly, due to electric charges, for instance, then we would have to include their potential energy as well.

The pressure satisfies the following relation:

$$p = -\left(\frac{\partial F}{\partial V}\right)_{T,n} = -\left(\frac{\partial U}{\partial V}\right)_{T,n} + T\left(\frac{\partial S}{\partial V}\right)_{T,n} = -\frac{a}{v^2} + T\left(\frac{\partial S}{\partial V}\right)_{T,n}. \tag{6.46}$$

From the comparison with (6.44), it follows that

$$\left(\frac{\partial S}{\partial V}\right)_{T,n} = \left(\frac{\partial s}{\partial v}\right)_{T} = \frac{R}{v-b}, \tag{6.47}$$

which means that the first term in the van der Waals equation is related to the entropy, hence

$$s(T,v) = R\ln\frac{v-b}{v_0-b} + \chi(T) \tag{6.48}$$

where v_0 is the molar volume of a reference state. To determine the function $\chi(T)$ we use the relation

$$\left(\frac{\partial u}{\partial T}\right)_{v} = T\left(\frac{\partial s}{\partial T}\right)_{v} = T\frac{\mathrm{d}\chi}{\mathrm{d}T}, \tag{6.49}$$

where $(\partial u/\partial T)_v = fR/2$, hence, $\chi(T) = s_0 + (fR/2)\ln(T/T_0)$ and

$$s(T,v) = s_0 + \frac{1}{2}fR\ln\frac{T}{T_0} + R\ln\frac{v-b}{v_0-b}. \tag{6.50}$$

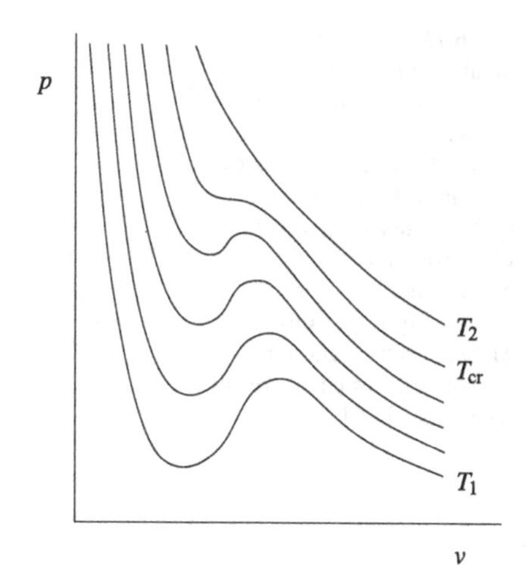

Fig. 6.12 Schematic picture of the isotherms obtained from the van der Waals equation of state for several values of T between $T_1 < T_{cr}$ and $T_2 > T_{cr}$. The part of an isotherm with a positive slope is unphysical

For $b = 0$, we recover the ideal gas entropy as a function of T and v. However, the thermodynamic potential whose natural variables are T and v is the molar Helmholtz free energy, $\phi = F/n = u - Ts$. Using relations (6.45) and (6.50), we obtain

$$\phi(T, v) = \left(\frac{1}{2} f R - s_0\right) T - \frac{a}{v} - \frac{1}{2} f RT \ln \frac{T}{T_0} - RT \ln \frac{v - b}{v_0 - b}. \quad (6.51)$$

6.7.1 Maxwell Construction

The first term in Eq. (6.44) gives a higher pressure than the ideal gas equation of state for the same values of T and v, whereas the presence of attractive interactions lowers the pressure. For a very dilute gas, the van der Waals equation of state gives similar results as the equation of state of the ideal gas. When v decreases the second term becomes important and the predictions of the two equations become very different.

Schematic picture of the isotherms obtained from Eq. (6.44) is shown in Fig. 6.12. Comparing Figs. 6.10 and 6.12, we notice an important difference for temperatures $T < T_{cr}$. The van der Waals isotherms do not have the horizontal segment corresponding to the two-phase region. Moreover, the part of an isotherm between the local minimum and local maximum has a positive slope, which corresponds to negative isothermal compressibility. Since a system with negative compressibility is not mechanically stable, the liquid–vapour transition must occur at a pressure between the local minimum and local maximum of the isotherm, which is shown in Fig. 6.13. To determine the pressure p_{coex} at the liquid–vapour coexistence from the van der Waals equation of state, we integrate the Gibbs–Duhem equation (see (6.3)): $d\mu = -s dT + v dp$, along the isotherm from the point A to B, hence

$$\mu_B - \mu_A = \int_A^B v dp = \int_A^B d(pv) - \int_A^B p dv = \int_{v_A}^{v_B} \left[p_{coex} - p(T, v)\right] dv. \quad (6.52)$$

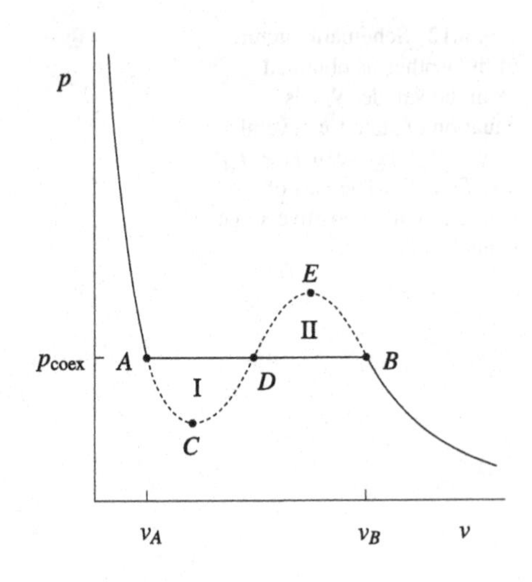

Fig. 6.13 Single van der Waals isotherm for $T < T_{cr}$. The phase transition occurs at the pressure $p_{coex}(T)$. The metastable states (*fragments* AC *and* EB) and the unstable states (*fragment* CDE) are drawn with a broken line. The pressure $p_{coex}(T)$ follows from the Maxwell construction, i.e., the condition of equal area of the regions I and II

From the condition of two-phase coexistence, $\mu_A = \mu_B$, we get

$$\int_{v_A}^{v_B} \big[p_{coex} - p(T, v)\big] dv = 0, \tag{6.53}$$

where $p(T, v)$ denotes the van der Waals isotherm for the given temperature T, and the molar volumes v_A and v_B satisfy the equation $p(T, v) = p_{coex}$. Therefore, condition (6.53) can be treated as an equation for p_{coex}. It is shown in Fig. 6.13 that the solution of (6.53) corresponds to the condition of equal area of the regions I and II, and $v_A = v^l(T)$ and $v_B = v^g(T)$ are the molar volumes of the coexisting liquid and gaseous phases at the temperature T. The construction shown in Fig. 6.13 is called the *Maxwell construction*. It replaces a part of the van der Waals isotherm (broken line in Fig. 6.13) with a horizontal segment, in such a way that the regions I and II between the isotherm and the horizontal segment have the same area. In other words, the Maxwell construction allows to transform the van der Waals isotherm into another isotherm which describes also the liquid–vapour two-phase region.

The fragments of the van der Waals isotherm with $\kappa_T > 0$, marked as AC and EB in Fig. 6.13, have a simple physical meaning. They correspond to metastable states: a super-heated liquid (AC) and a super-saturated vapour (EB). Metastable states can be observed experimentally if the change of one phase into another phase proceeds sufficiently slowly and the substance is free of impurities. The fragment CDE of the isotherm corresponds to unstable states for which $\kappa_T < 0$. In this range of molar volume, the system separates into two phases irrespective of the speed of the process and purity of the substance.

6.7.2 Principle of Corresponding States

From the van der Waals equation of state, we can determine the position of the critical point, that is, we can express T_{cr}, p_{cr} and v_{cr} in terms of the constants a and b

which characterize a given substance. To do this, we first rewrite the van der Waals equation, using the *reduced variables*: $\bar{T} = T/T_{cr}$, $\bar{p} = p/p_{cr}$ and $\bar{v} = v/v_{cr}$. In terms of the reduced variables, it adopts the same functional form as Eq. (6.44), i.e.,

$$\bar{p} = \frac{\bar{R}\bar{T}}{\bar{v} - \bar{b}} - \frac{\bar{a}}{\bar{v}^2}, \tag{6.54}$$

where $\bar{R} = RT_{cr}/(p_{cr}v_{cr})$, $\bar{a} = a/(p_{cr}v_{cr}^2)$ and $\bar{b} = b/v_{cr}$. Then we express (6.54) in the form of a third-order equation for the reduced molar volume $\bar{v}$:

$$\bar{p}\bar{v}^3 - (\bar{p}\bar{b} + \bar{R}\bar{T})\bar{v}^2 + \bar{a}\bar{v} - \bar{a}\bar{b} = 0. \tag{6.55}$$

From the shape of the isotherms shown in Fig. 6.12, it follows that Eq. (6.55) has one real root for $\bar{T} > 1$, whereas for $\bar{T} < 1$ it may have one or three real roots, depending on the value of the reduced pressure $\bar{p}$. In particular, for $\bar{p} = p_{coex}/p_{cr}$, Eq. (6.55) has three real roots corresponding to the points A, B and C in Fig. 6.13. When $\bar{T} \to 1$ from below, the roots approach one another, and at the critical temperature, they merge into one point corresponding to $v = v_{cr}$, i.e., $\bar{v} = 1$. It means that at $\bar{T} = 1$ and $\bar{p} = 1$, Eq. (6.55) must have one triple root at $\bar{v} = 1$, hence

$$\bar{v}^3 - (\bar{b} + \bar{R})\bar{v}^2 + \bar{a}\bar{v} - \bar{a}\bar{b} = (\bar{v} - 1)^3 = 0. \tag{6.56}$$

Comparing the coefficients at different powers of $\bar{v}$, we get

$$\bar{a} = 3, \qquad \bar{b} = \frac{1}{3}, \qquad \bar{R} = \frac{8}{3}, \tag{6.57}$$

hence, we can express the critical parameters in terms of a and b as follows:

$$v_{cr} = 3b, \qquad p_{cr} = \frac{a}{27b^2}, \qquad T_{cr} = \frac{8a}{27Rb}. \tag{6.58}$$

Then the van der Waals equation of state, expressed in terms of the reduced variables, adopts the following form:

$$\bar{p} = \frac{8\bar{T}}{3\bar{v} - 1} - \frac{3}{\bar{v}^2}. \tag{6.59}$$

Note that now it does not contain any quantities characterizing the substance. Therefore, we can formulate the following conclusion called the *principle of corresponding states*.

Corollary 6.2 *If two different substances have the same values of two reduced variables then the value of the third reduced variable must also be the same.*

We have just shown that this principle holds in the case of substances to which the van der Waals equation of state can be applied. However, it is more general than the van der Waals equation. Indeed, any equation of state that can be expressed only in terms of the reduced variables, i.e.,

$$f(\bar{p}, \bar{T}, \bar{v}) = 0, \tag{6.60}$$

does not depend on any quantities characterizing the substance, thus, it satisfies the principle of corresponding states. In reality, this principle is approximately satisfied only for gases of non-polar molecules, i.e, without electric dipole moments, whose shape resembles a sphere.

Finally, we note that any equation of state which involves only two constants characterizing the substance, a and b, can be expressed in a reduced form independent of the substance. Obviously the principle of corresponding states must hold for such equations of state. For instance, the *Dieterici equation of state*:

$$p = \frac{RT \exp(-a/RTv)}{v-b}, \tag{6.61}$$

in terms of the reduced variables, adopts the following form:

$$\bar{p} = \frac{\mathrm{e}^2 \bar{T} \exp(-2/\bar{T}\bar{v})}{2\bar{v}-1}, \tag{6.62}$$

where $\mathrm{e} = \exp(1)$ and $v_{\mathrm{cr}} = 2b$, $p_{\mathrm{cr}} = a/(2\mathrm{e}b)^2$, $T_{\mathrm{cr}} = a/(4bR)$.

6.8 Exercises

6.1 Using the relations $\mu = (\partial F/\partial n)_{T,V}$ and $p = -(\partial F/\partial V)_{T,n}$, we can express the chemical potential and pressure as functions of temperature and molar volume. Show that if P phases coexist in a pure substance then the system is described by $f = 3 - P$ independent intensive parameters.

6.2 Derive an equation analogous to the Clapeyron equation for the slope of the two-phase coexistence line in the $T\mu$ plane, i.e., the derivative $\mathrm{d}\mu_{\mathrm{coex}}/\mathrm{d}T$, where $\mu_{\mathrm{coex}} = \mu_{\mathrm{coex}}(T)$ denotes the chemical potential on the coexistence line.

6.3 Calculate the chemical potential $\mu_{\mathrm{coex}}(T)$ for the liquid–vapour coexistence, assuming that the vapour can be approximated by the monatomic ideal gas and the enthalpy of transition, Δh, does not depend on temperature.

6.4 We have n mol of a pure substance which undergoes a phase transition from the phase α to the phase β at the pressure p and temperature T. To change α into β we start to supply heat to the system and then the system is thermally insulated. What is the proportion of the two phases if the molar enthalpy of the transition amounts to Δh, and the heat supplied to the system from the beginning of the transition to the moment of the system insulation amounts to Q?

6.5 In a thermally insulated system, there are n mol of a pure substance. Initially, the phase α, of the molar volume v^α, is in equilibrium with the phase β, of the molar volume v^β. Then the heat Q is supplied to the system in a reversible process at constant temperature and pressure, which causes partial transformation of the phase α into β. At the end of this process the amount of α in the system is equal to n_f^α. Determine the average molar volume of the system at the beginning and at the end of the process. The enthalpy of transition amounts to Δh.

6.6 For the transition from ice to liquid water at the temperature $T_0 = 273.15$ K and pressure of 1 bar, the change in the molar volume and enthalpy of transition amount to $\Delta v = -1.7\ \mathrm{cm^3\,mol^{-1}}$ and $\Delta h = 6.01\ \mathrm{kJ\,mol^{-1}}$, respectively. Calculate the pressure at which the transition occurs at -10 °C. Neglect the dependence of Δv and Δh on temperature.

6.7 A certain substance undergoes the solid–liquid transition at the pressure of 1 bar and temperature of 350 K. The molar volume of the solid and liquid phases amounts to $v^s = 160\ \mathrm{cm^3\,mol^{-1}}$ and $v^l = 163\ \mathrm{cm^3\,mol^{-1}}$, respectively. The melting point at a pressure of 100 bar amounts to 351 K. Determine the enthalpy of melting.

6.8 The enthalpy of evaporation of a certain liquid at 180 K and the pressure of 1 bar amounts to $14.4\ \mathrm{kJ\,mol^{-1}}$. Find the temperature at which the liquid and vapour coexist at a pressure of 2 bar. Neglect the dependence of the enthalpy of evaporation on temperature.

6.9 Close to the triple point of ammonia the vapour pressure above the liquid and solid changes with temperature as

$$\ln(p_{\mathrm{coex}}/1\ \mathrm{atm}) = 15.16 - 3063\ \mathrm{K}/T,$$

and

$$\ln(p_{\mathrm{coex}}/1\ \mathrm{atm}) = 18.70 - 3754\ \mathrm{K}/T,$$

respectively. Determine the temperature and pressure at the triple point and the enthalpy of evaporation, sublimation and melting.

6.10 The temperature of air on a dry winter morning amounts to -5 °C. The pressure of water vapour is equal to 2 torr. Is frost going to disappear from the car glass? The pressure at the triple point of water is equal to 0.006 bar, and the enthalpy of sublimation amounts to $51\ \mathrm{kJ\,mol^{-1}}$. At what pressure of water vapour can frost remain?

6.11 A certain liquid boils at the temperature $T_{\mathcal{H}} = 368$ K at the peak of a mountain whose elevation above sea-level amounts to $\mathcal{H}$, and at the foot, it boils at $T_0 = 378$ K. The enthalpy of evaporation amounts to $\Delta h = 45\ \mathrm{kJ\,mol^{-1}}$. Estimate the height of the mountain, assuming that the temperature of air is equal to 20 °C.

6.12 Derive a formula for the derivative of the molar entropy difference of two phases, $\Delta s = s^\beta - s^\gamma$, with respect to temperature, along the coexistence line.

6.13 Making use of the result of Exercise 6.12, determine the enthalpy of transition, Δh, as a function of temperature in the range $T_0 \le T \le T_1$, assuming that for both phases c_p is a linear function of temperature of the form: $c_p^i = a^i(T - T_0) + b^i$, $i = \beta, \gamma$, where a^i and b^i are some constants. Neglect the effect of thermal expansion.

6.14 A vessel of the volume V contains n mol of a pure substance. Initially the phases α and β coexist in the system at the temperature T_0. Then heat is supplied to the system in a reversible process at constant n and V. In the final state, at the temperature T_1, the phases α and β still coexist and only their proportion is different. Derive a relation between the heat absorbed by the system and the pressure $p_{coex}(T)$ and chemical potential $\mu_{coex}(T)$ of the coexisting phases. Then apply this relation to the liquid–vapour coexistence. Assume that vapour can be approximated by the monatomic ideal gas, and the enthalpy of transition does not depend on temperature (see Exercise 6.3).

6.15 Determine T_{cr}, p_{cr} and v_{cr} for a gas to which the van der Waals equation of state applies, assuming $a = 0.15\ \mathrm{J\,m^3\,mol^{-2}}$ and $b = 4 \times 10^{-5}\ \mathrm{m^3\,mol^{-1}}$.

6.16 The internal energy of a gas which satisfies the van der Waals equation of state

$$p = \frac{RT}{v - b} - \frac{a}{v^2} \tag{6.63}$$

is given by

$$U = \frac{f}{2} nRT - \frac{an^2}{V},$$

where f is the number of degrees of freedom per molecule, and a and b are constants characterizing the gas. Express the molar internal energy at the liquid–vapour critical point, u_{cr}, in terms of a and b, and then express $u = U/n$ in terms of the reduced variables: $\bar{u} = u/u_{cr}$, $\bar{T} = T/T_{cr}$ and $\bar{v} = v/v_{cr}$.

6.17 Express the chemical potential μ of a gas to which the van der Waals equation of state applies as a function of temperature and pressure. Then find the limit:

$$\mu_0(T) = \lim_{p \to 0} \left[\mu(T, p) - RT \ln \frac{p}{p_0} \right],$$

where p_0 denotes the pressure of a reference state.

Chapter 7
Mixtures

7.1 Basic Concepts and Relations

7.1.1 Definitions

So far we have been mainly occupied with pure substances. In this chapter, we extend the formalism of thermodynamics to *mixtures*, i.e., to systems that contain more than one component. By a *component* we mean a substance of a definite chemical composition, i.e., a chemical compound. Here we consider only such processes in which all components maintain their identity, which means that no chemical reactions occur in a system. Application of thermodynamics to chemical reactions is discussed in Part III of this book.

Imagine a vessel isolated from the surroundings and consisting of two parts separated by a diathermal wall. The parts are occupied by two different gases. For simplicity, we assume that both gases are very dilute so they can be considered ideal gases. Each gas is in thermodynamic equilibrium. We also assume that the temperature and pressure in both parts are the same. Then we remove the dividing wall. What will happen? Although the temperature and pressure will not change, according to the second law of thermodynamics, the system will reach a new equilibrium state of higher entropy. Why does the entropy of the system increase? We know from experience that the gases will not stay in their parts of the vessel but will mix. Due to the process of mixing each component fills the whole volume of the vessel. The process is irreversible, hence, the entropy of the system must increase. The new equilibrium state corresponds to a homogeneous gas in which both components are mixed up on a molecular scale.

And what will happen if we use liquids instead of gases? We know from experiment that some liquids mix in arbitrary proportion, as gases do, although the process of mixing is slower than in gases. Other liquids mix only in certain proportion. Besides, mixing of liquids involves some effects that are absent in gases. For instance, when two liquids mix at constant temperature and pressure their total volume before and after the mixing is different, in general. A good example is

R. Hołyst, A. Poniewierski, *Thermodynamics for Chemists, Physicists and Engineers*,

DOI 10.1007/978-94-007-2999-5_7,

the mixture of liquid water and ethanol. The molar volume of water amounts to 18 $cm^3\,mol^{-1}$. If we mix one mole of water with a much larger amount of ethanol, of the volume V, then the volume of the mixture amounts to about $V + 14\,cm^3$, instead of $V + 18\,cm^3$. This is because in strong dilution almost each water molecule is surrounded by ethanol molecules. Due to different structure of both molecules the effective volume occupied by one water molecule surrounded by ethanol is smaller than in pure water.

If components are mixed up on a molecular scale the mixture is a homogeneous system, which means that it forms a phase. Mixtures, as pure substances, can undergo phase transitions. The number of possible phases usually increases with the number of components. For instance, in a mixture of two components, two liquid phases of different composition may exist. It occurs when two liquids do not mix in all proportion. Then, in a certain range of composition, the mixture cannot exist as a homogeneous liquid but it separates into two liquid phases of different concentration of each component.

Many phenomena that occur in liquid or solid mixtures can be explained by means of a certain idealization of a real system, called the *ideal mixture* or *ideal solution*. In the ideal mixture, as in the mixture of ideal gases, the only effect of mixing is an increase in the entropy of the system caused by an increase in the volume available to each component. When can we apply the ideal mixture approximation to real mixtures? Usually we can do so if the molecules of different components are similar in respect of size and intermolecular interactions. By a similarity of interactions we understand that the interaction of two molecules does not depend much on whether they are molecules of the same component or different components. Then, each molecule is always in a similar environment irrespective of whether it is a pure substance or a mixture of similar components.

The composition of a mixture can be defined in different ways. In this book, we use mainly the *molar fraction*. In the case of solutions, the *molar concentration* (molarity) or *molality* are often used.

Definition 7.1 *Molar fraction* is the ratio of the mole number of a component to the total mole number of all components. The sum of all molar fractions is equal to 1.

Definition 7.2 *Solution* is a mixture in a liquid or solid phase, composed of two or more substances, one of which—called the *solvent*—is treated differently from the other components—called the *solutes*. If the sum of molar fractions of all solutes is much smaller than unity we talk about a *dilute solution*.

Definition 7.3 *Molar concentration* is the amount of a component measured in moles divided by the volume of the mixture.

A commonly used unit of the molar concentration is *mol per litre* ($mol\,L^{-1}$). A solution of the molar concentration of 1 $mol\,L^{-1}$ is called the *1-mol solution* and is denoted by 1 M.

Definition 7.4 *Molality* is the amount of a solute measured in moles divided by the mass of the solvent.

A commonly used unit of the molality is *mol per kilogram* ($\text{mol}\,\text{kg}^{-1}$).

7.1.2 Internal Energy

We recall that in one-component systems, the internal energy U is a function of three extensive parameters: the entropy S, volume V and mole number n. The relation between these extensive parameters: $U = U(S, V, n)$, is called the fundamental relation. It contains complete information about a given system in thermodynamic equilibrium (see Sect. 4.3.1).

Extension of the fundamental relation to mixtures is simple. Instead of one mole number n we have the mole numbers: $n_1, \dots, n_C$, of the C components, hence

$$U = U(S, V, n_1, \dots, n_C), \tag{7.1}$$

$$\mathrm{d}U = T\mathrm{d}S - p\mathrm{d}V + \sum_{i=1}^{C} \mu_i \mathrm{d}n_i, \tag{7.2}$$

$$U = TS - pV + \sum_{i=1}^{C} \mu_i n_i, \tag{7.3}$$

where μ_i denotes the chemical potential of the ith component. Equation (7.1) is the *fundamental relation* for a mixture. The physical meaning of all terms in (7.2) is the same as in the case of a pure substance. From the form of $\mathrm{d}U$, the following relations follow:

$$\left(\frac{\partial U}{\partial S}\right)_{V,n_i} = T, \qquad \left(\frac{\partial U}{\partial V}\right)_{S,n_i} = -p, \qquad \left(\frac{\partial U}{\partial n_i}\right)_{S,V,n_{j\neq i}} = \mu_i, \tag{7.4}$$

where the differentiation at constant n_i means that all mole numbers are fixed, and the differentiation at constant $n_{j\neq i}$ means that all n_j except $j = i$ are fixed. Identity (7.3) is the *Euler relation* (cf. (4.60)), which results from the extensiveness of all parameters in (7.1). Mathematically this fact is expressed by the identity:

$$U(mS, mV, mn_1, \dots, mn_C) = mU(S, V, n_1, \dots, n_C), \tag{7.5}$$

where $m > 0$. Differentiating both sides of (7.5) with respect to m and then putting $m = 1$, we obtain

$$U = \left[\frac{\mathrm{d}}{\mathrm{d}m} U(mS, mV, mn_1, \dots, mn_C)\right]_{m=1} = TS - pV + \sum_{i=1}^{C} \mu_i n_i, \tag{7.6}$$

where we have used relations (7.4). Thus, we have derived the Euler relation for mixtures.

7.1.3 Thermodynamic Potentials

The thermodynamic potentials for mixtures are defined in the same way as for pure substances. Here we restrict ourselves to the three most often used potentials, i.e., the Helmholtz free energy, enthalpy, and Gibbs free energy. All of them are presented below as functions of their natural variables.

Helmholtz free energy: $F = F(T, V, n_1, \ldots, n_C)$
The Legendre transform of $U(S, V, n_1, \ldots, n_C)$ with respect to S;

$$F = U - TS, \qquad \mathrm{d}F = -S\mathrm{d}T - p\mathrm{d}V + \sum_{i=1}^{C} \mu_i \mathrm{d}n_i, \tag{7.7}$$

$$\left(\frac{\partial F}{\partial T}\right)_{V,n_i} = -S, \qquad \left(\frac{\partial F}{\partial V}\right)_{T,n_i} = -p, \qquad \left(\frac{\partial F}{\partial n_i}\right)_{T,V,n_{j\neq i}} = \mu_i. \tag{7.8}$$

Enthalpy: $H = H(S, p, n_1, \ldots, n_C)$
The Legendre transform of $U(S, V, n_1, \ldots, n_C)$ with respect to V;

$$H = U + pV, \qquad \mathrm{d}H = T\mathrm{d}S + V\mathrm{d}p + \sum_{i=1}^{C} \mu_i \mathrm{d}n_i, \tag{7.9}$$

$$\left(\frac{\partial H}{\partial S}\right)_{p,n_i} = T, \qquad \left(\frac{\partial H}{\partial p}\right)_{S,n_i} = V, \qquad \left(\frac{\partial H}{\partial n_i}\right)_{S,p,n_{j\neq i}} = \mu_i. \tag{7.10}$$

Gibbs free energy: $G = G(T, p, n_1, \ldots, n_C)$
The Legendre transform of $U(S, V, n_1, \ldots, n_C)$ with respect to S and V;

$$G = U - TS + pV, \qquad \mathrm{d}G = -S\mathrm{d}T + V\mathrm{d}p + \sum_{i=1}^{C} \mu_i \mathrm{d}n_i, \tag{7.11}$$

$$\left(\frac{\partial G}{\partial T}\right)_{p,n_i} = -S, \qquad \left(\frac{\partial G}{\partial p}\right)_{T,n_i} = V, \qquad \left(\frac{\partial G}{\partial n_i}\right)_{T,p,n_{j\neq i}} = \mu_i. \tag{7.12}$$

The Gibbs free energy is of great importance to chemical thermodynamics because an equilibrium state at constant temperature and pressure corresponds to a minimum of G (see Corollary 5.3 and 5.4). Equations (7.11) are called the *fundamental equations of chemical thermodynamics*.

In the case of pure substances, the molar Gibbs free energy, $g = G/n$, is equal to the chemical potential. In the case of mixtures, $n = \sum_{i=1}^{C} n_i$ is the total number of moles of all components in a mixture, and $x_i = n_i/n$ is the *molar fraction* of the ith component. The molar fractions satisfies the identity

$$\sum_{i=1}^{C} x_i = 1. \tag{7.13}$$

From the extensiveness of G, it follows that

$$g = \frac{1}{n} G(T, p, n_1, \ldots, n_C) = G(T, p, x_1, \ldots, x_C), \tag{7.14}$$

thus, the molar Gibbs free energy is a function of T, p and $C - 1$ independent molar fractions, e.g., $g = g(T, p, x_1, \ldots, x_{C-1})$. Dividing $\mathrm{d}G$ in (7.11) by n and substituting $\mathrm{d}x_C = -\sum_{i=1}^{C-1} \mathrm{d}x_i$, we get

$$\mathrm{d}g = -s\mathrm{d}T + v\mathrm{d}p + \sum_{i=1}^{C-1} (\mu_i - \mu_C)\mathrm{d}x_i, \tag{7.15}$$

hence for $i < C$, we have

$$\left(\frac{\partial g}{\partial x_i}\right)_{T,p,x_{j\neq i}} = \mu_i - \mu_C. \tag{7.16}$$

It follows from the definition of G and from the Euler relation that

$$G = \sum_{i=1}^{C} n_i \mu_i, \tag{7.17}$$

hence

$$g = \sum_{i=1}^{C} x_i \mu_i. \tag{7.18}$$

Gibbs–Duhem Equation The differential of G obtained from relation (7.17), i.e.,

$$\mathrm{d}G = \sum_{i=1}^{C} (\mu_i \mathrm{d}n_i + n_i \mathrm{d}\mu_i), \tag{7.19}$$

is equal to $\mathrm{d}G$ in (7.11), provided that the *Gibbs–Duhem equation*:

$$S\mathrm{d}T - V\mathrm{d}p + \sum_{i=1}^{C} n_i \mathrm{d}\mu_i = 0, \tag{7.20}$$

holds. Dividing both sides by n, we get

$$s\mathrm{d}T - v\mathrm{d}p + \sum_{i=1}^{C} x_i \mathrm{d}\mu_i = 0, \tag{7.21}$$

where $s = S/n$ and $v = V/n$. In the case of one component, we recover relation (5.30). The Gibbs–Duhem equation is a relation between $C + 2$ intensive parameters, thus, $C + 1$ of them can be varied independently. For instance, the pressure can be treated as a function of temperature and chemical potentials of all components, i.e.,

$$\mathrm{d}p = -\frac{s}{v}\mathrm{d}T + \sum_{i=1}^{C} \varrho_i \mathrm{d}\mu_i, \tag{7.22}$$

where s/v is the entropy of the system per unit volume, and $\varrho_i = n_i/V$ denotes the molar concentration of the ith component. Another possible choice of independent variables is T, p and the chemical potentials of $C-1$ components. If we study processes at constant temperature and pressure then the Gibbs–Duhem equation reduces to a relation between chemical potentials:

$$\sum_{i=1}^{C} x_i \mathrm{d}\mu_i = 0, \quad \text{at } T = const \quad \text{and} \quad p = const. \tag{7.23}$$

7.2 Intrinsic Stability of a Mixture

In Sect. 5.5, we derived the conditions of intrinsic stability for one-component systems, i.e., $c_v > 0$ and $\kappa_T > 0$, using the entropy maximum principle. For mixtures, these two conditions have the same form because

$$c_v = T\left(\frac{\partial S}{\partial T}\right)_{V,n_i}, \qquad \kappa_T = -\frac{1}{V}\left(\frac{\partial V}{\partial p}\right)_{T,n_i}, \tag{7.24}$$

are calculated at constant mole numbers of all components. Moreover, we showed that the condition $\kappa_T > 0$ can also be derived from the minimum principle for the Helmholtz free energy, for systems at constant temperature.

In the case of mixtures, there are additional conditions of intrinsic stability related to the possibility of flow of different components between subsystems. To derive these new conditions, we apply the minimum principle for the Gibbs free energy (see Sect. 5.3.2) to a closed system at constant temperature and pressure. Using the same reasoning as in Sect. 5.5, we separate a small subsystem from a given system. Then the Gibbs free energy of the whole system adopts the following form:

$$G_{\mathrm{tot}} = ng + n'g' = ng(T, p, x_1, \ldots, x_{C-1}) + n'g\big(T, p, x_1', \ldots, x_{C-1}'\big), \tag{7.25}$$

where n and g correspond to the small subsystem, and n', g' correspond to the complementary subsystem. The molar Gibbs free energy of both subsystems has the same functional form because they are parts of the same system. We also assume that $n \ll n'$. The whole system is closed, hence

$$n\mathrm{d}x_i + n'\mathrm{d}x_i' = \mathrm{d}\big(n_i + n_i'\big) = 0, \tag{7.26}$$

for $i = 1, \ldots, C-1$. The equilibrium state of the system corresponds to the minimum of G_{tot}. From the necessary condition for a minimum of G_{tot} and conditions (7.26), we obtain

$$\mathrm{d}G_{\mathrm{tot}} = n\mathrm{d}g + n'\mathrm{d}g' = n\sum_{i=1}^{C-1}\left[\left(\frac{\partial g}{\partial x_i}\right)_{T,p,x_{j\neq i}} - \left(\frac{\partial g}{\partial x_i'}\right)_{T,p,x_{j\neq i}'}\right]\mathrm{d}x_i = 0, \tag{7.27}$$

hence

$$\mu_i - \mu_C = \mu_i' - \mu_C', \tag{7.28}$$

where we have used (7.16). Equation (7.28) resembles the condition of equilibrium with respect to matter flow but here we have difference between the chemical potentials: $\mu_i - \mu_C = \tilde{\mu}_i$ and $\mu_i' - \mu_C' = \tilde{\mu}_i'$. We note, however, that using (7.13), we can transform the Gibbs–Duhem equation (see (7.21)) as follows:

$$\mathrm{d}\mu_C = -s\mathrm{d}T + v\mathrm{d}p - \sum_{i=1}^{C-1} x_i \mathrm{d}\tilde{\mu}_i, \tag{7.29}$$

hence, we conclude that $\mu_C = \mu_C(T, p, \tilde{\mu}_1, \ldots, \tilde{\mu}_{C-1})$. Thus, from the equality $\tilde{\mu}_i = \tilde{\mu}_i'$, for $i = 1, \ldots, C-1$, it follows that $\mu_C = \mu_C'$, which means that $\mu_i = \mu_i'$, for $i = 1, \ldots, C$, in accord with the condition of equilibrium with respect to matter flow.

Then we use the sufficient condition for the minimum of G_{tot}:

$$\mathrm{d}^2 G_{\text{tot}} = n\mathrm{d}^2 g + n'\mathrm{d}^2 g' > 0, \tag{7.30}$$

where

$$\mathrm{d}^2 g = \frac{1}{2} \sum_{i=1}^{C-1} \sum_{j=1}^{C-1} g_{ij} \mathrm{d}x_i \mathrm{d}x_j, \tag{7.31}$$

$$g_{ij} = \left(\frac{\partial^2 g}{\partial x_i \partial x_j} \right)_{T,p} \tag{7.32}$$

and an analogous expression can be written for $\mathrm{d}^2 g'$. Repeating the reasoning presented in Sect. 5.5, one shows that the term $n'\mathrm{d}^2 g'$ in (7.30) can be neglected if $n \ll n'$, thus, it suffices to study the condition $\mathrm{d}^2 g > 0$. To proceed with it, it is convenient to present $\mathrm{d}^2 g$ as a sum of quadratic terms. We show below how to do it in a systematic way.

In the first step, we express $\mathrm{d}^2 g$ in the following form:

$$\mathrm{d}^2 g = \frac{1}{2} g_{11} (\mathrm{d}x_1)^2 + \sum_{j=2}^{C-1} g_{1j} \mathrm{d}x_1 \mathrm{d}x_j + \frac{1}{2} \sum_{i=2}^{C-1} \sum_{j=2}^{C-1} g_{ij} \mathrm{d}x_i \mathrm{d}x_j. \tag{7.33}$$

We recall (see (7.15)) that

$$\mathrm{d}g = \sum_{i=1}^{C-1} (\mu_i - \mu_C) \mathrm{d}x_i = \sum_{i=1}^{C-1} \tilde{\mu}_i \mathrm{d}x_i, \tag{7.34}$$

for constant T and p, hence $\tilde{\mu}_i$ is a function of $C-1$ molar fractions. Substituting $\mathrm{d}x_1$ obtained from the condition

$$\mathrm{d}\tilde{\mu}_1 = g_{11} \mathrm{d}x_1 + \sum_{i=2}^{C-1} g_{1i} \mathrm{d}x_i, \tag{7.35}$$

into (7.33), we get

$$\mathrm{d}^2 g = \frac{(\mathrm{d}\tilde{\mu}_1)^2}{2g_{11}} + \frac{1}{2} \sum_{i=2}^{C-1} \sum_{j=2}^{C-1} \left(g_{ij} - \frac{g_{1i} g_{1j}}{g_{11}} \right) \mathrm{d}x_i \mathrm{d}x_j. \tag{7.36}$$

From (7.35), it follows that

$$\left(\frac{\partial x_1}{\partial x_j}\right)_{T,p,\tilde{\mu}_1} = -\frac{g_{1j}}{g_{11}}, \tag{7.37}$$

hence, making use of the relation $\partial g/\partial x_i = \tilde{\mu}_i$, we obtain

$$g_{ij} - \frac{g_{1i}g_{1j}}{g_{11}} = g_{ij} + g_{1i}\left(\frac{\partial x_1}{\partial x_j}\right)_{T,p,\tilde{\mu}_1} = \left(\frac{\partial \tilde{\mu}_i}{\partial x_j}\right)_{T,p,\tilde{\mu}_1}. \tag{7.38}$$

To simplify notation, we have suppressed all the variables $x_{k\neq j}$. Now (7.36) can be presented as follows:

$$\mathrm{d}^2 g = \frac{(\mathrm{d}\tilde{\mu}_1)^2}{2g_{11}} + \frac{1}{2}\sum_{i=2}^{C-1}\sum_{j=2}^{C-1}\left(\frac{\partial \tilde{\mu}_i}{\partial x_j}\right)_{T,p,\tilde{\mu}_1} \mathrm{d}x_i \mathrm{d}x_j. \tag{7.39}$$

Then we introduce the partial Legendre transform of the function g with respect to x_1, i.e.

$$g[\tilde{\mu}_1] = g - \tilde{\mu}_1 x_1. \tag{7.40}$$

Thus, we get

$$\left(\frac{\partial \tilde{\mu}_i}{\partial x_j}\right)_{T,p,\tilde{\mu}_1} = \left(\frac{\partial^2 g[\tilde{\mu}_1]}{\partial x_i \partial x_j}\right)_{T,p,\tilde{\mu}_1}, \tag{7.41}$$

which follows from the observation that at constant T and p we have

$$\mathrm{d}g[\tilde{\mu}_1] = -x_1 \mathrm{d}\tilde{\mu}_1 + \sum_{i=2}^{C-1} \tilde{\mu}_i \mathrm{d}x_i, \tag{7.42}$$

where

$$\tilde{\mu}_i = \left(\frac{\partial g[\tilde{\mu}_1]}{\partial x_i}\right)_{T,p,\tilde{\mu}_1}. \tag{7.43}$$

If also $\tilde{\mu}_1$ is constant then

$$\mathrm{d}g[\tilde{\mu}_1] = \sum_{i=2}^{C-1} \tilde{\mu}_i \mathrm{d}x_i \tag{7.44}$$

has the same form as in (7.34) but the number of variables has been reduced by one. Here, however, $\tilde{\mu}_i = \tilde{\mu}_i(\tilde{\mu}_1, x_2, \ldots, x_{C-1})$ for $i = 2, \ldots, C-1$. Finally, $\mathrm{d}^2 g$ can be expressed in the following form:

$$\mathrm{d}^2 g = \frac{(\mathrm{d}\tilde{\mu}_1)^2}{2g_{11}} + \frac{1}{2}\sum_{i=2}^{C-1}\sum_{j=2}^{C-1} g[\tilde{\mu}_1]_{ij} \mathrm{d}x_i \mathrm{d}x_j, \tag{7.45}$$

where $g[\tilde{\mu}_1]_{ij} = (\partial^2 g[\tilde{\mu}_1]/\partial x_i \partial x_j)_{T,p,\tilde{\mu}_1}$.

We proceed in the same way as described above with the second term on the right-hand side of (7.45). Having performed $C-1$ steps, we obtain

$$\mathrm{d}^2 g = \frac{(\mathrm{d}\tilde{\mu}_1)^2}{2g_{11}} + \frac{1}{2}\sum_{k=2}^{C-1} \frac{(\mathrm{d}\tilde{\mu}_k)^2}{g[\tilde{\mu}_1, \ldots, \tilde{\mu}_{k-1}]_{kk}}, \tag{7.46}$$

where

$$g[\tilde{\mu}_1, \ldots, \tilde{\mu}_{k-1}] = g - \sum_{j=1}^{k-1} \tilde{\mu}_j x_j, \tag{7.47}$$

is the partial Legendre transform of g with respect to $x_1, \ldots, x_{k-1}$, and

$$g[\tilde{\mu}_1, \ldots, \tilde{\mu}_{k-1}]_{kk} = \left(\frac{\partial^2 g[\tilde{\mu}_1, \ldots, \tilde{\mu}_{k-1}]}{\partial x_k^2}\right)_{T,p,\tilde{\mu}_1,\ldots,\tilde{\mu}_{k-1}}. \tag{7.48}$$

The differentials $\mathrm{d}g$ and $\mathrm{d}\tilde{\mu}_k$ at constant T, p and $\tilde{\mu}_1, \ldots, \tilde{\mu}_{k-1}$ amount to

$$\mathrm{d}g[\tilde{\mu}_1, \ldots, \tilde{\mu}_{k-1}] = \sum_{j=k}^{C-1} \tilde{\mu}_j \mathrm{d}x_j \tag{7.49}$$

and

$$\mathrm{d}\tilde{\mu}_k = g[\tilde{\mu}_1, \ldots, \tilde{\mu}_{k-1}]_{kk}\mathrm{d}x_k + \sum_{j=k+1}^{C-1} g[\tilde{\mu}_1, \ldots, \tilde{\mu}_{k-1}]_{kj}\mathrm{d}x_j, \tag{7.50}$$

respectively, where

$$\tilde{\mu}_j = \left(\frac{\partial g[\tilde{\mu}_1, \ldots, \tilde{\mu}_{k-1}]}{\partial x_j}\right)_{T,p,\tilde{\mu}_1,\ldots,\tilde{\mu}_{k-1}}, \tag{7.51}$$

for $j = k, \ldots, C-1$. In particular, for $k = C-1$ we have

$$\mathrm{d}g[\tilde{\mu}_1, \ldots, \tilde{\mu}_{C-2}] = \tilde{\mu}_{C-1}\mathrm{d}x_{C-1}, \tag{7.52}$$

$$\mathrm{d}\tilde{\mu}_{C-1} = g[\tilde{\mu}_1, \ldots, \tilde{\mu}_{C-2}]_{C-1,C-1}\mathrm{d}x_{C-1}. \tag{7.53}$$

Thus, the condition $\mathrm{d}^2 g > 0$ is satisfied if all coefficients at $(\mathrm{d}\tilde{\mu}_k)^2$ are positive, i.e.,

$$\left(\frac{\partial^2 g[\tilde{\mu}_1, \ldots, \tilde{\mu}_{k-1}]}{\partial x_k^2}\right)_{T,p,\tilde{\mu}_1,\ldots,\tilde{\mu}_{k-1}} = \left(\frac{\partial \tilde{\mu}_k}{\partial x_k}\right)_{T,p,\tilde{\mu}_1,\ldots,\tilde{\mu}_{k-1}} > 0, \tag{7.54}$$

for $k = 1, \ldots, C-1$, and for $k = 1$, we have $g[\tilde{\mu}_1, \ldots, \tilde{\mu}_{k-1}] = g$.

Example 7.1 For $C = 2$, we have only one independent molar fraction, e.g., x_1.

Then

$$d^2g = \frac{1}{2}g_{11}(dx_1)^2 = \frac{(d\tilde{\mu}_1)^2}{2g_{11}}, \tag{7.55}$$

where we have used the relation: $d\tilde{\mu}_1 = g_{11}dx_1$. Thus, the condition of intrinsic stability has the following form:

$$\left(\frac{\partial^2 g}{\partial x_1^2}\right)_{T,p} = \left(\frac{\partial \tilde{\mu}_1}{\partial x_1}\right)_{T,p} > 0. \tag{7.56}$$

From the condition $x_2 = 1 - x_1$ and relations (7.29) and (7.56), we conclude that

$$\left(\frac{\partial \mu_2}{\partial x_2}\right)_{T,p} = -\left(\frac{\partial \mu_2}{\partial x_1}\right)_{T,p} = x_1\left(\frac{\partial \tilde{\mu}_1}{\partial x_1}\right)_{T,p} > 0. \tag{7.57}$$

The same inequality holds for the first component, since from Eq. (7.23), we get

$$x_1\left(\frac{\partial \mu_1}{\partial x_1}\right)_{T,p} = x_2\left(\frac{\partial \mu_2}{\partial x_2}\right)_{T,p} > 0. \tag{7.58}$$

This means that for intrinsically stable two-component systems, the chemical potential of a given component is an increasing function of its molar fraction at constant T and p.

7.3 Partial Molar Quantities and Functions of Mixing

7.3.1 Partial Molar Quantities

First, we give a formal definition of a partial molar quantity and then present a few examples of such quantities.

We consider an extensive quantity Y which is a state function of the state parameters: T, p and $n_1, \ldots, n_C$, e.g., the entropy or Gibbs free energy.

Definition 7.5 *Partial molar quantity* is a change in a given extensive quantity due to addition of a small amount of one component at constant temperature, pressure and amounts of all other components, divided by the amount added.

Although it sounds complicated the formula is very simple. For the extensive function of state $Y(T, p, n_1, \ldots, n_C)$, the partial molar quantity related to the ith component is defined by the following formula:

$$y_i = \left(\frac{\partial Y}{\partial n_i}\right)_{T,p,n_{j\neq i}}. \tag{7.59}$$

By definition partial molar quantities are intensive parameters.

For the function Y, we can derive an equation analogous to the Euler relation, using the identity

$$Y(T, p, mn_1, \ldots, mn_C) = mY(T, p, n_1, \ldots, n_C), \tag{7.60}$$

where m is an arbitrary positive number. This identity expresses extensiveness of Y in the language of mathematics. Differentiating both sides of (7.60) with respect to m and putting $m = 1$ at the end, we arrive at the following equation:

$$\sum_{i=1}^{C} n_i \left(\frac{\partial Y}{\partial n_i}\right)_{T,p,n_{j\neq i}} = Y, \tag{7.61}$$

and because of (7.59) we have

$$Y = \sum_{i=1}^{C} n_i y_i. \tag{7.62}$$

Relation (7.62) simply means that y_i is a partial contribution to the quantity Y per one mole of the ith component, which justifies its name.

Example 7.2 First, we substitute the Gibbs free energy for Y. From (7.12), we conclude that the partial molar quantities associated with G are:

$$\left(\frac{\partial G}{\partial n_i}\right)_{T,p,n_{j\neq i}} = \mu_i, \tag{7.63}$$

i.e., the chemical potentials of individual components, and (7.62) takes on the form of relation (7.17):

$$G = \sum_{i=1}^{C} n_i \mu_i. \tag{7.64}$$

Example 7.3 We substitute for Y the total volume V occupied by the mixture. Since V is the derivative of G with respect to p (see (7.11)) we have $V = V(T, p, n_1, \ldots, n_C)$, and

$$v_i = \left(\frac{\partial V}{\partial n_i}\right)_{T,p,n_{j\neq i}} \tag{7.65}$$

is the partial molar volume of the ith component. It is understood as the effective volume occupied by molecules of the ith component in the mixture. It depends on the temperature and pressure and also on the mixture composition. Relation (7.62) takes on the following form:

$$V = \sum_{i=1}^{C} n_i v_i. \tag{7.66}$$

7.3.2 Relations Between Partial Molar Quantities

There exist useful relations between some partial molar quantities. For instance, from (7.11) and (7.65), we obtain a relation between v_i and μ_i:

$$v_i = \left(\frac{\partial \mu_i}{\partial p}\right)_{T,n_j}, \tag{7.67}$$

were we have used the equality of the partial second derivatives of G with respect to p and n_i. In an analogous way, a relation between μ_i and the partial molar entropy s_i is derived, where

$$s_i = \left(\frac{\partial S}{\partial n_i}\right)_{T,p,n_{j\neq i}}. \tag{7.68}$$

From the equality of the second derivatives of G with respect to T and n_i, we get

$$s_i = -\left(\frac{\partial \mu_i}{\partial T}\right)_{p,n_j}. \tag{7.69}$$

Example 7.4 Since the chemical potential is an intensive parameter we can divide $n_1, \ldots, n_C$ by the total mole number n and treat μ_i as a function of T, p and $C-1$ independent molar fractions, e.g., $x_1, \ldots, x_{C-1}$. Then, making use of relations (7.67) and (7.69), we obtain

$$\mathrm{d}\mu_i = -s_i \mathrm{d}T + v_i \mathrm{d}p + \sum_{j=1}^{C-1} \left(\frac{\partial \mu_i}{\partial x_j}\right)_{T,p,x_{k\neq j}} \mathrm{d}x_j. \tag{7.70}$$

For $C = 1$, relation (7.70) reduces to the Gibbs–Duhem equation for a pure substance.

A relation between partial molar quantities, analogous to the Gibbs–Duhem equation at constant temperature and pressure, cane be derived. It follows from definition (7.59) that the differential of Y at constant T and p is given by the formula

$$\mathrm{d}Y = \sum_{i=1}^{C} y_i \mathrm{d}n_i. \tag{7.71}$$

The same differential can also be calculated from relation (7.62), hence

$$\mathrm{d}Y = \sum_{i=1}^{C} (y_i \mathrm{d}n_i + n_i \mathrm{d}y_i). \tag{7.72}$$

From the comparison of (7.71) and (7.72), we obtain the relation:

$$\sum_{i=1}^{C} n_i \mathrm{d}y_i = 0, \tag{7.73}$$

which means that amongst C intensive quantities y_i there are $C-1$ independent ones.

Example 7.5 Substituting for y_i the chemical potentials in relation (7.73), we get

$$\sum_{i=1}^{C} n_i \mathrm{d}\mu_i = 0, \tag{7.74}$$

which is the Gibbs–Duhem equation at constant T and p (cf. Eq. (7.20)).

Example 7.6 The substitution of v_i for y_i in (7.73) leads to the relation

$$\sum_{i=1}^{C} n_i \mathrm{d}v_i = 0, \tag{7.75}$$

which shows that the partial molar volumes are not independent of one another.

7.3.3 Functions of Mixing

Definition 7.6 *Function of mixing* is a change in a given extensive state function of a mixture due to the transition of the components from the pure state to the mixture, at constant temperature and pressure.

It is to be understood as follows. In the initial state, the system consists of C subsystems, where C denotes the number of components, and each subsystem contains a different component. Between the subsystems, there are walls that prevent the components from mixing but allow to reach the thermal and mechanical equilibrium. Thus, the temperature and pressure have the same values in all subsystems equal to the temperature and pressure of the surroundings. We denote the value of the function Y in the initial state by Y^*. Since Y is an extensive quantity we have

$$Y^*(T, p, n_1, \ldots, n_C) = \sum_{i=1}^{C} n_i y_i^*(T, p), \tag{7.76}$$

where y_i^* is the molar quantity Y for the pure substance i. For instance, if $Y = V$ then v_i^* is the molar volume of the pure substance i. When all internal constraints are removed the components start to mix and the system reaches a new equilibrium state for which $Y = Y(T, p, n_1, \ldots, n_C)$. The function of mixing for the quantity Y is denoted by $\Delta_M Y$ and according to Definition 7.6 it is equal to $\Delta_M Y = Y - Y^*$. Making use of relations (7.62) and (7.76), we can express $\Delta_M Y$ in the following form:

$$\Delta_M Y = \sum_{i=1}^{C} n_i \left(y_i - y_i^*\right), \tag{7.77}$$

and since $y_i \neq y_i^*$, in general, also $\Delta_M Y \neq 0$.

The *Gibbs free energy of mixing* is of fundamental importance. It is defined by

$$\Delta_M G = \sum_{i=1}^{C} n_i \left(\mu_i - \mu_i^*\right), \tag{7.78}$$

where μ_i^* is the chemical potential of the i-component in the pure state, i.e., for the pure substance. Differentiating $\Delta_M G$ with respect to p at constant temperature and composition, we arrive at the *volume of mixing*, $\Delta_M V$:

$$\left(\frac{\partial \Delta_M G}{\partial p}\right)_{T,n_j} = \sum_{i=1}^{C} n_i \left(v_i - v_i^*\right) = \Delta_M V, \tag{7.79}$$

where we have used relation (7.67), which applies both to mixtures and to pure substances. In a similar way, using (7.69), we derive the following relation:

$$\left(\frac{\partial \Delta_M G}{\partial T}\right)_{p,n_j} = -\sum_{i=1}^{C} n_i \left(s_i - s_i^*\right) = -\Delta_M S, \tag{7.80}$$

where $\Delta_M S$ denotes the *entropy of mixing*. Then we use the relation $H = G + TS$, to define the *enthalpy of mixing*, $\Delta_M H$:

$$\Delta_M H = \Delta_M G + T \Delta_M S. \tag{7.81}$$

Finally, the *internal energy of mixing*, $\Delta_M U$, is determined from the relation $U = H - pV$:

$$\Delta_M U = \Delta_M H - p \Delta_M V. \tag{7.82}$$

We can see that the knowledge of $\Delta_M G$ is a crucial question, as we can derive from $\Delta_M G$ other functions of mixing. We will show that it has a particularly simple form for a mixture of ideal gases.

7.4 Mixture of Ideal Gases

With the concept of the ideal gas, we associate the lack of interactions between molecules, except collisions, which are necessary to reach thermodynamic equilibrium. However, we neglect the size of molecules, treating them as point objects, and also the attractive interactions. Therefore, the internal energy of the ideal gas is the sum of energy of individual molecules.

7.4.1 Dalton's Law

From the assumption that molecules in the ideal gas do not interact, we conclude that a mixture of ideal gases must also be an ideal gas, and each gaseous component can be treated as if it filled separately the whole vessel. Thus, the total force exerted by all molecules on the walls of the vessel must be equal to the sum of the forces exerted by molecules of individual components. This means that the individual components and the mixture satisfy the equation of state of the ideal gas, i.e.,

$$p_i V = n_i RT, \tag{7.83}$$

$$pV = nRT, \tag{7.84}$$

where $i = 1, \ldots, C$, $n = \sum_{i=1}^{C} n_i$, and p_i and p denote the pressure of the ith component and the total pressure of the mixture, respectively. From Eqs. (7.83) and (7.84), it follows that

$$\sum_{i=1}^{C} p_i = p, \tag{7.85}$$

where $p_i = x_i p$. Relation (7.85) is called *Dalton's law.*

Corollary 7.1 *The total pressure exerted by a mixture of gases on the vessel walls is the sum of the pressures exerted separately by each gas.*

The pressure $p_i = x_i p$ is called the *partial pressure* of the ith component. It has sense for any mixture because the total pressure p and the molar fractions are well defined quantities. The sum of all partial pressures is equal to the total pressure since

$$\sum_{i=1}^{C} p_i = p \sum_{i=1}^{C} x_i = p. \tag{7.86}$$

However, if gases are not ideal the partial pressure p_i differs, in general, from the pressure of the ith gas filling separately the whole vessel, thus, Dalton's law is not satisfied in that case.

7.4.2 Chemical Potential of a Component

In Sect. 5.2.2, we derived the expression for the chemical potential of the ideal gas (see (5.52)). Here we are mainly interested in the dependence of μ on pressure, therefore, it is convenient to express it in the following form:

$$\mu(T, p) = \mu^0(T) + RT \ln \frac{p}{p^0}, \tag{7.87}$$

where p^0 denotes the pressure of a reference state, and $\mu^0(T)$ is the chemical potential for $p = p^0$. In principle, the reference state can be arbitrarily chosen, however, the following convention is used in practice.

Definition 7.7 *Standard pressure* denoted by p^0 is the pressure of 10^5 Pa = 1 bar.[1]

[1]Before 1982 the value 101 325 Pa (= 1 atm) was used.

Definition 7.8 *Standard state* is the state of a system chosen as standard for reference by convention.

Definition 7.9 *Standard state of a gas* is the hypothetical state of the pure substance in the gaseous phase at the standard pressure p^0, assuming the ideal gas behaviour.

Thus, $\mu^0(T)$ in (7.87) is the chemical potential of the ideal gas in the standard state. We note, however, that Definition 7.9 of the standard state applies not only to ideal gases, and the meaning of *hypothetical state* is explained in the section devoted to real gases.

Now, we can apply expression (7.87) to the ith component at the partial pressure p_i, which gives

$$\mu_i = \mu_i^0(T) + RT \ln \frac{p_i}{p^0}. \tag{7.88}$$

Substituting $p_i = x_i p$, we arrive at the expression for the chemical potential of the ith component in a mixture of ideal gases at the temperature T and total pressure p:

$$\mu_i = \mu_i^*(T, p) + RT \ln x_i, \tag{7.89}$$

where

$$\mu_i^*(T, p) = \mu_i^0(T) + RT \ln \frac{p}{p^0} \tag{7.90}$$

is the chemical potential of the ith gas in the pure state, at the temperature T and pressure p.

7.4.3 Functions of Mixing for Ideal Gases

Substituting (7.89) into expression (7.78), we get

$$\Delta_M G = RT \sum_{i=1}^{C} n_i \ln x_i = nRT \sum_{i=1}^{C} x_i \ln x_i, \tag{7.91}$$

hence, the entropy of mixing amounts to (see (7.80))

$$\Delta_M S = -nR \sum_{i=1}^{C} x_i \ln x_i. \tag{7.92}$$

Since the molar fractions satisfy the inequality $x_i < 1$, we have $\Delta_M G < 0$ and $\Delta_M S > 0$. Therefore, the process of mixing is irreversible because the Gibbs free energy decreases at constant temperature and pressure. Then we find that the volume of mixing

$$\Delta_M V = 0, \tag{7.93}$$

since $\Delta_M G$ does not depend on pressure. From relations (7.81), (7.91) and (7.92), we obtain the enthalpy of mixing

$$\Delta_M H = 0 \tag{7.94}$$

and the internal energy of mixing

$$\Delta_M U = \Delta_M H - p\Delta_M V = 0. \tag{7.95}$$

We can summarize these results as follows.

Corollary 7.2 *Mixing of ideal gases does not cause any thermal effect or change in the volume. The only effect of mixing is an increase in the entropy of the system.*

7.5 Ideal Mixture

If we mix different liquids whose molecules are so similar that the differences in the intermolecular interactions can be neglected then we do not observe, as in the case of ideal gases, any thermal effect in the system or change in its volume. The reason of such behaviour is that for similar molecules it does not really matter if a given molecule interacts with molecules of the same or a different component. Nevertheless, molecules of different components are distinguishable, and the entropy of the system increases in the process of mixing. As in the case of ideal gases, this is caused by an increase in the volume available to molecules of each component. Therefore, our consideration that was restricted originally to mixtures of ideal gases can now be extended to the case of interacting molecules, provided that these interactions do not differ much for different components.

Definition 7.10 *Ideal mixture* is a mixture of the components $A, B, \ldots$ such that the chemical potential of each component i is given by the formula

$$\mu_i = \mu_i^*(T, p) + RT \ln x_i, \tag{7.96}$$

where μ_i^* is the chemical potential of the ith component in the pure state.

The ideal mixture in the liquid or solid phase is also called the *ideal solution*. Expression (7.96) has the same form as (7.89). It should be remembered, however, that the chemical potential μ_i^* for a liquid or solid is not given by expression (7.90), which applies to the ideal gas only. From relation (7.96), it follows that $\Delta_M V = 0$ and $\Delta_M H = 0$, and the entropy of mixing is given by formula (7.92), i.e.,

$$\Delta_M S = -nR \sum_{i=1}^{C} x_i \ln x_i.$$

Corollary 7.3 *In the ideal mixture, the properties of a component are not affected by the presence of other components, and the only effect of mixing is the dissolution of each component in the others.*

7.6 Real Mixtures

A non-ideal mixture is called a real mixture. The expression for the chemical potential of a component in a real mixture has to be modified. However, we want to do it in such a way that it should resemble expressions (7.96) or (7.88) as much as possible.

7.6.1 Fugacity

Fugacity of a Pure Gas First, we consider a pure substance in the gaseous phase. As we know, the chemical potential of the ideal gas, here denoted μ^{id}, is given by the formula

$$\mu^{\mathrm{id}}(T, p) = \mu^0(T) + RT \ln \frac{p}{p^0}. \tag{7.97}$$

To express the chemical potential of a real gas, we use an analogous formula, replacing only the pressure with a certain function of T and p.

Definition 7.11 *Fugacity* f is an intensive quantity of the dimension of pressure, defined by the relation

$$\mu(T, p) = \mu^0(T) + RT \ln \frac{f(T, p)}{p^0}, \tag{7.98}$$

where μ denotes the chemical potential of the gas. The fugacity of the ideal gas is equal to its pressure.

The advantage of expressing the chemical potential in terms of the fugacity may seem illusive, since we simply express one unknown quantity with another one. We know, however, that in the case of dilute gases $f \approx p$, as then the ideal gas is a good approximation, therefore

$$\lim_{p \to 0} \frac{f(T, p)}{p} = 1. \tag{7.99}$$

Thus, $f = f^{\mathrm{id}} = p$ for the ideal gas, and in the case of real gases, the fugacity can be interpreted as a *corrected* pressure. Besides, various relations derived for mixtures of ideal gases remain valid also for real gases if the partial pressure of a component is replaced by its fugacity.

The fugacity can be related to the difference between the molar volume of a real gas and the ideal gas, at the same temperature and pressure. From (7.98), it follows that for $T = const$,

$$RT \mathrm{d} \ln f = \left(\frac{\partial \mu}{\partial p}\right)_T \mathrm{d}p = v \mathrm{d}p, \tag{7.100}$$

and for the ideal gas,

$$RT\,\mathrm{d}\ln p = \frac{RT}{p}\mathrm{d}p = v^{\mathrm{id}}\mathrm{d}p. \tag{7.101}$$

Subtracting (7.101) from (7.100) and integrating over pressure, we get

$$\ln\frac{f(T,p)}{p} = \frac{1}{RT}\int_0^p \left[v(T,p') - v^{\mathrm{id}}(T,p')\right]\mathrm{d}p', \tag{7.102}$$

which satisfies condition (7.99).

We return once more to the definition of the standard state (see Definition 7.9). It follows from Eqs. (7.97) and (7.98) that the standard chemical potential $\mu^0(T)$ is equal to $\mu^{\mathrm{id}}(T,p^0)$, and not to $\mu(T,p^0)$, since $f(T,p^0) \neq p^0$ for real gases, in general. The chemical potential μ^{id} represents here the ideal gas to which a given real gas tends in the limit $p \to 0$. This means that the standard state is not a state of the real gas but a *hypothetical* state of the ideal gas at the pressure p^0. We shall see that an analogous procedure is used to define the standard state of a solute in a dilute solution.

Fugacity of a Gaseous Component in a Mixture The fugacity of a gaseous component in a mixture of real gases is defined in a similar way as for a pure substance. The starting point is expression (7.88) for the chemical potential of a component in the mixture of ideal gases, denoted here μ_i^{id}, i.e.,

$$\mu_i^{\mathrm{id}} = \mu_i^0(T) + RT\ln\frac{p_i}{p^0}, \tag{7.103}$$

where $p_i = x_i p$ denotes the partial pressure. In the case of real gases, the partial pressure of the ith component is replaced by its fugacity, f_i, i.e.,

$$\mu_i = \mu_i^0(T) + RT\ln\frac{f_i}{p^0}, \tag{7.104}$$

where f_i is a function of temperature, pressure and composition of the mixture. Taking the differential of both sides of (7.104) at constant temperature and at constant mole numbers $n_1,\ldots,n_C$, we get

$$RT\,\mathrm{d}\ln f_i = v_i\mathrm{d}p, \tag{7.105}$$

where v_i is the partial molar volume at the pressure p. The molar volume of the ideal gas at the pressure p is equal to $v^{\mathrm{id}} = RT/p$, hence

$$RT\,\mathrm{d}\ln\frac{f_i}{p} = \left(v_i - v^{\mathrm{id}}\right)\mathrm{d}p. \tag{7.106}$$

Integrating the last relation from zero to p, we obtain

$$\ln\frac{f_i}{p} = \frac{1}{RT}\int_0^p \left(v_i - v^{\mathrm{id}}\right)\mathrm{d}p' + c_i, \tag{7.107}$$

where c_i is an integration constant. It must be chosen in such a way that f_i tends to p_i in the limit $p \to 0$. This requirement is satisfied if $c_i = \ln x_i$, since then

$$\ln\frac{f_i}{x_i p} = \frac{1}{RT}\int_0^p \left(v_i - v^{\mathrm{id}}\right)\mathrm{d}p'. \tag{7.108}$$

Definition 7.12 *Fugacity coefficient* Φ_i is the ratio of the fugacity to the partial pressure of the gaseous component:

$$\Phi_i = \frac{f_i}{x_i p}. \tag{7.109}$$

For the ideal gas, $\Phi_i = 1$. Now we can express the difference $\mu_i - \mu_i^{\text{id}}$ in terms of the fugacity coefficient, i.e.,

$$\mu_i - \mu_i^{\text{id}} = RT \ln \Phi_i. \tag{7.110}$$

Thus, the fugacity coefficient is a measure of the deviation of a real gas from the ideal gas in a mixture at the same temperature, pressure and composition.

7.6.2 Activity

The activity is a quantity related to the fugacity. Usually it is used in the context of liquids or solids.

Definition 7.13 *Relative activity* is a dimensionless quantity given by the formula[2]

$$a = \exp\left(\frac{\mu - \mu^0}{RT}\right), \tag{7.111}$$

where μ^0 is the standard chemical potential whose exact definition depends on the choice of the standard state

Definition 7.14 *Standard state of a pure substance or solvent* in the liquid or solid phase is the state of the pure substance in the liquid or solid phase at the standard pressure p^0.

Here, the standard state is a real state of the pure substance at the pressure p^0, and $\mu^0 = \mu(T, p^0)$ depends only on temperature. The activity of the standard state is equal to unity by definition. Formally, the activity of a pure substance depends on both pressure and temperature. In practice, however, it is often assumed that

$$a \approx 1 \tag{7.112}$$

in the liquid or solid state if the pressure p does not differ too much from p^0.

[2]The concept of *absolute activity*, $\lambda = \exp(\mu/RT)$, is also used but in this book we always mean the relative activity a.

Explanation The liquid and solid phases are not very compressible,[3] which means that their molar volume weakly depends on pressure. Therefore, a change in the chemical potential caused by a change in pressure can be approximated by the first term of the Taylor expansion around $p = p^0$, i.e.,

$$\mu(T, p) - \mu^0(T) \approx v(T, p^0)(p - p^0), \tag{7.113}$$

where v is the molar volume of the liquid or solid phase. Using (7.113) and the ideal gas equation of state: $v^{\text{id}}(T, p^0) = RT/p^0$, we obtain

$$\frac{\mu(T, p) - \mu^0(T)}{RT} \approx \frac{v(T, p^0)}{v^{\text{id}}(T, p^0)}\left(\frac{p}{p^0} - 1\right). \tag{7.114}$$

The typical value of the ratio v/v^{id} is of the order 0.1 %, hence, the difference $(\mu - \mu^0)/RT$ is small and the activity $a \approx 1$, provided that p and p^0 are not very different.

Activity of a Substance in a Mixture The chemical potential of a component of the ideal mixture, denoted μ_i^{id}, is given by expression (7.96):

$$\mu_i^{\text{id}} = \mu_i^*(T, p) + RT \ln x_i. \tag{7.115}$$

We want to modify this expression that it could be applied to real mixtures as well. Using definition (7.13), we can express the chemical potential of the ith component in terms of its activity:

$$\mu_i = \mu_i^0(T) + RT \ln a_i. \tag{7.116}$$

We can do the same with the difference between the chemical potential of the pure substance i and its standard chemical potential, defined as $\mu_i^0(T) = \mu_i^*(T, p^0)$, i.e.,

$$\mu_i^*(T, p) - \mu_i^0(T) = RT \ln a_i^*, \tag{7.117}$$

where a_i^* denotes the activity of the pure substance. Eliminating μ_i^0 from (7.116) and (7.117), we get

$$\mu_i = \mu_i^*(T, p) + RT \ln \frac{a_i}{a_i^*}. \tag{7.118}$$

As we have already shown, the dependence of the chemical potential of a pure liquid or solid on pressure can often be neglected and the approximation $a_i^* = 1$ can be used. For formal reasons, however, we prefer to include a_i^* in thermodynamic relations. From the comparison of (7.115) with (7.118), it follows that the ratio a_i/a_i^* replaces the molar fraction x_i in a similar way as the fugacity replaces pressure in the case of real gases. It is convenient to express this ratio in the following form:

$$\frac{a_i}{a_i^*} = \gamma_i x_i, \tag{7.119}$$

where γ_i is called the activity coefficient.

[3]The compressibility of a liquid becomes large only near the critical point.

Definition 7.15 *Activity coefficient* γ_i is a dimensionless quantity defined by the formula

$$RT \ln(x_i \gamma_i) = \mu_i(T, p, x_1, x_2, \ldots) - \mu_i^*(T, p), \tag{7.120}$$

where μ_i is the chemical potential of the ith component in a mixture and μ_i^* is the chemical potential of that component in the pure state, at the temperature and pressure of the mixture.

It follows from the definition that the activity coefficient is a function of T, p and the composition, i.e.,

$$\gamma_i = \gamma_i(T, p, x_1, x_2, \ldots), \tag{7.121}$$

and that $\gamma_i = 1$ for the pure substance ($x_i = 1$). In the case of the ideal mixture, $\gamma_i = 1$ for all components. Similarly to the fugacity coefficient w have

$$\mu_i - \mu_i^{\text{id}} = RT \ln \gamma_i, \tag{7.122}$$

where $\mu_i^{\text{id}} = \mu_i^* + RT \ln x_i$ denotes the chemical potential of the ith component in a hypothetical ideal mixture, at the temperature, pressure and composition of the real mixture. Thus, all deviations from the ideal behaviour are contained in the activity coefficient γ_i.

7.6.3 Dilute Solutions

Chemical Potential of a Solute We consider now a dilute solution. The index $i = 1$ is assigned to the solvent and the indices $i \geq 2$ are assigned to the solutes, and we assume also that

$$\sum_{i=2}^{C} x_i = 1 - x_1 \ll 1. \tag{7.123}$$

When the sum of all molar fractions of the solutes is small, the solute molecules practically do not interact with one another but only with the solvent molecules. Therefore, we can assume that $\mu_i = \mu_i(T, p, x_i)$ for $i \geq 2$. This assumption means that the chemical potential of the ith component depends on the molar fraction of that component, as well as on the solvent, but it does not depend on the molar fractions of other components.

We express the chemical potential of the solute i in terms of its activity (see (7.116)):

$$\mu_i = \mu_i^0(T) + RT \ln a_i, \tag{7.124}$$

without specifying at the moment the standard chemical potential μ_i^0. We assume also that a_i can be expanded in the Taylor series around $x_i = 0$, which means that

a_i becomes proportional to x_i in the limit $x_i \to 0$. Since the logarithm of the proportionality coefficient, multiplied by RT, can be added to μ_i^0, μ_i can be expressed in the following form for small x_i:

$$\mu_i = \mu_i^\infty(T, p) + RT \ln x_i. \tag{7.125}$$

The quantity μ_i^∞, which denotes the chemical potential of a reference state, differs from the chemical potential of the pure substance, μ_i^*, in general. The infinity symbol refers to the limit of infinite dilution, $x_i \to 0$. Note that (7.125) has the same form as in the case of ideal solutions, but the chemical potential μ_i^∞ depends not only on temperature and pressure but also on the solvent, and expression (7.125) can be used only when x_i is small. However, it can be extended to less dilute solutions if we introduce the activity coefficient in a similar way as in Definition 7.15, i.e.,

$$RT \ln(x_i \gamma_i) = \mu_i(T, p, x_1, x_2, \ldots) - \mu_i^\infty(T, p). \tag{7.126}$$

In the limit of infinite dilution, $\gamma_i \to 1$. For the standard chemical potential of the solute in a dilute solution we assume

$$\mu_i^0(T) = \mu_i^\infty(T, p^0). \tag{7.127}$$

The difference $\mu_i^\infty - \mu_i^0$ can be expressed in terms of the activity a_i^∞ as follows:

$$\mu_i^\infty(T, p) - \mu_i^0(T) = RT \ln a_i^\infty, \tag{7.128}$$

hence, after the substitution into (7.126), we get

$$\mu_i = \mu_i^0(T) + RT \ln\left(x_i \gamma_i a_i^\infty\right). \tag{7.129}$$

From the comparison of the last relation with (7.124), we obtain an expression analogous to (7.119), i.e.,

$$\frac{a_i}{a_i^\infty} = \gamma_i x_i. \tag{7.130}$$

At first glance it may seem strange that we use different reference states for the solute and for the solvent. If the mixture was formed only by similar substances, then the state of the pure substance at the temperature and pressure of the mixture should be a good reference state for each component. However, the situation becomes qualitatively different, for instance, when a gas is dissolved in the liquid or solid phase, e.g., oxygen dissolved in liquid water. In the solution, it exists in the liquid phase, whereas as a pure substance at the temperature and pressure of the mixture it exists in the gaseous phase. In such a case, the state of the pure substance is not a good reference state, as it corresponds to a different phase. Then it is just the state of the solute in an infinitely dilute solution which is a good reference state of that solute in less dilute solutions.

Apart from the molar fraction, we can use the molality m or molar concentration c (see Definitions 7.3 and 7.4), to specify the amount of the solute in a solution. Relation (7.126) can be transformed to have a dimensionless quantity, either m/m^0 or c/c^0, as an argument of the logarithm, where m^0 and c^0 denote the standard molality and standard molar concentration, respectively. It is usually assumed that

$m^0 = 1 \text{ mol kg}^{-1}$ and $c^0 = 1 \text{ mol L}^{-1}$. For instance, if the molality is used then the activity coefficient of the solute is defined by the formula

$$RT \ln\left(\frac{m_i \gamma_i}{m^0}\right) = \mu_i - \mu_i^0, \tag{7.131}$$

where the standard chemical potential is defined as follows:

$$\mu_i^0(T, p) = \lim_{m_i \to 0}\left(\mu_i - RT \ln \frac{m_i}{m^0}\right). \tag{7.132}$$

Thus, in the limit of infinite dilution, γ_i tends to 1. In a similar way, we define the activity coefficient and the standard state for the molar concentration. From the comparison of (7.126) with (7.132), it follows that the value of the activity coefficient and the definition of the standard state depend on the way the composition of the solution is specified. Now we can formulate the following definition of the standard state of a solute.

Definition 7.16 *Standard state of a solute* in a solution is a hypothetical state of the solute, at the standard molality m^0 or standard pressure p^0 or standard molar concentration c^0, which behaves as in an infinitely dilute solution.

We simply want to assign a certain state of the solute to the chemical potential μ_i^0. It is not a state of the pure substance but a state of the solute in a solution of a specified composition. From Eqs. (7.126) and (7.131), it follows that we can do it by assuming $\gamma_i = 1$, which corresponds to an infinitely dilute solution. However, the values of thermodynamic parameters for which $\mu_i = \mu_i^0$, do not correspond to infinite dilution. For instance, $x_i = 1$ and $p = p^0$ are to be substituted into Eq. (7.126), to get μ^0, and in Eq. (7.131), we have to substitute $m_i = m^0$. For this reason, we talk about a *hypothetical* state, since it is not a real state of the solute. We note also that since the standard state corresponds to a specified composition, μ^0 depends only on temperature for $p = p^0$. However, if the standard state is defined by the condition $m_i = m^0$ or $c_i = c^0$ then μ^0 depends on both temperature and pressure.

Chemical Potential of the Solvent We recall the Gibbs–Duhem equation at constant T and p (see (7.23)):

$$\sum_{i=1}^{C} x_i \mathrm{d}\mu_i = 0. \tag{7.133}$$

Using this equation, we can easily determine the chemical potential of the solvent, μ_1, in the limit of infinite dilution. Substituting (7.125) into (7.133), we get

$$x_1 \mathrm{d}\mu_1 + \sum_{i=2}^{C} x_i \mathrm{d}\mu_i = x_1 \mathrm{d}\mu_1 + RT \sum_{i=2}^{C} \mathrm{d}x_i = 0, \tag{7.134}$$

hence, $x_1 \mathrm{d}\mu_1 = RT \mathrm{d}x_1$ because $\sum_{i=2}^{C} \mathrm{d}x_i = -\mathrm{d}x_1$. Therefore,

$$\mu_1 = \mu_1^*(T, p) + RT \ln x_1, \tag{7.135}$$

where we have taken the chemical potential of the pure solvent as the integration constant. Note that μ_1 has exactly the same form as in the case of ideal mixtures, but relation (7.135) holds only in the limit $x_1 \to 1$, in general.

7.6.4 Excess Functions

Real mixtures differ from the ideal mixture and this difference can be conveniently expressed in terms of excess functions.

Definition 7.17 *Excess function* expresses the deviation of a state function of a mixture from the value of that function in a hypothetical ideal mixture, at the temperature, pressure and composition of the given mixture.

We mark the excess functions with the index E. For instance, the excess chemical potential of the ith component, μ_i^E, is defined as

$$\mu_i^E = \mu_i(T, p, x_1, x_2, \ldots) - \mu_i^{\mathrm{id}}(T, p, x_i). \tag{7.136}$$

Comparing (7.136) with (7.122), we find that

$$\mu_i^E = RT \ln \gamma_i. \tag{7.137}$$

In the case of an extensive quantity $Y = Y(T, p, n_1, \ldots, n_C)$, we have

$$Y^E = Y - Y^{\mathrm{id}} = \left[Y - \sum_{i=1}^{C} n_i y_i^*(T, p) \right] - \left[Y^{\mathrm{id}} - \sum_{i=1}^{C} n_i y_i^*(T, p) \right], \tag{7.138}$$

where y_i^* denotes the molar counterpart of Y for the pure substance i. We know from Definition 7.6 that the expressions in brackets are the functions of mixing for Y and Y^{id}, respectively, hence, Y^E can also be expressed as

$$Y^E = \Delta_M Y - \Delta_M Y^{\mathrm{id}}. \tag{7.139}$$

If $\Delta_M Y^{\mathrm{id}} = 0$ then $Y^E = \Delta_M Y$. For instance, this is the case of the enthalpy of mixing, $\Delta_M H$, and the volume of mixing, $\Delta_M V$, which vanish in the ideal mixture. For the Gibbs free energy, we have

$$G^E = \sum_{i=1}^{C} n_i \mu_i^E = RT \sum_{i=1}^{C} n_i \ln \gamma_i. \tag{7.140}$$

If we know the dependence of the activity coefficients on temperature and pressure we can determine the excess entropy S^E and excess volume V^E from the following relations:

$$S^E = -\left(\frac{\partial G^E}{\partial T} \right)_{p,n_i}, \tag{7.141}$$

$$V^E = \left(\frac{\partial G^E}{\partial p} \right)_{T,n_i}. \tag{7.142}$$

7.7 Phase Rule

In Chap. 6, we derived the condition of phase coexistence in the case of a pure substance. To generalize this condition to mixtures, we use the necessary condition for a minimum of the Gibbs free energy: $\mathrm{d}G = 0$, at constant T and p (see (5.75)). Each phase forms a separate homogeneous subsystem. When temperature and pressure are the same in all subsystems the whole system is in thermodynamic equilibrium if also the condition of equilibrium with respect to matter flow is satisfied. This requires that the chemical potential of each component has the same value in all phases (subsystems).

Corollary 7.4 *The condition of coexistence of the phases $\alpha, \beta, \gamma, \ldots$ is the equality of the chemical potentials: $\mu_i^\alpha = \mu_i^\beta = \mu_i^\gamma = \cdots$, for each component i.*

We assume now that P phases coexist in a mixture of C components. The components are numbered with the index $i = 1, \ldots, C$ and the phases are numbered with the index $\phi = 1, \ldots, P$. The temperature and pressure have the same values in all phases. For each phase, the Gibbs–Duhem equation (see (7.21)) holds, i.e.,

$$s^\phi \mathrm{d}T - v^\phi \mathrm{d}p + \sum_{i=1}^{C} x_i^\phi \mathrm{d}\mu_i^\phi = 0, \tag{7.143}$$

where s^ϕ and v^ϕ denote the molar entropy and molar volume of the phase ϕ, and x_i^ϕ and μ_i^ϕ denote the molar fraction and chemical potential of the ith component in this phase, respectively. The Gibbs–Duhem equation is a relation between $C + 2$ intensive parameters: T, p, and $\mu_1^\phi, \ldots, \mu_C^\phi$, hence, $C + 1$ of them can be varied independently. Since T and p have the same values in all phases, there are $C - 1$ independent chemical potentials for each phase. Therefore, the number of independent intensive parameters in the system, including T and p, amounts to $P(C - 1) + 2$. Then we have to take into account the condition of phase coexistence. It has the form of $P - 1$ independent equations for each component i, for instance,

$$\mu_i^\phi = \mu_i^1, \tag{7.144}$$

for $\phi = 2, \ldots, P$, which gives $C(P - 1)$ independent equations altogether. Subtracting the number of equations from the number of independent intensive variables, we find the *number of degrees of freedom* f for the system:[4]

$$f = C - P + 2. \tag{7.145}$$

It specifies the number of independent intensive parameters in the system that can be changed without violation of the P-phase coexistence. Relation (7.145) is called the *phase rule* (or the Gibbs phase rule). From the inequality $f \geq 0$, it follows that

$$P \leq C + 2. \tag{7.146}$$

[4]It should not be confused with the number of degrees of freedom of a molecule and with the fugacity, for which we have used the same symbol.

The case $P = C + 2$ corresponds to the maximum number of phases that can coexist in a C-component system.

Example 7.7 For a pure substance, we have $C = 1$ and $f = 3 - P$, hence $P \leq 3$. For one phase ($P = 1$), there are two degrees of freedom ($f = 2$) since T and p can be varied independently. On a phase diagram, a single phase is represented by a two-dimensional region in the Tp plane. For the two-phase coexistence ($P = 2$), there is one degree of freedom ($f = 1$) since either T or p can be varied independently. On a phase diagram, the two-phase coexistence is represented by a line. In the case of three phases ($P = 3$), we have $f = 0$, which means that T and p have definite values. This corresponds to the triple point on a phase diagram.

Example 7.8 In the case of two components, $f = 4 - P$, hence $P \leq 4$. Thus, the coexistence of four phases is possible. For a single phase ($P = 1$) we have $f = 3$. As independent intensive parameters, we can choose, for instance, T, p and the chemical potential of the first component, μ_1. For the two-phase coexistence ($P = 2$), there are two degrees of freedom ($f = 2$) and usually T and p are used as independent parameters. On a three-dimensional phase diagram, the coexistence of two phases can be represented as the surface $\mu_1 = \mu_1(T, p)$. In practice, the molar fraction x_1 rather than μ_1 is used as the third parameter. Then the coexistence of the phases α and β is represented by two surfaces: $x_1 = x_1^\alpha(T, p)$ and $x_1 = x_1^\beta(T, p)$, since the molar fractions, as the molar entropy and volume, are discontinuous at a first-order phase transition. Usually one of the parameters, e.g., pressure, is fixed and the lines $x_1 = x_1^\alpha(T)$ and $x_1 = x_1^\beta(T)$ are drawn. Inverting the relation between T and x_1, we obtain the *composition lines* for the two phases, i.e., $T = T^\alpha(x_1)$ and $T = T^\beta(x_1)$. The compositions of the coexisting phases at the given temperature T_0 correspond to the intersections of the line $T = T_0$ with the composition lines (see Chaps. 8 and 9).

7.8 Exercises

7.1 In a solution, the partial molar volume of the component A, of the molar mass $M_A = 58\ \mathrm{g\,mol^{-1}}$, amounts to $v_A = 74\ \mathrm{cm^3\,mol^{-1}}$, and the partial molar volume of the component B, of the molar mass $M_B = 118\ \mathrm{g\,mol^{-1}}$, amounts to $v_B = 80\ \mathrm{cm^3\,mol^{-1}}$. The molar fraction of B is equal to $x_B = 0.45$, and the mass of the solution amounts to 0.85 kg. Determine the volume of the solution.

7.2 Show that the partial molar enthalpy of the ith component, h_i, satisfies the relation

$$\left(\frac{\partial h_i}{\partial p}\right)_{T,x} = -T^2\left(\frac{\partial v_i/T}{\partial T}\right)_{p,x},$$

where the index x means that we differentiate at constant composition.

7.3 Prove that if the chemical potential of each component in a mixture has the form: $\mu_i = \mu_i^* + RT \ln x_i$, where μ_i^* is a function of temperature and pressure, then the Gibbs–Duhem equation at constant T and p is always satisfied.

7.4 In a two-component mixture $A + B$, the chemical potentials are assumed to have the following form:

$$\mu_A = \mu_A^* + RT \ln x_A + W(x_B),$$
$$\mu_B = \mu_B^* + RT \ln x_B + W(x_A),$$

where $W(x)$ is a polynomial and $W(0) = 0$. The coefficients of the polynomial, as well as μ_A^* and μ_B^*, are some functions of temperature and pressure. What form should the polynomial $W(x)$ have, to satisfy the Gibbs–Duhem equation at constant T and p, i.e., $x_A \mathrm{d}\mu_A + x_B \mathrm{d}\mu_B = 0$?

7.5 Find the Gibbs free energy of mixing per one mole of the mixture $A+B$ defined in Exercise 7.4.

7.6 Assuming that the partial molar volumes of the components in the mixture $A+B$ are analytic functions of the molar fraction x_A, show that $(\partial v_B/\partial x_A)_{T,p} = 0$, for $x_A = 0$, and $(\partial v_A/\partial x_B)_{T,p} = 0$, for $x_B = 0$. Find the form of the Taylor expansion of the function $v_A(x_B)$ around $x_B = 0$ and $v_B(x_A)$ around $x_A = 0$.

7.7 The partial molar volumes of the components in the two-component mixture $A+B$ are given by the following expressions:

$$v_A = v_A^* + a x_B^2 - \frac{2}{3}(a-b)x_B^3,$$
$$v_B = v_B^* + b x_A^2 + \frac{2}{3}(a-b)x_A^3,$$

where v_A^*, v_B^*, a and b are functions of temperature and pressure. Show that v_A and v_B satisfy the equation $x_A \mathrm{d}v_A + x_B \mathrm{d}v_B = 0$ at constant T and p, and then find the volume of mixing per one mole of the mixture.

7.8 Assuming that air is an ideal mixture of the composition: $x_{N_2} = 0.781$, $x_{O_2} = 0.210$, $x_{Ar} = 0.009$, calculate its entropy of mixing per mole.

7.9 Calculate the partial pressures and total pressure for the mixture of gases that form air, at the temperature of 0 °C. The molar volume of the mixture amounts to 22.4 $\mathrm{L\,mol^{-1}}$. Assume that air is a mixture of ideal gases whose composition is specified in Exercise 7.8.

7.10 The entropy of mixing of a certain mixture is the same as in the case of ideal mixtures, and the enthalpy of mixing does not vanish, but it is independent of pressure. What can be said about the volume of mixing and internal energy of mixing?

7.11 Show that if the volume of mixing does not depend on temperature then the entropy of mixing does not depend on pressure.

7.12 Calculate the fugacity $f(T, p)$ of the gas characterized by the following equation of state:

$$\frac{pv}{RT} = 1 + B(T)p + C(T)p^2.$$

7.13 Derive expressions for the coefficients $B(T)$ and $C(T)$ in Exercise 7.12 for a gas which satisfies the van der Waals equation of state, and then compute them, assuming $a = 0.15\ \mathrm{J\,m^3\,mol^{-2}}$, $b = 4 \times 10^{-5}\ \mathrm{m^3\,mol^{-1}}$ and $T = 273.15$ K. Calculate also the fugacity coefficient $\Phi = f/p$ at the pressure $p = 5$ bar and temperature $T = 273.15$ K.

7.14 Using the solution of Exercise 6.17, express the fugacity of the van der Waals gas as a function of the temperature T and molar volume v. Then calculate the fugacity coefficient at $T = T_{cr}$ and $p = p_{cr}$.

7.15 What is the dimension of a region representing three-phase coexistence in a four-component system?

7.16 What is the minimal number of components in a system in which five phases can coexist?

Chapter 8
Phase Equilibrium in Ideal Mixtures

In this chapter, we use the concept of ideal mixture, to study phase transitions in multi-component systems, i.e., we assume that the chemical potential of a substance in mixture has the following simple form:

$$\mu_i(\alpha, T, p, x_i) = \mu_i^*(\alpha, T, p) + RT \ln x_i, \tag{8.1}$$

where α can refer to the gaseous (g), liquid (l) or solid (s) phase and μ_i^* denotes the chemical potential of pure substance ($x_i = 1$). To denote the dependence of the chemical potential on the phase, we will often use the notation $\mu_i(\alpha)$, which is common in physical chemistry. In the case of mixtures, it is more convenient than the notation μ_i^α, since the latter leads to rather cumbersome symbols like $\mu_i^{*\alpha}$. The same convention is also applied to other state functions.

Phase diagrams for systems composed of more than two components can be quite complex. Therefore, we restrict ourselves only to two-component mixtures, whose components are labeled with the capital letters A and B. We consider the liquid–gas and liquid–solid equilibrium, and also a sort of liquid–liquid coexistence known as *osmotic equilibrium*.

8.1 Liquid–Gas Equilibrium

8.1.1 Raoult's Law

Raoult's law applies to ideal solutions in the liquid phase, in equilibrium (coexistence) with the gaseous phase treated as a mixture of ideal gases. Raoult's law states that the partial pressure of a given component in the vapour above the solution is proportional to the molar fraction of that component in the solution, i.e.,

$$p_A = p_A^* x_A, \tag{8.2}$$

$$p_B = p_B^* x_B, \tag{8.3}$$

R. Hołyst, A. Poniewierski, *Thermodynamics for Chemists, Physicists and Engineers*,
DOI 10.1007/978-94-007-2999-5_8,

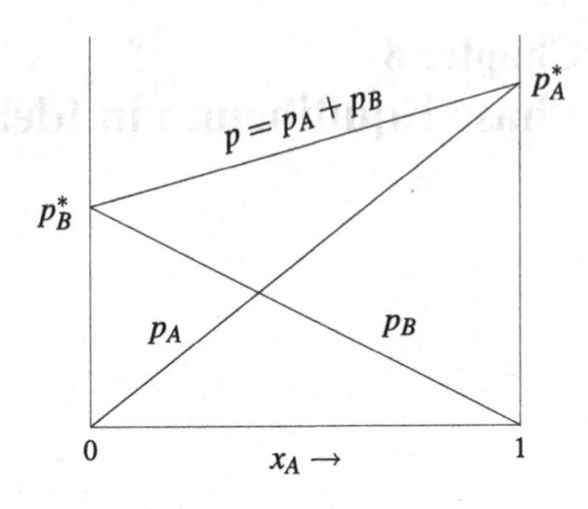

Fig. 8.1 Partial pressures p_A and p_B, and the total pressure of the vapour above the solution, p, against the molar fraction of the component A, for a solution to which Raoult's law applies. The vertical lines $x_A = 0$ and $x_A = 1$ correspond to pure components B and A, respectively

where $p_A^* = p_A^*(T)$ and $p_B^* = p_B^*(T)$ denote the vapour pressures at the liquid–vapour coexistence for the pure A and B, at the temperature of the solution. Historically, Raoult's law was formulated as a result of observation of similar liquids, whose mixing is well described by the ideal mixture approximation.

Because of the relation $x_A + x_B = 1$, the composition of the solution is defined by one molar fraction, which we usually choose to be x_A. Using Dalton's law (see (7.85)), we can determine the total vapour pressure above the solution:

$$p = p_A + p_B = p_B^* + (p_A^* - p_B^*)x_A. \tag{8.4}$$

The plot of the partial and total pressures against the solution composition is shown in Fig. 8.1. Note that the partial pressures can also be expressed as functions of the total pressure p. To show this, we first determine x_A from (8.4):

$$x_A = \frac{p - p_B^*}{p_A^* - p_B^*}, \tag{8.5}$$

and then substitute it into (8.2) and (8.3), hence

$$p_A = \frac{p_A^*(p - p_B^*)}{p_A^* - p_B^*}, \tag{8.6}$$

$$p_B = \frac{p_B^*(p_A^* - p)}{p_A^* - p_B^*}. \tag{8.7}$$

Derivation of Raoult's Law We focus on the component A since the same reasoning can be repeated for the second component. For the liquid phase (l) and gaseous phase (g), we have, respectively:

$$\mu_A(c, T, p, x_A) = \mu_A^*(c, T, p) + RT \ln x_A, \tag{8.8}$$

$$\mu_A(g, T, p_A) = \mu_A^0(T) + RT \ln \frac{p_A}{p^0}, \tag{8.9}$$

where we have used relation (7.88) for the ideal gas. At the liquid–vapour coexistence for the pure substance A, the following equality holds:

$$\mu_A^*(c, T, p_A^*) = \mu_A^*(g, T, p_A^*) = \mu_A^0(T) + RT \ln \frac{p_A^*}{p^0}. \tag{8.10}$$

Using Eqs. (8.8), (8.9) and (8.10), and the condition that μ_A must have the same value in the solution and in the vapour above the solution, we get

$$\mu_A^*(c, T, p) + RT \ln x_A - \mu_A^*(c, T, p_A^*) = RT \ln \frac{p_A}{p_A^*}, \tag{8.11}$$

hence

$$\ln \frac{p_A}{p_A^* x_A} = \frac{\mu_A^*(c, T, p) - \mu_A^*(c, T, p_A^*)}{RT}. \tag{8.12}$$

Raoult's law (8.2) is satisfied if the right-hand side of (8.12) equals zero. It is not true, in general, but we have already mentioned (see (7.114)) that a change in the chemical potential of the liquid or solid phase due to a small change in pressure is small compared to RT. Here the pressure of the liquid–vapour equilibrium has a value between p_A^* and p_B^*. If the components A and B do not differ much from each other then the pressure p_A^* is rather close to p_B^*. In what follows, we neglect a small difference between them and assume that relations (8.2) and (8.3) are satisfied for all compositions of the solution.

8.1.2 Liquid–Vapour Phase Diagram at Constant Temperature

It follows from the phase rule (7.145) that a two-component system in which two phases coexist has two degrees of freedom. This means that temperature and pressure can be changed independently from each other. Phase diagrams for a binary mixture are usually presented in the $x_A p$ plane, at constant temperature, or in the $x_A T$ plane, at constant pressure. Here we consider the case $T = const$.

The phase diagram of the ideal solution is shown in Fig. 8.2. The upper line, called the *liquid composition line*, defines the relation between the pressure and composition of the solution in equilibrium with the vapour. We know from Raoult's law that it is a straight line given by formula (8.4). To emphasize that the dependence of p on the molar fraction in the liquid phase is concerned, we write:

$$p(l, x_A) = p_B^* + (p_A^* - p_B^*) x_A. \tag{8.13}$$

The lower line, called the *vapour composition line*, defines the relation between the pressure and composition of the gaseous phase in equilibrium with the solution. To determine the vapour composition line, we use the relation between the partial pressure and the molar fraction of A in the gaseous phase, which is assumed to be a mixture of ideal gases, hence

$$x_A = \frac{p_A}{p}. \tag{8.14}$$

Substituting p_A given by formula (8.6) into (8.14) and solving the linear equation for p, we obtain the vapour composition line:

$$p(g, x_A) = \frac{p_A^* p_B^*}{p_A^* + (p_B^* - p_A^*) x_A}. \tag{8.15}$$

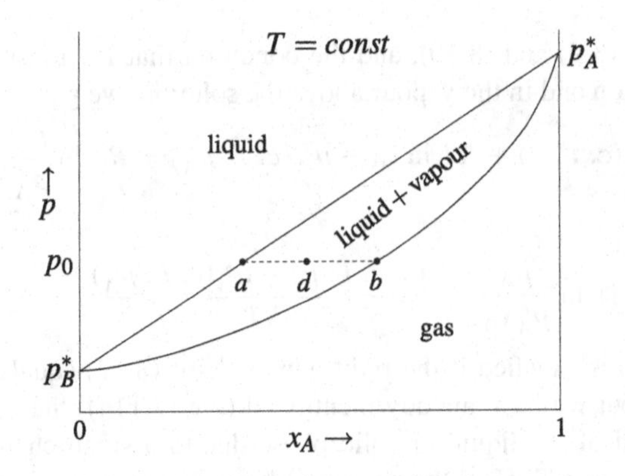

Fig. 8.2 Liquid–vapour equilibrium for the ideal solution at constant temperature. The *upper* and *lower* lines represent the liquid and vapour composition lines, respectively. The two-phase region in between corresponds to the liquid–vapour coexistence. The horizontal (*dashed*) line shows a certain value of pressure, $p = p_0$. The composition of the liquid and vapour which coexist at p_0 corresponds to the points a and b, respectively. A mixture whose composition represented by the point d is in the two-phase region, separates into the liquid and gaseous phases in the proportion given by the lever rule (see (8.23))

From expressions (8.13) and (8.15), the inequality

$$p(l, x_A) \geq p(g, x_A), \tag{8.16}$$

follows, where $p(l, 0) = p(g, 0) = p_B^*$ and $p(l, 1) = p(g, 1) = p_A^*$, which means that for $0 < x_A < 1$, the liquid composition line is above the vapour one as shown in Fig. 8.2.

To determine the composition of the liquid and gaseous phases at coexistence, we have to solve the equations:

$$p_0 = p(l, x_A), \tag{8.17}$$

$$p_0 = p(g, x_A), \tag{8.18}$$

for the given value of pressure, p_0, which is done graphically in Fig. 8.2. The intersection of the horizontal line $p = p_0$ with the liquid composition line (Eq. (8.17)) gives the liquid composition x_A. Similarly, the intersection of the horizontal line with the vapour composition line (Eq. (8.18)) gives the vapour composition, which is denoted y_A, to distinguish it from the liquid composition. The molar fractions x_A and y_A depend on pressure and temperature. The vapour is richer in the more volatile component, i.e., that one whose vapour pressure at the liquid–vapour coexistence of a pure substance is higher than for the other component at the same temperature. In the case considered, it is the component A ($p_A^* > p_B^*$), therefore, $y_A > x_A$.

8.1.3 Lever Rule

A mixture whose composition $\bar{x}_A$ is between x_A and y_A, at given values of temperature and pressure, does not exist as a single phase, but separates into the liquid and gaseous phases. To determine the amount of each phase, we use the two obvious relations:

$$n_A = n_A(l) + n_A(g), \tag{8.19}$$

$$n_B = n_B(l) + n_B(g), \tag{8.20}$$

where n_A, $n_A(l)$ and $n_A(g)$ denote the total mole number of the component A and the mole numbers of A in the liquid and gaseous phases, respectively, and analogously for the component B. Adding (8.19) to (8.20), we get

$$n = n_A + n_B = n(l) + n(g), \tag{8.21}$$

where $n(l) = n_A(l) + n_B(l)$ and $n(g) = n_A(g) + n_B(g)$. From the definition of the molar fraction, we derive the following relations: $n_A = \bar{x}_A n$, $n_A(l) = x_A n(l)$ and $n_A(g) = y_A n(g)$, hence

$$\bar{x}_A\big[n(l) + n(g)\big] = n_A = x_A n(l) + y_A n(g). \tag{8.22}$$

Finally, we write the last relation in the form called the *lever rule*:

$$n(l)(\bar{x}_A - x_A) = n(g)(y_A - \bar{x}_A), \tag{8.23}$$

which allows to determine the proportion of the two phases for the given composition $\bar{x}_A$. Note that (8.23) has a similar meaning as the lever rule for a pure substance (see (6.43)), but instead of the molar volume we have the molar fraction.

8.1.4 Liquid–Vapour Phase Diagram at Constant Pressure

The liquid and vapour composition lines defined by relations (8.13) and (8.15) depend also on temperature through $p_A^*(T)$ and $p_B^*(T)$ (see relation (6.42)). Therefore, for a given value of pressure, we obtain the liquid and vapour composition lines in the $x_A T$ plane, i.e., the functions $T(l, x_A)$ and $T(g, x_A)$, respectively, solving Eqs. (8.17) and (8.18) with respect to temperature. In this case, however, we usually cannot derive the explicit form of these functions. Both lines are shown schematically in Fig. 8.3. Now the liquid corresponds to the lower part of the diagram and the vapour corresponds to the upper part. We assume that atmospheric pressure is exerted on the solution, because this is the most common situation in practice. The boiling points of pure substances, T_A^* and T_B^*, are different, in general, and the more volatile component boils at a lower temperature. In the case considered, it is the component A ($T_A^* < T_B^*$). The boiling point of a solution is the point on the liquid composition line at a given composition of the solution. In the case of ideal solutions, the boiling point of a solution lies always between the boiling points of pure components.

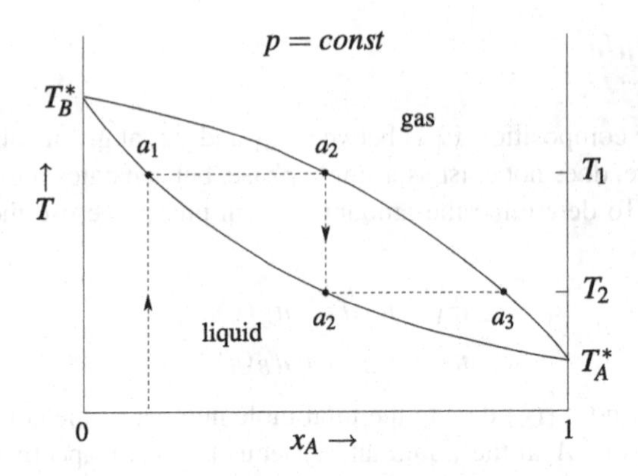

Fig. 8.3 Liquid–vapour equilibrium for the ideal solution at constant pressure. The liquid–vapour two-phase region is between the liquid composition line (*lower line*) and the vapour composition line (*upper line*). The horizontal segments (*dashed line*) show selected values of temperature, and their end-points correspond to the composition of coexisting phases. *Idea of distillation*. A solution of the composition a_1 is heated at constant pressure to the boiling point T_1. The vapour of the composition a_2 is removed from the system and then cooled down to the temperature T_2. Then the solution formed is again brought to boiling at $T_2 < T_1$. The vapour of the composition a_3 at $T = T_2$ is again removed from the system. The process repeats until practically pure components are obtained

The difference between the liquid and vapour composition at the liquid–vapour coexistence can be used to separate components of a mixture in the *distillation process*. The idea of distillation is presented in Fig. 8.3. Suppose that we start with a solution, of the composition a_1, rich in the less volatile component B. The liquid heated at constant pressure begins to boil at the temperature T_1. The vapour in equilibrium with the liquid at $T = T_1$ has the composition $a_2 > a_1$, i.e., it is richer in the more volatile component A. Then the vapour of the composition a_2 is taken away from the system and cooled down to the temperature T_2, at which it changes into the liquid phase. The solution formed in this way contains more component A than the original solution. In the next step, the liquid of the composition a_2 boils at the temperature T_2, giving a vapour of the composition $a_3 > a_2$, which is again removed from the system. Repeating this process, we obtain a solution of increasingly high content of the component A and lower and lower boiling point, whereas the liquid remaining in the system becomes increasingly rich in the component B. Since the boiling point of the ideal solution changes monotonically with the composition, the distillation process allows to separate the components practically with an arbitrary accuracy. We will see later that components of some real solutions can be separated by distillation only up to a definite composition.

8.1.5 Boiling Point of a Solution

We are going to investigate how the boiling point of the pure solvent A changes when a small amount of the substance B is dissolved in it at constant external pressure. We assume that the mixture forms an ideal solution. To maximally simplify the problem, we assume also that the substance B is non-volatile. This means that the amount of the component B in the vapour is negligible compared to the component A, hence we can put $p_B^* = 0$.

Due to the presence of the non-volatile component the boiling point of the solution must exceed the boiling point of the pure solvent, T_A^*. According to Raoult's law the vapour pressure above the solution satisfies the inequality

$$p_A = p_A^* x_A < p_A^*, \tag{8.24}$$

where x_A is the molar fraction of the solvent in the solution. Boiling occurs when the pressure of the vapour above the solution equals the external pressure p. For the pure solvent, we have

$$p_A^*(T_A^*) = p. \tag{8.25}$$

The pressure p_A^* increases with temperature, hence the equality

$$p_A(T) = p_A^*(T) x_A = p \tag{8.26}$$

is satisfied at a temperature T higher than T_A^*.

To determine quantitatively the elevation of the boiling point, we use the equality of the chemical potential of the solvent in the liquid and gaseous phases, taking into account that only the component A is present in the vapour, i.e.,

$$\mu_A(l) = \mu_A^*(l, T, p) + RT \ln x_A = \mu_A^*(g, T, p). \tag{8.27}$$

At a given pressure and composition, the temperature at which Eq. (8.27) is satisfied is the boiling point of the solution, T_b; for $x_A = 1$, $T_b = T_A^*$. Equation (8.27) is solved graphically in Fig. 8.4. Note that the lines of temperature dependence of $\mu_A^*(l)$ and $\mu_A^*(g)$ extend beyond their intersection corresponding to the liquid–vapour coexistence in the pure solvent. This is related to the existence of metastable states: a superheated liquid and supersaturated vapour. A metastable state corresponds to a local minimum of the Gibbs free energy, whereas a stable state corresponds to the absolute minimum of G, at the given temperature and pressure. The existence of metastable states is crucial for our consideration, because to solve Eq. (8.27), we have to assume that $\mu_A^*(l)$ is well defined also for temperatures slightly higher than T_A^*, for which the stable state of the system is the gaseous phase.

Denoting by $\Delta\mu_A^* = \mu_A^*(g) - \mu_A^*(l)$ the difference between the chemical potential of the pure solvent in a stable gaseous phase and in a metastable liquid phase, we can express (8.27) as follows:

$$\ln x_A = \frac{\Delta\mu_A^*}{RT}. \tag{8.28}$$

Then we use the *Gibbs–Helmholtz relation*:

$$\left(\frac{\partial G/T}{\partial T}\right)_{p,n} = -\frac{H}{T^2}, \tag{8.29}$$

the derivation of which is presented at the end of this section. As we know, the chemical potential of a pure substance is equal to its molar Gibbs free energy, hence,

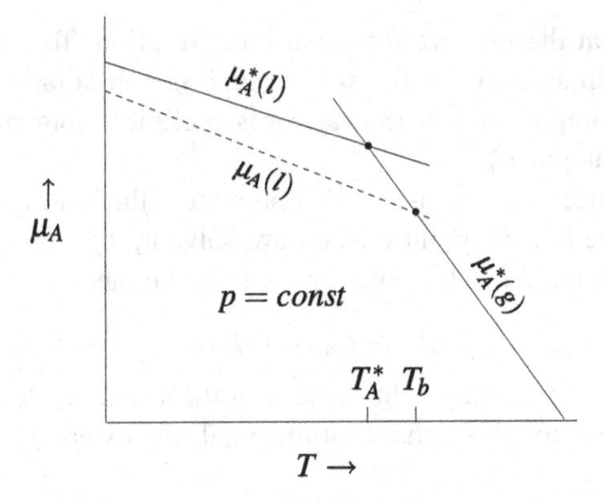

Fig. 8.4 Boiling point elevation in a solution of the substance A (solvent) with a non-volatile substance B (solute). The chemical potentials $\mu_A^*(l)$ and $\mu_A^*(g)$ refer to the liquid and gaseous phases of the pure solvent (*solid line*), respectively, and $\mu_A(l)$ denotes the chemical potential of the solvent in the solution (*dashed line*). The intersection of $\mu_A^*(l)$ with $\mu_A^*(g)$ corresponds to the boiling point of the pure solvent, T_A^*, and the intersection of $\mu_A(l)$ with $\mu_A^*(g)$ corresponds to the boiling point of the solution, $T_b > T_A^*$

differentiating both sides of (8.28) with respect to T at constant pressure and using (8.29), we get

$$\left(\frac{\partial \ln x_A}{\partial T}\right)_p = -\frac{\Delta h_A^*}{RT^2}, \tag{8.30}$$

where Δh_A^* denotes the enthalpy of evaporation of the pure solvent at the boiling point of the solution. The elevation of the boiling point is small, provided that the amount of the solute B is small, which we assumed at the very beginning. Therefore, we can approximate Δh_A^* by the enthalpy of evaporation of the pure solvent at its boiling point T_A^*. Then we integrate (8.30) over T from $T = T_A^*$ to $T = T_b$, hence

$$\ln x_A = \frac{\Delta h_A^*}{R}\left(\frac{1}{T_b} - \frac{1}{T_A^*}\right). \tag{8.31}$$

Expression (8.31) can be simplified, because for small x_B we have

$$\ln x_A = \ln(1 - x_B) \approx -x_B \approx -\frac{n_B}{n_A}, \tag{8.32}$$

$$\frac{1}{T_b} \approx \frac{1}{T_A^*} - \frac{\Delta T_b}{(T_A^*)^2}, \tag{8.33}$$

where $\Delta T_b = T_b - T_A^*$. Finally, we arrive at the following relation between the boiling point of the solution and the amount of the solute:

$$\Delta T_b = \frac{R(T_A^*)^2}{\Delta h_A^*}\frac{n_B}{n_A}. \tag{8.34}$$

In the case of dilute solutions, a convenient measure of the composition is the molality of the solute (see Definition 7.4), expressed usually in the units mole per kilogram. Denoting the molality of the solute by m_B, we can write (8.34) as follows:

$$\Delta T_b = K_b m_B, \tag{8.35}$$

where the proportionality coefficient K_b is called the *ebullioscopic constant*. Recall that we calculate m_B, dividing the mole number of the solute by the mass of the solvent:

$$m_B = \frac{n_B}{n_A M_A}, \tag{8.36}$$

where M_A is the molar mass of the solvent. From Eqs. (8.34), (8.35) and (8.36), we obtain the following formula for the ebullioscopic constant:

$$K_b = \frac{M_A R (T_A^*)^2}{\Delta h_A^*}, \tag{8.37}$$

whose unit is $\mathrm{K\,kg\,mol^{-1}}$. It can be inferred from (8.37) that K_b depends only on the solvent. Therefore, relation (8.35) can be used to determine the molar mass of an unknown non-volatile substance. If we know the mass of the solvent and the constant K_b then by measurement of the boiling point of the solution, we determine the molality m_B and hence also the mole number of the solute. Then, dividing the mass of the solute by the mole number, we obtain its molar mass.

Derivation of the Gibbs–Helmholtz Relation

$$\left(\frac{\partial G/T}{\partial T}\right)_{p,n} = \frac{1}{T}\left(\frac{\partial G}{\partial T}\right)_{p,n} - \frac{G}{T^2} = -\frac{TS+G}{T^2}, \tag{8.38}$$

where we have used the relation $(\partial G/\partial T)_{p,n} = -S$. From the definition of the Gibbs free energy and enthalpy, we have $G = U - TS + pV = H - TS$, hence $G + TS = H$, which ends the proof of relation (8.29).

8.1.6 Solubility of Gases in Liquids. Henry's Law

When we open a bottle of carbonated water the number of gas bubbles increases rapidly. Carbonated water is simply a solution of liquid water as a solvent and CO_2 as a solute. In a closed bottle, CO_2 in the gaseous state is in equilibrium with CO_2 in the solution at a slightly higher pressure than atmospheric pressure. When we open the bottle the gas pressure decreases rapidly, which causes the excess of CO_2 dissolved in water to escape from the solution. This means that the amount of gas which can be dissolved in a liquid at a given temperature depends on the gas pressure.

We assume that a small amount of the substance B, which in the pure state exists in the gaseous phase in the temperature range of interest, is dissolved in the liquid solvent A. The solution and the gas phase are in equilibrium with each other. When the molar fraction of the solute B, x_B, tends to zero, the partial pressure of the gas above the solution, p_B, also tends to zero. The limit of the ratio p_B/x_B, i.e.,

$$k_B = \lim_{x_B \to 0} \frac{p_B}{x_B}, \tag{8.39}$$

is called the *Henry constant*. Note that $x_B = 0$ corresponds to the liquid–vapour equilibrium for the pure solvent, which depends on temperature, therefore, k_B is also a function of temperature. The Henry constant depends also on both the solute and solvent. This is not surprising because behaviour of the solute molecules in a very dilute solution results mainly from their interaction with the solvent molecules. From (8.39), it follows that for small values of x_B *Henry's law* holds:

$$p_B = k_B x_B. \tag{8.40}$$

Definition 8.1 *Ideal dilute solution* is such a dilute solution that Henry's law applies to the solute.

Note that although relation (8.40) resembles Raoult's law (8.3), the proportionality coefficient between p_B and x_B is usually different. In fact, the quantity p_B^* may not even exist, as it refers to the liquid–vapour equilibrium for the pure solute, whereas the temperature of the solution is usually higher than the critical temperature of the gas dissolved. Nevertheless, Henry's law applies also to solutions of two liquids. We will return to this problem in the next chapter, where we discuss deviations from Raoult's law in real solutions.

8.1.7 Ostwald Absorption Coefficient

According to Henry's law, the molar fraction of a solute in a dilute solution is related to the partial pressure of that solute in the gaseous phase coexisting with the solution by formula (8.40). If the solvent A can be considered a non-volatile substance then the total pressure above the solution amounts to $p = p_B$. Thus, the solubility of a gas in a non-volatile liquid is proportional to the gas pressure above the solution, i.e.,

$$x_B = \frac{p}{k_B}. \tag{8.41}$$

In a dilute solution, $x_B \ll 1$, hence

$$x_B \approx \frac{n_B}{n_A} = \frac{V_B}{v_B^*} \frac{v_A^*}{V_A}, \tag{8.42}$$

where V_A and V_B denote the volume occupied by the solvent and gas, respectively, in their pure states, and v_A^* and v_B^* are their molar volumes. If the gas pressure is low we can use the ideal gas equation of state: $pv_B^* = RT$, which together with relations (8.41) and (8.42) gives

$$\frac{V_B}{V_A} = \frac{RT}{k_B v_A^*}. \tag{8.43}$$

The ratio of the volume of the gas dissolved to the volume of the liquid solvent, V_B/V_A, is called the *Ostwald absorption coefficient.* From relation (8.43), it follows that it is a function of temperature, however, due to small compressibility of liquids it is almost independent of pressure. In practice, Ostwald absorption coefficient is determined as the volume of the gas dissolved in the unit volume of the liquid, at a given temperature and the pressure of 1 bar.

We conclude from relation (8.41) that if the Henry constant increases with temperature then the solubility of the gas in the liquid decreases. For instance, this is the case of ammonia dissolved in water. It is not a general rule, however. For the solution of hydrogen in hexane, for instance, the Henry constant is a decreasing function of temperature. In general, the dependence of the Henry constant on temperature does not have to be monotonic. Such a non-monotonic behaviour of the Henry constant is observed, e.g., in the solutions of oxygen and nitrogen in water. For temperatures below about 360 K, the Henry constant increases with temperature, whereas at higher temperatures, it is a decreasing function of temperature. This means that at lower temperatures, the solubility of oxygen and nitrogen in water decreases with increasing temperature, whereas at higher temperatures it begins to increase.

If a liquid solution is in equilibrium with a mixture of gases, then Henry's law applies to each gas separately, provided that the solution is dilute. Because of different values of the Henry constant for different gases, their proportion in the solution is usually different than in the gaseous phase. For instance, at a temperature of 18 °C the proportion of oxygen in air dissolved in water amounts to 34.1 %, whereas the proportion of oxygen in the atmosphere amounts to 21.1 %.

8.2 Liquid–Solid Equilibrium

8.2.1 Freezing Point of a Solution

We are going to show that a solution freezes at a lower temperature than the pure solvent. This phenomenon is analogous to the boiling point elevation discussed in Sect. 8.1.5.

We consider a solution of a small amount of the substance B, e.g., common salt, dissolved in the liquid solvent A, e.g., water. We are interested in the effect of the solute on the freezing point of the solution. For simplicity, we assume that A does not form a solid solution with B, i.e., the solution crystallizes in the form of the pure substance A. It is a similar assumption to that the molecules of the solute are absent from the vapour above the solution, made in Sect. 8.1.5. The condition of liquid–solid equilibrium for the ideal solution has the following form:

$$\mu_A(l) = \mu_A^*(l, T, p) + RT \ln x_A = \mu_A^*(s, T, p). \tag{8.44}$$

We have simply replaced in formula (8.27) the chemical potential of the solvent in the gaseous phase, $\mu_A^*(g)$, with its chemical potential in the solid phase, $\mu_A^*(s)$.

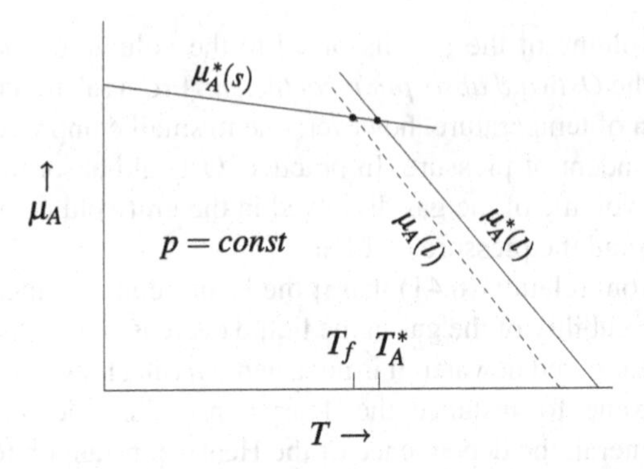

Fig. 8.5 Freezing point depression in a solution of the solvent A with the solute B; A and B do not form a solid solution. The chemical potentials $\mu_A^*(l)$ and $\mu_A^*(s)$ refer to the pure solvent in the liquid and solid phases (*solid line*), respectively, and $\mu_A(l)$ denotes the chemical potential of the solvent in the liquid solution (*dashed line*). The intersection of $\mu_A^*(l)$ with $\mu_A^*(s)$ corresponds to the freezing point of the pure solvent, T_A^*. The intersection of $\mu_A(l)$ with $\mu_A^*(s)$ corresponds to the freezing point of the solution, $T_f < T_A^*$

Condition (8.44) is satisfied only if $\mu_A^*(l) > \mu_A^*(s)$ at the freezing point of the solution, T_f, which is shown in Fig. 8.5. Thus, the temperature T_f must be slightly lower than the freezing point of the pure solvent, T_A^*. At the temperature T_f, the pure solvent has a lower chemical potential in the solid phase than in a metastable supercooled liquid phase.

The formula for T_f is derived in the same way as in Sect. 8.1.5. Therefore, we can use Eq. (8.31), replacing only T_b with T_f and the enthalpy of evaporation with the enthalpy of freezing with the minus sign, since freezing is the reverse of melting, hence

$$\ln x_A = \frac{\Delta h_A^*}{R}\left(\frac{1}{T_A^*} - \frac{1}{T_f}\right). \tag{8.45}$$

For a dilute solution, we can also repeat the reasoning presented in Sect. 8.1.5. Finally, we obtain the following expression for the freezing point depression:

$$\Delta T_f = -K_f m_B, \tag{8.46}$$

where $\Delta T_f = T_f - T_A^*$, m_B is the molality of the solute, and the proportionality coefficient

$$K_f = \frac{M_A R (T_A^*)^2}{\Delta h_A^*} \tag{8.47}$$

is called the *cryoscopic constant*. This phenomenon is commonly used in the winter when roads are sprinkled with salt, which causes that a liquid solution of water and salt, instead of ice, persists even at temperatures well below 0 °C. Note that the freezing point depression, as well as the boiling point elevation, depends only on the mole number of the solute and not on its nature.

8.2.2 Solubility of Solids in Liquids

What happens when we add the solid substance B to the liquid solvent A? A small amount of B added to the solvent dissolves in it, i.e., it undergoes a transition to the liquid phase. At a given temperature, a definite amount of the substance B can be dissolved in the unit mass of the solvent A. It means that there exists a maximum concentration of B in the solution, which is then called the *saturated solution*. Adding more B, we simply observe equilibrium between the solution in the solid and liquid phases. Moreover, we assume that the solid phase is the pure substance B. It is a realistic assumption because different molecules of the solvent and solute cannot, in general, form a common crystalline structure too easily. Note that we made the same assumption in Sect. 8.2.1 in relation to the solvent. Also the condition of phase equilibrium has a form analogous to condition (8.44), but in this case it concerns the solute instead of the solvent, i.e.,

$$\mu_B(l) = \mu_B^*(l, T, p) + RT \ln x_B = \mu_B^*(s, T, p). \tag{8.48}$$

The chemical potential $\mu_B^*(l)$ refers to a metastable liquid phase, i.e., the supercooled liquid of the substance B. We transform Eq. (8.48) into the following form:

$$\ln x_B = -\frac{\Delta\mu_B^*}{RT}, \tag{8.49}$$

where $\Delta\mu_B^* = \mu_B^*(l) - \mu_B^*(s)$, which expresses the composition of the saturated solution as a function of temperature and pressure. At the melting point of the pure B, T_B^*, we have $\Delta\mu_B^* = 0$ and $x_B = 1$.

Then we differentiate (8.49) with respect to T at constant pressure and use the Gibbs–Helmholtz relation (8.29), to get

$$\left(\frac{\partial \ln x_B}{\partial T}\right)_p = \frac{\Delta h_B^*}{RT^2}, \tag{8.50}$$

where Δh_B^* denotes the enthalpy of melting for the pure B. If we assume that Δh_B^* does not depend on temperature in the range of temperature of our interest then we can integrate (8.50) from T to T_B^*, which gives

$$\ln x_B = \frac{\Delta h_B^*}{R}\left(\frac{1}{T_B^*} - \frac{1}{T}\right). \tag{8.51}$$

Relation (8.51) defines the line of the solid–liquid equilibrium in the solution, which is called the *solubility line*. In the case of ideal solutions, we talk about *ideal solubility* of a solid in a liquid, which does not depend on the solvent nature. Note that because $x_B \leq 1$ and $\Delta h_B^* > 0$, expression (8.51) makes sense only when temperature is lower than the melting point T_B^*. Figure 8.6 shows the solubility line versus T^{-1}, which is a straight line in the case of ideal solubility. For instance, the solubility line for a solution of naphthalene in benzene is well described by relation (8.51). This relation also allows to draw the following general conclusions.

1. Solubility increases with increasing temperature.

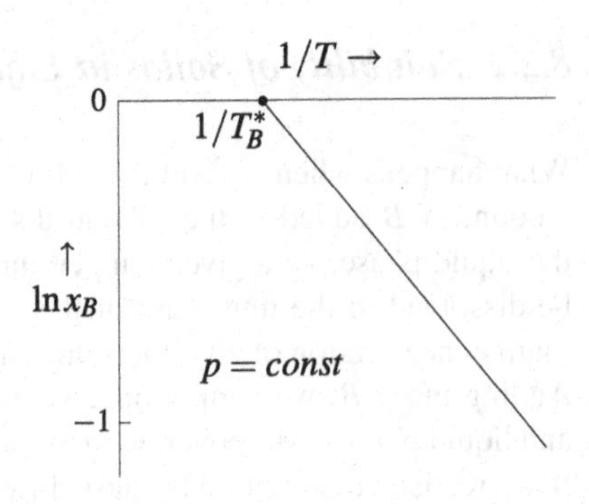

Fig. 8.6 Line of ideal solubility of the solid B in a liquid solvent. T_B^* denotes the melting point of the pure substance

2. If two solids have similar melting points then the more soluble one is the solid with a smaller enthalpy of melting.
3. If two solids have similar enthalpy of melting then the more soluble one is the solid with a lower melting point.

8.2.3 Simple Eutectic

A distinction between the line of freezing, given by (8.45), and the line of solubility, given by (8.51), is a matter of convention, since they are both expressed by the same formula. Usually we talk about freezing when the solid phase that forms is composed of the solvent molecules, and about solubility when it is composed of the solute molecules.

We consider now two substances: A and B, which are completely miscible in the liquid phase and completely non-miscible in the solid phase, which means that the solution crystallizes in the form of pure substances. Such a mixture is called a *simple eutectic*. As we have already seen, addition of a small amount of the substance B to the solvent A causes depression of the freezing point in relation to the freezing point of the pure solvent, T_A^*. Similarly, addition of the substance A to the pure substance B shifts the freezing point below T_B^*. Thus, depending on the composition, a liquid solution can be in equilibrium with the solid phase of either A or B.

Suppose that the liquid phase is an ideal solution. Then the composition of the liquid phase in equilibrium with the solid phase of the pure A is given by relation (8.45) and an analogous relation holds for the component B. Therefore, we can draw two lines of the liquid–solid equilibrium in the $x_A T$ plane (see Fig. 8.7):

$$\ln x_A = \frac{\Delta h_A^*}{R}\left(\frac{1}{T_A^*} - \frac{1}{T}\right), \tag{8.52}$$

$$\ln x_B = \frac{\Delta h_B^*}{R}\left(\frac{1}{T_B^*} - \frac{1}{T}\right), \tag{8.53}$$

where T_A^* and T_B^* denote the freezing points of the pure A and B, respectively, and $x_B = 1 - x_A$. The intersection of these lines is called the *eutectic point*. It is

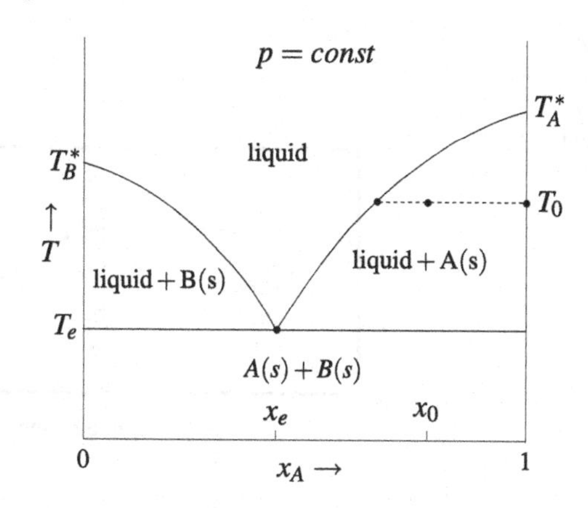

Fig. 8.7 Phase diagram of a simple eutectic. The freezing lines meet at the eutectic point, of the composition x_e and temperature T_e. Meaning of symbols: 'liquid + A(s)' and 'liquid + B(s)' refer to the liquid–solid two-phase regions, and '$A(s) + B(s)$' refers to the coexistence of pure solid phases A and B. The point on the *dashed line* represents a mixture composition in the 'liquid + A(s)' region, at the given temperature $T = T_0$. For instance, a mixture of the composition x_0 separates into the solid phase A and the liquid phase whose composition corresponds to the intersection of the *dashed line* with the freezing line

defined by the composition $x_A = x_e$ and temperature $T = T_e$. Note that only the parts of the lines above the eutectic point have physical meaning, because T_e is the lowest temperature at which the solution can exist in the liquid phase. Below the line $T = T_e$, the solid phase of pure A coexists with the solid phase of pure B, because their molecules do not form a common crystalline structure. At the temperature T_e, three phases are in equilibrium with one another, i.e., the liquid solution, of the composition x_e, and the solid phases A and B.

At the eutectic composition x_e, the liquid solution crystallizes like a pure substance, i.e., its temperature remains constant and equal to T_e until the whole liquid changes into an inhomogeneous mixture of small crystals of pure substances A and B, whose average composition amounts to x_e. If the initial composition of the solution differs from x_e then, depending on the composition, either the solid phase A or B starts to separate from the solution at a certain temperature. Let us consider for instance, the composition $x_0 > x_e$, i.e., a solution richer in the component A, which is shown in Fig. 8.7. At a certain temperature below T_A^*, the solid A begins to separate from the solution. Further cooling of the solution causes more solid A to separate, while the liquid phase becomes poorer in the component A, i.e., its composition approaches the eutectic point from the right-hand side. At the given temperature T_0, the composition of the solution in equilibrium with the solid A is determined by the intersection of the line $T = T_0$ with the freezing line of the solution. When the composition reaches the eutectic point, the solution crystallizes like a pure substance at the temperature T_e. If the initial composition of the solution is smaller than x_e we observe an analogous behaviour, but then the solid B begins to separate from the solution at a certain temperature below T_B^*.

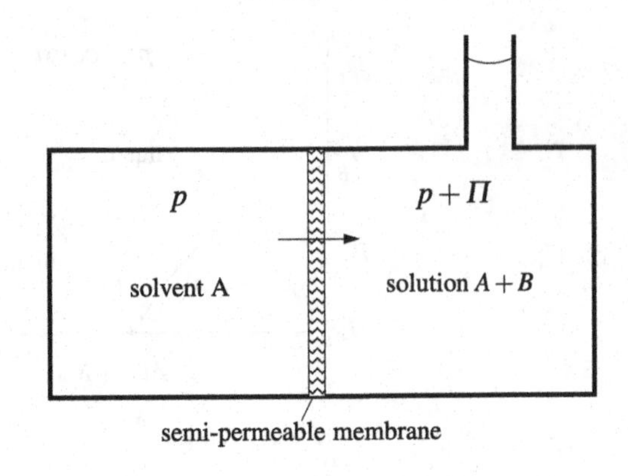

Fig. 8.8 In the state of osmotic equilibrium, there is a pressure difference Π between the pure solvent and the solution, called the *osmotic pressure*

8.3 Osmotic Equilibrium

Substances that form ideal mixtures can mix in arbitrary proportion. The phenomenon of separation into distinct liquid phases do not occur in this case. However, in some situations, mixing of different components is impossible or difficult because of internal constraints imposed on the system.

Imagine a vessel divided by a semi-permeable membrane. One part of the vessel is occupied by the pure substance A in the liquid phase, whereas in the second part, the same substance is a solvent in a solution with the substance B (see Fig. 8.8). The membrane is permeable to the solvent molecules and impermeable to the solute molecules. Such a phenomenon is called *osmosis*. In this case, we consider only the equilibrium condition with respect to the flow of the component A, i.e.,

$$\mu_A^*(T,p) = \mu_A(T,p+\Pi,x_A), \tag{8.54}$$

where $p+\Pi$ denotes the pressure in the solution. On the right-hand side of (8.54), we substitute the expression for the chemical potential of a component in the ideal solution, hence

$$\mu_A^*(T,p) = \mu_A^*(T,p+\Pi) + RT\ln x_A. \tag{8.55}$$

As we can see, the pressure in both parts of the vessel is the same only if $x_A = 1$. When $x_A < 1$ there exists a pressure difference Π called the *osmotic pressure*. We can simplify expression (8.55), expanding $\mu_A^*(T,p+\Pi)$ around p. For a small osmotic pressure, it leads to the following relation:

$$\Pi = -\frac{RT\ln x_A}{v_A^*}, \tag{8.56}$$

where v_A^* is the molar volume of pure solvent at the pressure p. If the solution is dilute then $\ln x_A = \ln(1-x_B) \approx -x_B$. Moreover, $x_B \approx n_B/n_A$ and $v_A^* \approx V/n_A$, where V is the total volume occupied by the solution. Then we can express the osmotic pressure as follows:

$$\Pi = \frac{RTn_B}{V}. \tag{8.57}$$

Note that the osmotic pressure depends only on the amount of the substance dissolved, and not on the nature of that substance. Measurement of the osmotic pressure allows to determine the molar mass of the solute if we know its mass in the solution.

8.4 Colligative Properties

We know already that some properties of solutions do not depend on the nature of the solute but only on its amount in the solution. Such properties of solutions are referred to as *colligative properties*. Here we recall the phenomena in which the colligative properties are manifested.

8.4.1 Vapour Pressure Depression

Addition of a small amount of the non-volatile substance B to the liquid solvent A causes depression of the vapour pressure in equilibrium with the liquid phase. From Raoult's law, it follows that

$$p \approx p_A = p_A^* x_A = p_A^* - p_A^* x_B. \tag{8.58}$$

In the case of dilute solutions, the depression of the vapour pressure below its value for the pure solvent depends only on the mole number of the substance B, i.e.,

$$p - p_A^* \approx -p_A^* \frac{n_B}{n_A}. \tag{8.59}$$

8.4.2 Boiling Point Elevation

In Sect. 8.1.5, we showed that addition of a small amount of the non-volatile substance B to a liquid solvent causes elevation of the boiling point of the solution in relation to the boiling point of the pure solvent, T_A^*. The boiling point elevation is given by the following formula:

$$\Delta T_b = K_b m_B, \tag{8.60}$$

where the ebullioscopic constant K_b characterizes the solvent, hence, ΔT_b depends only on the mole number of the solute.

8.4.3 Freezing Point Depression

In Sect. 8.2.1, we considered a solution of a small amount of the substance B dissolved in the liquid solvent A, assuming that A and B do not form a solid solution. We showed that such a solution freezes at a lower temperature than the freezing point of the pure solvent, T_A^*. The depression of the freezing point is given by the following formula:

$$\Delta T_f = -K_f m_B, \tag{8.61}$$

where the cryoscopic constant K_f characterizes the solvent, hence, ΔT_f depends only on the mole number of the solute.

8.4.4 *Osmotic Pressure*

In Sect. 8.3, we studied equilibrium between the pure solvent A and the solvent A in a solution with a small amount of the substance B, in the presence of a membrane permeable only to the solvent molecules. The pressure difference between the pure solvent and the solution:

$$\Pi = \frac{RTn_B}{V}, \tag{8.62}$$

depends only on the mole number of the solute, but not on its nature.

8.5 Exercises

8.1 At a given temperature, the vapour pressure above the liquid solution $A + B$ amounts to $p = (p_A^* + p_B^*)/2$. Using Raoult's law, find the composition of the liquid and vapour at this pressure.

8.2 Two liquids, A and B, form an ideal solution at the external pressure p. Using Raoult's law and the Clausius–Clapeyron equation, determine the liquid and vapour composition lines in the form of a relation between the composition and temperature. Assume that the enthalpy of melting of both components does not depend on temperature.

8.3 The boiling point of pure liquids A and B at the pressure $p = 1$ bar amounts to $T_A^* = 340$ K and $T_B^* = 360$ K, respectively. The enthalpy of evaporation amounts to $\Delta h_A^* = 20$ kJ/mol and $\Delta h_B^* = 25$ kJ/mol, respectively. Find the pressure of the liquid–vapour equilibrium for pure substances A and B at the temperature $T = 350$ K. Then find the liquid and vapour composition for the solution $A + B$ whose boiling point at the pressure of 1 bar amounts to 350 K. Assume that A and B form an ideal solution.

8.4 For very dilute solutions, Henry's law:

$$p_B = k_B x_B,$$

holds, where p_B is the partial pressure of the solute B in the gaseous phase, and x_B is its molar fraction in the solution. In particular, Henry's law applies to gases dissolved in liquids. For instance, the concentration of oxygen in water, necessary to sustain life, amounts to 4 $\mathrm{mg\,L^{-1}}$, and the Henry constant for oxygen in water at the temperature of 25 °C amounts to 3.3×10^7 torr. Calculate the partial pressure of oxygen above the surface of water needed to maintain the required molar concentration of oxygen in water. By comparison, the partial pressure of oxygen above sea surface amounts to 160 torr.

8.5 The gas B and liquid A form an ideal dilute solution, where A is treated as a non-volatile substance. Assuming A to be water and B to be oxygen, calculate the ratio of the volume that oxygen dissolved in water would occupy in the gaseous state to the volume of water, at the temperature of 25 °C. Assume also that the concentration of oxygen in water is small and the gas above the solution is ideal.

8.6 The mass m of polyethylglicol, of the molar mass M, is dissolved in water of the volume V. Then the solution is connected to a narrow capillary filled with pure water through a semi-permeable membrane, impermeable to the polymer. Assuming dilute solution, find the elevation of water level in the capillary, h, at a given temperature T.

8.7 The boiling point of pure benzene at atmospheric pressure amounts to 353.2 K and its ebullioscopic constant $K_b = 2.53\ \mathrm{K\,kg\,mol^{-1}}$. When a small amount of a non-volatile substance B, of the molar mass $M_B = 64\ \mathrm{g\,mol^{-1}}$, is added to benzene, a solution of the mass $m = 100$ g forms. What is the mass of the non-volatile component if the boiling point of the solution amounts to 355 K?

8.8 After the addition of 24 g of ethanol (C_2H_5OH) to 1 kg of water the freezing point of the solution amounts to −0.97 °C. Calculate the cryoscopic constant of water.

8.9 The cryoscopic constant of acetic acid (CH_3COOH) amounts to $K_f = 3.70\ \mathrm{K\,kg\,mol^{-1}}$. What amount of acetone ($(CH_3)_2CO$) should be dissolved in 1.5 kg of acetic acid, to lower the freezing point of the solution by 0.5 K below the freezing point of the pure solvent?

8.10 Naphthalene melts at the temperature of 352.3 K, and its enthalpy of melting amounts to 19.0 $\mathrm{kJ\,mol^{-1}}$. Calculate the solubility of naphthalene in benzene, i.e., the molar fraction at which naphthalene in the solution coexists with the solid phase, at the temperature of 298 K. Assume that: (1) the solid precipitated from the solution is pure naphthalene, (2) the solution of naphthalene in benzene is an ideal solution, (3) the enthalpy of melting of naphthalene does not depend on temperature.

8.11 A mixture of substances A and B forms a simple eutectic at the composition $x_e = 0.4$. The temperature of the eutectic point, T_e, equals 90 % of the freezing point of pure A, T_A^*, and 84 % of the freezing point of pure B, T_B^*. Calculate the ratio of the enthalpy of melting of pure components, $\Delta h_A^* / \Delta h_B^*$, assuming that A and B form an ideal solution in the liquid phase and that the enthalpy of melting of pure components does not depend on temperature.

Chapter 9
Phase Equilibrium in Real Mixtures

There are many mixtures to which predictions of the ideal solution model, discussed in the previous chapter, do not apply. For instance, the liquid and vapour composition lines do not have to be monotonic. They may have a minimum or maximum at which liquid and vapour have the same composition. A mixture of such a composition is called an *azeotrope*. Another phenomenon, not observed in ideal mixtures, is partial miscibility of components. This means that there exists a certain range of composition in which the components do not form a single liquid or solid phase. A mixture whose composition is in this range separates into two liquid or solid phases of different composition. In such a case, the mixture is said to have a *miscibility gap*. As in the previous chapter, we consider only two-component systems.

9.1 Liquid–Vapour Equilibrium

9.1.1 Deviations From Raoult's Law

In real solutions, Raoult's law is satisfied for any composition only if similar liquids are mixed. For other mixtures, the closer to unity the molar fraction of the solvent, x_A, is, the better Raoult's law is satisfied. In the case of the solute, denoted B, we learned in Sect. 8.1.6 that in a dilute solution it satisfies Henry's law. Thus, we have

$$p_A = p_A^* x_A, \tag{9.1}$$

$$p_B = k_B x_B, \tag{9.2}$$

for $x_B \to 0$ ($x_A \to 1$), where p_A^* is the vapour pressure of the pure solvent in liquid–vapour equilibrium, and k_B is the Henry constant of the solute B in the solvent A. Interchanging A with B, we get

$$p_B = p_B^* x_B, \tag{9.3}$$

$$p_A = k_A x_A, \tag{9.4}$$

R. Hołyst, A. Poniewierski, *Thermodynamics for Chemists, Physicists and Engineers*,
DOI 10.1007/978-94-007-2999-5_9,

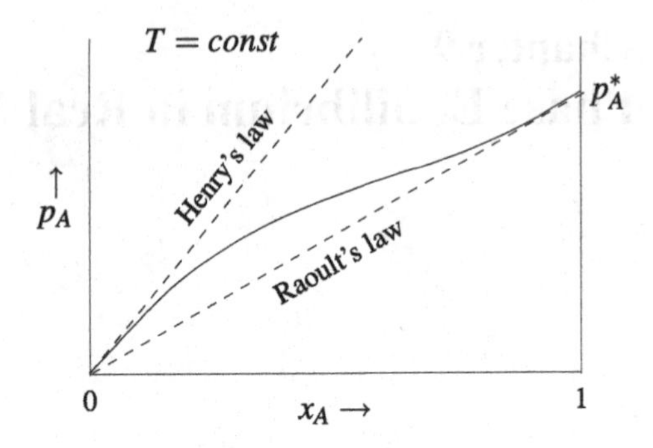

Fig. 9.1 Schematic plot of the partial vapour pressure of the component A vs its molar fraction in the solution. Raoult's law (9.1) is satisfied in the limit $x_A \to 1$, whereas in the limit $x_A \to 0$, Henry's law (9.4) holds

for $x_A \to 0$ ($x_B \to 1$). The dependence of the partial pressure p_A on x_A is shown schematically in Fig. 9.1. Raoult's and Henry's laws concern the slope of the function $p_A(x_A)$ at $x_A = 1$ and $x_A = 0$, respectively. Beyond these two limits, p_A is not a linear function of x_A, in general. A similar drawing can be made for the component B.

If there are no specific interactions between components, such as hydrogen bonds for instance, then *positive deviations* from Raoult's law are usually observed (see Fig. 9.2a). It means that molecules exhibit a stronger tendency to escape from the liquid phase than in the case of the ideal solution. The case of a *negative deviation* from Raoult's law is shown in Fig. 9.2b. An example of such a behaviour is the mixture of acetone and chloroform.

9.1.2 Simple Solutions

To take into account deviations from Raoult's law, we express the chemical potential of a component in its general form (see (7.120)), e.g.,

$$\mu_A = \mu_A^* + RT \ln(x_A \gamma_A), \tag{9.5}$$

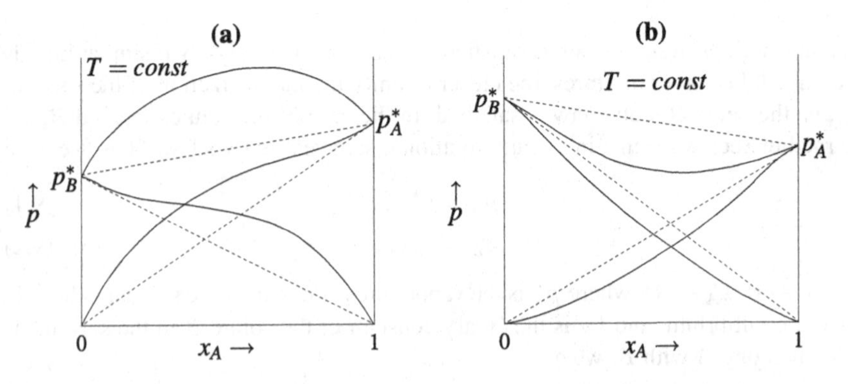

Fig. 9.2 Partial vapour pressure of the components and the total pressure, for a solution showing: (**a**) positive deviations from Raoult's law, (**b**) negative deviations from Raoult's law. The *dashed lines* correspond to the ideal solution

where μ_A^* is the chemical potential of the pure substance, γ_A is the activity coefficient and x_A is the molar fraction of the component A in the solution. A similar relation holds for the component B. From the definition of the activity coefficient, it follows that $\gamma_A = 1$ for the pure substance, and in the case of ideal solutions, $\gamma_A = 1$ for any composition. In general, γ_A is a function of temperature, pressure and composition of the solution. The activity coefficient simply shows how the behaviour of a given component in a real solution deviates from the behaviour of that component in a hypothetical ideal solution.

To derive a relation between the partial pressure p_A and the composition of the solution, we need to repeat the reasoning presented in Sect. 8.1.1. However, it is easy to notice that we only need to replace the molar fraction x_A with the product $\gamma_A x_A$ in the formula for the chemical potential. The same replacement needs to be done in the expression for p_A, and we proceed analogously with the component B. In this way, we obtain a generalized form of Raoult's law:[1]

$$p_A = p_A^* \gamma_A x_A, \tag{9.6}$$

$$p_B = p_B^* \gamma_B x_B. \tag{9.7}$$

The total vapour pressure above the solution amounts to $p = p_A + p_B$, with $p_A = p y_A$ and $p_B = p y_B$, where y_A and y_B are the molar fractions of the components in the gaseous phase. If we know the composition of both the solution and vapour we can determine the activity coefficients, measuring the vapour pressure.

We consider now a model of a real solution, called the *simple solution*, which allows us to express the activity coefficients in terms of a single parameter. We recall first the Gibbs–Duhem equation at constant temperature and pressure (see (7.23)):

$$x_A \mathrm{d}\mu_A + x_B \mathrm{d}\mu_B = 0. \tag{9.8}$$

Since it is satisfied by the chemical potentials in the ideal solution, it must be also satisfied by the excess chemical potentials (see (7.137)), i.e.,

$$x_A \mathrm{d}\mu_A^E + x_B \mathrm{d}\mu_B^E = 0, \tag{9.9}$$

where $\mu_A^E = RT \ln \gamma_A$ and $\mu_B^E = RT \ln \gamma_B$. As we know, μ_A has the same form as in the ideal solution when $x_A \to 1$. Then $\mu_A^E \to 0$ and also $x_B = 1 - x_A \to 0$. Therefore, it is convenient to treat μ_A^E as a function of x_B and, analogously, μ_B^E is treated as a function of x_A. Since $\mathrm{d}x_B = -\mathrm{d}x_A$, we obtain the following equation:

$$-x_A \left(\frac{\partial \mu_A^E}{\partial x_B} \right)_{T,p} + x_B \left(\frac{\partial \mu_B^E}{\partial x_A} \right)_{T,p} = 0. \tag{9.10}$$

The simplest non-trivial solution of Eq. (9.10) is given by the functions

$$\mu_A^E = g_{AB} x_B^2, \tag{9.11}$$

$$\mu_B^E = g_{AB} x_A^2, \tag{9.12}$$

[1] As before, we treat the vapour as an ideal gas. In the general case, the pressure should replaced by the fugacity.

where g_{AB} is a certain function of temperature and pressure. Then the activity coefficients are given by the following expressions:

$$\gamma_A = \exp(g_{AB}x_B^2/RT), \tag{9.13}$$

$$\gamma_B = \exp(g_{AB}x_A^2/RT). \tag{9.14}$$

Substituting (9.11) and (9.12) into the expression for the excess molar Gibbs free energy (see (7.140)), i.e.,

$$g^E = \frac{G^E}{n} = x_A\mu_A^E + x_B\mu_B^E, \tag{9.15}$$

we obtain the following form of g^E for the simple solution:

$$g^E = g_{AB}x_Ax_B(x_A + x_B) = g_{AB}x_Ax_B. \tag{9.16}$$

We can see that the excess part of the Gibbs free energy is related to interactions between components of the solution.

Depending on the sign of g^E we talk about positive or negative deviations from the ideal behaviour. For the simple solution, g^E has the same sign as the coefficient g_{AB}. In the case of positive deviations ($g^E > 0$), both activity coefficients are greater than unity, and the partial vapour pressures lie above the straight lines predicted by Raoult's law, as shown in Fig. 9.2a. In the case of negative deviations ($g^E < 0$), $\gamma_A < 1$ and $\gamma_B < 1$, and the partial pressures lie below the values predicted by Raoult's law (see Fig. 9.2b). In both cases, the total vapour pressure above the solution deviates in the same direction as the partial pressures.

9.1.3 Zeotropic and Azeotropic Mixtures

The total vapour pressure above the solution is a sum of the partial pressures:

$$p = p_A + p_B = p_A^*\gamma_Ax_A + p_B^*\gamma_Bx_B. \tag{9.17}$$

The expressions for the liquid and vapour composition lines are derived in the same way as in the case of ideal solutions. Note that they follow directly from relations (8.13) and (8.15) if the products $p_A^*\gamma_A$ and $p_B^*\gamma_B$ are substituted for p_A^* and p_B^*, respectively, hence

$$p(l, x_A) = p_B^*\gamma_B + (p_A^*\gamma_A - p_B^*\gamma_B)x_A, \tag{9.18}$$

$$p(g, x_A) = \frac{p_A^*\gamma_A p_B^*\gamma_B}{p_A^*\gamma_A + (p_B^*\gamma_B - p_A^*\gamma_A)x_A}. \tag{9.19}$$

It should be emphasized, however, that now the right-hand side of relations (9.18) and (9.19) also depends on the pressure p through the activity coefficients. The liquid and vapour composition lines are also called the *bubble point curve* and *dew point curve*, respectively. At constant temperature, they are the bubble point and dew point isotherms, and at constant pressure, they are the bubble point and dew point isobars. In a pure substance, the bubble point and dew point are the same and are referred to as the boiling point. In respect of shape of the bubble point isotherm, mixtures are divided into *zeotropic* and *azeotropic* ones. Below we present equations for the bubble point and dew point isotherms and isobars. Their derivation is given at the end of this section.

Bubble point isotherm

$$\left(\frac{\partial p}{\partial x_A}\right)_T = \frac{(y_A - x_A)g_{xx}}{y_A \Delta v_A + y_B \Delta v_B}. \tag{9.20}$$

Dew point isotherm

$$\left(\frac{\partial p}{\partial y_A}\right)_T = \frac{(y_A - x_A)g_{yy}}{x_A \Delta v_A + x_B \Delta v_B}. \tag{9.21}$$

Bubble point isobar

$$\left(\frac{\partial T}{\partial x_A}\right)_p = -\frac{(y_A - x_A)g_{xx}}{y_A \Delta s_A + y_B \Delta s_B}. \tag{9.22}$$

Dew point isobar

$$\left(\frac{\partial T}{\partial y_A}\right)_p = -\frac{(y_A - x_A)g_{yy}}{x_A \Delta s_A + x_B \Delta s_B}. \tag{9.23}$$

The following symbols have been used: $x_A = 1 - x_B$ (liquid composition), $y_A = 1 - y_B$ (vapour composition), Δv_A, Δv_B (change in the partial molar volume at the liquid–vapour transition), Δs_A, Δs_B (change in the partial molar entropy at the liquid–vapour transition),

$$g_{xx} = \left(\frac{\partial^2 g}{\partial x_A^2}\right)_{T,p}, \tag{9.24}$$

$$g_{yy} = \left(\frac{\partial^2 g}{\partial y_A^2}\right)_{T,p}, \tag{9.25}$$

where g denotes the molar Gibbs free energy for the liquid or gaseous phase. At the liquid–vapour transition, Δv_A, Δv_B, Δs_A, and Δs_B are positive. Moreover, the condition of intrinsic stability must be satisfied (see Example 7.1), i.e.,

$$\left(\frac{\partial^2 g}{\partial x_A^2}\right)_{T,p} > 0. \tag{9.26}$$

Corollary 9.1 *The bubble point and dew point isotherms are inclined according to the sign of the difference between the vapour and liquid composition, $y_A - x_A$, whereas the bubble point and dew point isobars are inclined in the opposite direction.*

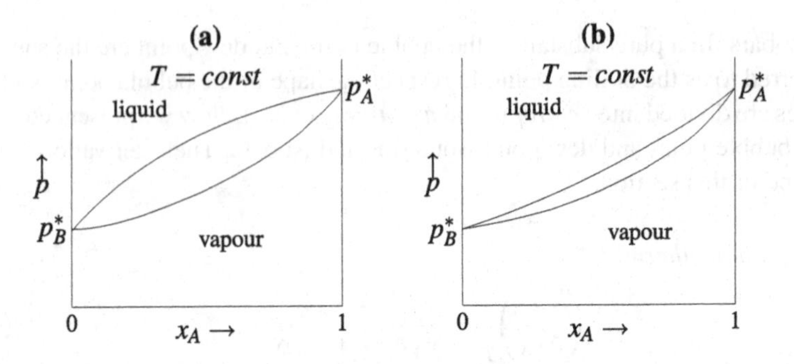

Fig. 9.3 Bubble point isotherm (*upper curve*) and the dew point isotherm (*lower curve*) for a zeotropic mixture, in the case of: (**a**) positive deviations from Raoult's law, (**b**) negative deviations from Raoult's law

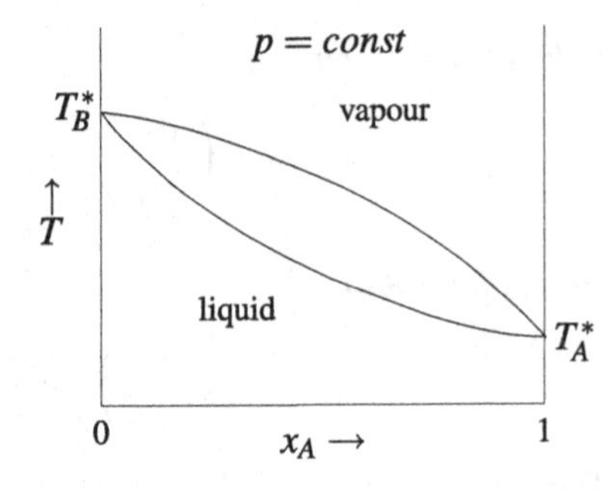

Fig. 9.4 Bubble point isobar (*lower curve*) and dew point isobar (*upper curve*) for a zeotropic mixture. The more volatile component A has the lower boiling point T_A^*

9.1.4 Zeotropic Mixtures

In a zeotropic mixture, the pressure on the bubble point and due point isotherms increases monotonically with the molar fraction of the more volatile component, which we assume to be the component A ($p_A^* > p_B^*$). Therefore, we infer from relations (9.20) and (9.21) that $y_A > x_A$, i.e., the vapour is richer in the more volatile component than the liquid. If γ_A and γ_B are greater than unity then $g^E > 0$ and the bubble point isotherm lies above the straight line predicted by Raoult's law (Fig. 9.3a). If γ_A and γ_B are smaller than unity then $g^E < 0$ and the bubble point isotherm is below that straight line (Fig. 9.3b).

The bubble point and dew point isobars are shown in Fig. 9.4. In this case there is no such a simple relation between the sign of deviation from Raoult's law and shape of the bubble point isobar as in the case of the isotherm. Therefore, only one diagram is shown in Fig. 9.4. According to relations (9.22) and (9.23), the temperature on the bubble point and dew point isobars decreases monotonically with an increasing molar fraction of the component A.

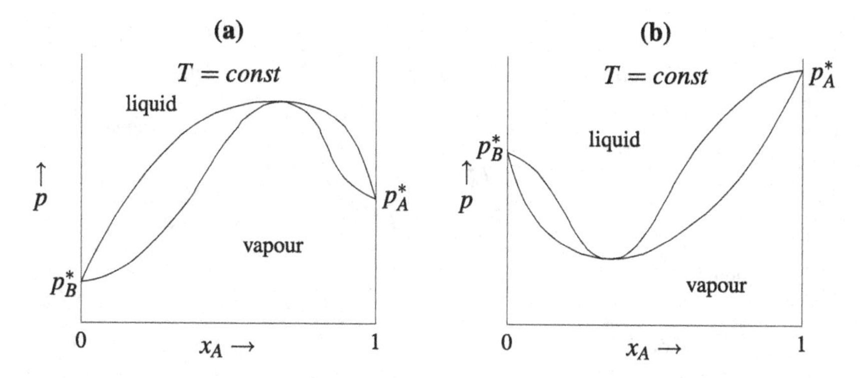

Fig. 9.5 Bubble point and dew point isotherms for a mixture with an azeotropic point: (**a**) a positive azeotrope, (**b**) a negative azeotrope

9.1.5 Azeotropic Mixtures

In the case of large deviation from Raoult's law, the bubble point isotherm has an extremum called the *azeotropic point*. From Eqs. (9.20) and (9.21), we infer that at the azeotropic point, the vapour composition y_A is equal to the liquid composition x_A; the dew point isotherm also have an extremum at that point. A mixture with a composition of the azeotropic point is called an *azeotropic mixture* or an *azeotrope*. Azeotrops boil, as do pure liquids, at constant temperature and pressure. A *positive azeotrope* corresponds to the maximum on the bubble point isotherm and a *negative azeotrope* corresponds to the minimum on that isotherm (Fig. 9.5). The bubble point and dew point isobars for a mixture with an azeotropic point are shown in Fig. 9.6. Contrary to the bubble point isotherms, a positive azeotrope has a minimum on the bubble point isobar and a negative azeotrope has a maximum (cf. (9.22) and (9.23)). A well known mixture that forms a positive azeotrope is a mixture of ethanol and water. At atmospheric pressure, the azeotrope boils at 78.2 °C, whereas the boiling points of the components amount to 100 °C (water) and 78.4 °C (ethanol). The azeotrope composition is equal to 95.63 % (by weight) of ethanol.

Distillation Process In the distillation process, the mixture is heated at constant pressure. The liquid begins to boil when the vapour pressure becomes equal to the external pressure. Distillation of a zeotropic mixture proceeds in a similar way as distillation of an ideal mixture (see Fig. 8.3 in Sect. 8.1.1), which means that the components can be separated to an arbitrary accuracy.

In the case of an azeotropic mixture, components cannot be completely separated by means of distillation. This is because in the distillation process, one makes use of the difference between the liquid and vapour composition in liquid–vapour equilibrium. Since this difference disappears at the azeotropic point, the components can be separated only up to that point by distillation. Let us consider, for instance, distillation of a positive azeotrope shown in Fig. 9.6a. If the initial composition of the liquid is to the left of the azeotropic point then, due to distillation, we remove vapour

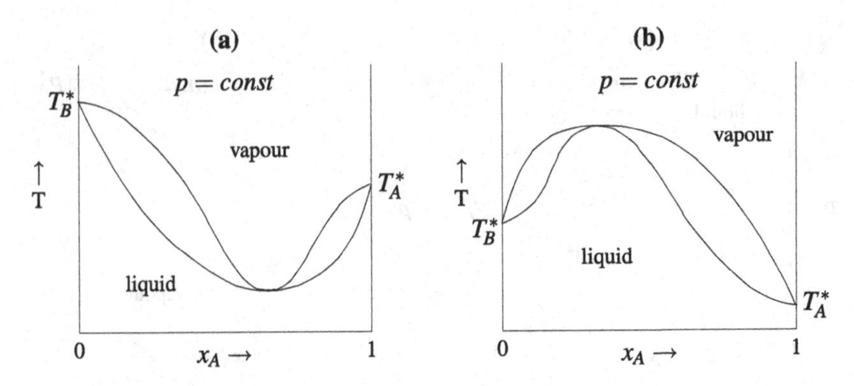

Fig. 9.6 Bubble point and dew point isobars for a mixture with an azeotropic point: (**a**) a positive azeotrope, (**b**) a negative azeotrope

richer and richer in the component A, whereas the liquid remaining in the vessel becomes increasingly rich in the component B with a higher boiling point. When the azeotropic point is reached the liquid and vapour have the same composition and the components cannot be further separated by distillation. If the initial composition is to the right of the azeotropic point then a liquid richer and richer in the component A stays in the vessel, whereas the vapour composition approaches the azeotropic point. A similar reasoning can be used for a negative azeotrope (Fig. 9.6b). In this case, we have an opposite situation, i.e., the vapour removed from the system becomes richer and richer in the component B or A, depending on the initial composition, whereas the composition of the liquid staying in the vessel approaches the azeotropic point.

9.1.6 Derivation of Equations for the Bubble Point and Dew Point Isotherms and Isobars

We derive here relations (9.20)–(9.23). In the liquid–vapour equilibrium, we have

$$\mu_A(l) = \mu_A(g), \tag{9.27}$$

$$\mu_B(l) = \mu_B(g). \tag{9.28}$$

We are interested in such changes of the state parameters of the system which do not influence the liquid–vapour equilibrium. For infinitesimal changes of the parameters, the following equations must be satisfied:

$$d\mu_A(l) = d\mu_A(g), \tag{9.29}$$

$$d\mu_B(l) = d\mu_B(g), \tag{9.30}$$

which lead to differential equations analogous to the Clapeyron equation for a pure substance.

Using (7.70), we obtain

$$d\mu_A = -s_A dT + v_A dp + \left(\frac{\partial \mu_A}{\partial x_A}\right)_{T,p,} dx_A, \tag{9.31}$$

$$d\mu_B = -s_B dT + v_B dp + \left(\frac{\partial \mu_B}{\partial x_A}\right)_{T,p,} dx_A, \tag{9.32}$$

where v_A, v_B, s_A and s_B denote the partial molar volume and partial molar entropy of the components, respectively. The derivative of the chemical potentials with respect to x_A can be expressed in terms of the second derivative of the molar Gibbs free energy g. Since

$$\left(\frac{\partial g}{\partial x_A}\right)_{T,p} = \mu_A - \mu_B \tag{9.33}$$

(see (7.16)) we have

$$\left(\frac{\partial^2 g}{\partial x_A^2}\right)_{T,p} = \left(\frac{\partial \mu_A}{\partial x_A}\right)_{T,p} - \left(\frac{\partial \mu_B}{\partial x_A}\right)_{T,p}. \tag{9.34}$$

The chemical potentials must satisfy the Gibbs–Duhem equation at constant T and p (see (7.23)): $x_A d\mu_A + x_B d\mu_B = 0$, hence

$$x_A \left(\frac{\partial \mu_A}{\partial x_A}\right)_{T,p} + x_B \left(\frac{\partial \mu_B}{\partial x_A}\right)_{T,p} = 0. \tag{9.35}$$

Form Eqs. (9.34) and (9.35), it follows that

$$\left(\frac{\partial \mu_A}{\partial x_A}\right)_{T,p} = x_B \left(\frac{\partial^2 g}{\partial x_A^2}\right)_{T,p}, \tag{9.36}$$

$$\left(\frac{\partial \mu_B}{\partial x_A}\right)_{T,p} = -x_A \left(\frac{\partial^2 g}{\partial x_A^2}\right)_{T,p}. \tag{9.37}$$

Substituting (9.36) into (9.31), and (9.37) into (9.32), we can write Eqs. (9.29) and (9.30) as follows:

$$-s_A^l dT + v_A^l dp + g_{xx} x_B dx_A = -s_A^g dT + v_A^g dp + g_{yy} y_B dy_A, \tag{9.38}$$

$$-s_B^l dT + v_B^l dp - g_{xx} x_A dx_A = -s_B^g dT + v_B^g dp - g_{yy} y_A dy_A. \tag{9.39}$$

Now x_A corresponds to the liquid phase (l) and y_A corresponds to the gaseous phase (g), hence, g_{xx} and g_{yy} denote the values of $\partial^2 g/\partial x_A^2$ in the liquid and gaseous phases, respectively. Finally, we obtain the following two equations:

$$-\Delta s_A dT + \Delta v_A dp - g_{xx} x_B dx_A + g_{yy} y_B dy_A = 0, \tag{9.40}$$

$$-\Delta s_B dT + \Delta v_B dp + g_{xx} x_A dx_A - g_{yy} y_A dy_A = 0, \tag{9.41}$$

where $\Delta s_A = s_A^g - s_A^l$, $\Delta v_A = v_A^g - v_A^l$, and analogously for the component B. To obtain equations for the isotherms, we put $dT = 0$ in (9.40) and (9.41). Then we multiply (9.40) by y_A and (9.41) by y_B and add up both sides, to eliminate the terms with $\mathrm{d}y_A$. Using the relations $x_B = 1 - x_A$ and $y_B = 1 - y_A$, we obtain equation (9.20) for the bubble point isotherm. In an analogous way, we derive equation (9.21) for the dew point isotherm. To obtain equations (9.22) and (9.23) for the isobars, we put $\mathrm{d}p = 0$ in (9.40) and (9.41). Then the derivation proceeds identically as for the isotherms.

9.2 Liquid Solutions with Miscibility Gap

So far we have been assuming that components can mix in any proportion. It is usually true if the molecules of different components have similar chemical structure. However, we know very well from everyday experience that some substances, e.g., water and oil, do not form a homogeneous liquid. For a wide range of composition, such a mixture exists in the form of two liquid phases of different content of each component. Big differences in the chemical structure of molecules can lead to separation of a solution into two liquid phases. Such a solution is said to have a *miscibility gap*. It is a range of composition for which a given solution cannot exist as a homogeneous liquid.

First, we discuss different types of a miscibility gap and then give thermodynamic description of this phenomenon based on the simple solution model. In particular, we show that a miscibility gap can occur only in solutions which exhibit positive deviation from the ideal behaviour.

9.2.1 Miscibility Curve and Critical Temperatures

The line separating the one-phase region from the two-phase region on a diagram showing the liquid–liquid equilibrium is called the *liquid–liquid miscibility curve*. Such a curve has a critical point or it terminates on a line of liquid–vapour equilibrium. Here we consider only the first possibility and the second one is discussed in Sect. 9.3.

Figure 9.7 presents a typical phase diagram for a solution with a miscibility gap. Above a certain temperature, called the *upper critical solution temperature*, T^u_{cr}, the liquids A and B are completely miscible. Below T^u_{cr}, the mixture exists as a single liquid phase, either α, rich in the component A, or β, rich in the component B, only in a certain range of composition. If the mixture composition is in the two-phase region then, at a given temperature, the phase α with the composition x^α_A and the phase β with the composition x^β_A coexist, and their proportion is determined by the lever rule. The liquid–vapour coexistence, which occur at higher temperatures, is not shown in Fig. 9.7.

Let us consider a mixture with the composition $x_A = x_{\mathrm{cr}}$ at a temperature $T < T^u_{\mathrm{cr}}$. When the temperature T approaches the critical temperature the composition of each liquid phase tends to the critical composition, and at the critical temperature it

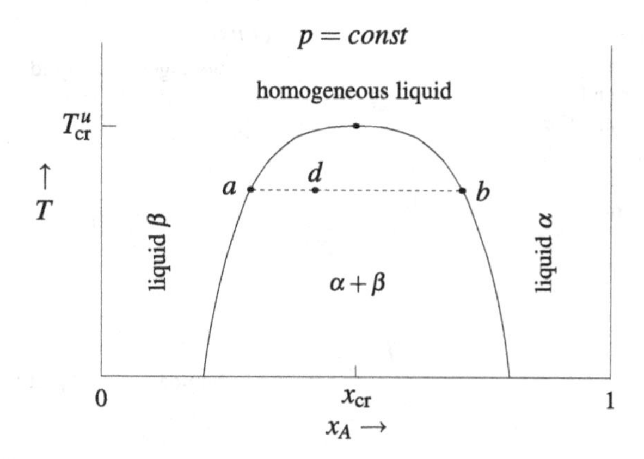

Fig. 9.7 Miscibility gap below the temperature T^u_{cr}; x_{cr} denotes the mixture composition at the critical point. For $T > T^u_{cr}$, the liquids A and B form a homogeneous mixture. For $T < T^u_{cr}$, there exist two liquid phases: α and β. A mixture whose composition corresponds to the point d separates into the phases α and β with the composition corresponding to the points a and b, respectively

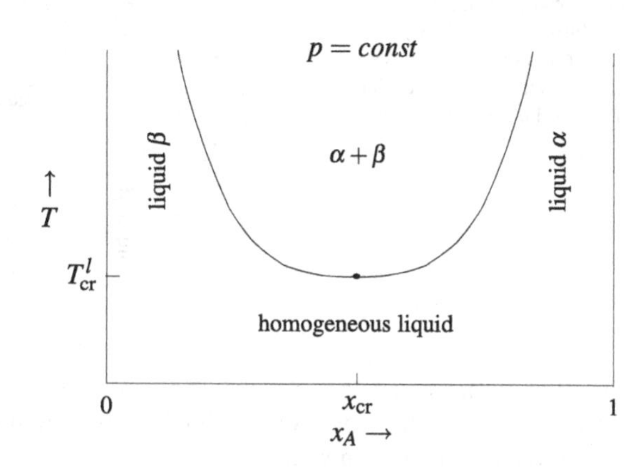

Fig. 9.8 Miscibility gap with a lower critical solution temperature T^l_{cr}. The liquids A and B form a homogeneous mixture when temperature is lower than T^l_{cr}. For $T > T^l_{cr}$, two liquid phases α and β exist

becomes equal to x_{cr}. At the critical point the difference between the phases α and β disappears, and above the critical point a single liquid phase exists.

It may also happen that two liquids are completely miscible at low temperatures and a miscibility gap occurs at higher temperatures. Such a situation is shown in Fig. 9.8. Below the *lower critical solution temperature*, T^l_{cr}, a mixture of the liquids A and B is homogeneous independently of its composition, and above T^l_{cr} two liquid phases α and β exist.

In Sect. 9.2.2, we show that the existence of a lower or upper critical solution temperature is related to the sign of the excess molar entropy h^E. If in the process of mixing the system absorbs heat, i.e., it is an *endothermic process*, then $h^E > 0$. When such a system is heated it is easier for the components to mix, therefore, the region of complete miscibility occurs at higher temperatures (Fig. 9.7). This is typical behaviour, characteristic of many systems, e.g., a mixture of methanol and carbon tetrachloride. In the case of mixtures with a lower critical solution tempera-

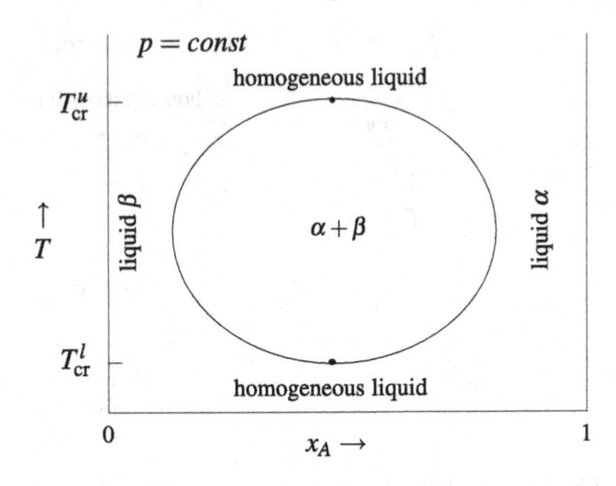

Fig. 9.9 Miscibility gap bound by a closed curve possesses an upper and lower critical solution points. For temperatures higher than T^u_{cr} or lower than T^l_{cr}, the components are completely miscible. For $T^l_{cr} < T < T^u_{cr}$, two liquid phases α and β exist

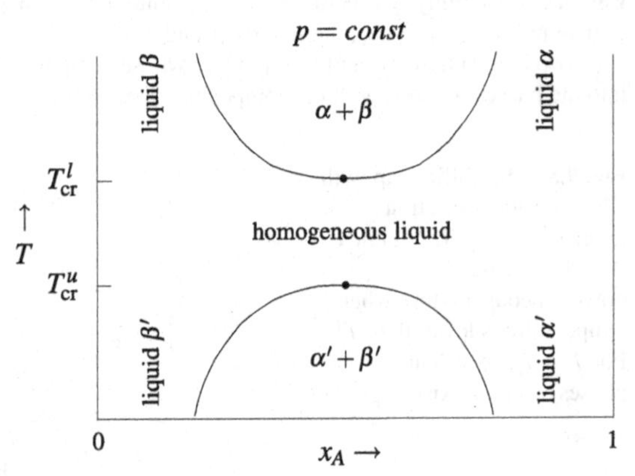

Fig. 9.10 Two separate miscibility gaps. For temperatures higher than T^u_{cr} and lower than T^l_{cr}, the mixture is completely miscible. For temperatures higher than T^l_{cr} or lower than T^u_{cr}, two liquid phases exist

ture (Fig. 9.8), we have $h^E < 0$, which means that mixing is an *exothermic process* (system gives off the heat). Associating liquids exhibit such behaviour, e.g., a mixture of diethylamine with water.

For some systems, the miscibility curve is closed (Fig. 9.9), for instance, a mixture of nicotine and water. In this case, complete miscibility occurs at temperatures higher than T^u_{cr} or lower than T^l_{cr}. Between the lower and upper critical solution temperature, the liquid phases α and β exist.

Two critical points can also occur in the case of two separate miscibility gaps: the upper gap, above the lower critical solution temperature T^l_{cr}, and the lower gap, below the upper critical solution temperature T^u_{cr} (see Fig. 9.10). Note that then $T^l_{cr} > T^u_{cr}$, therefore, the region of complete miscibility occurs between T^u_{cr} and T^l_{cr}. In this case, h^E changes sign when temperature varies between the upper and lower critical solution temperature.

Finally, it should be emphasized that we have presented only schematic pictures of miscibility curves. In the next point, we discuss the simple solution model, for which the molar Gibbs free energy is symmetric with respect to interchange of components, i.e., $g(x_A) = g(1 - x_A)$, provided that the chemical potentials of pure components are equal. In general, $\mu^*_A \neq \mu^*_B$ and experimental curves exhibit some asymmetry. Besides, the simple solution is only an approximation of real solutions, and the excess molar Gibbs free energy used, g^E, is symmetric with respect to interchange of components. However, a more general form of g^E without this symmetry can also be considered.

9.2.2 Miscibility Gap in Simple Solutions

We are going to show now that the miscibility gap occurs in solutions with positive deviation from the ideal behaviour. To do this, we consider a simple solution (see (9.16)), therefore, the molar Gibbs free energy is given by

$$g = g^{\mathrm{id}} + g^E, \tag{9.42}$$

where

$$g^{\mathrm{id}} = \mu_A^* x_A + \mu_B^* x_B + RT(x_A \ln x_A + x_B \ln x_B), \tag{9.43}$$

$$g^E = g_{AB} x_A x_B. \tag{9.44}$$

The first term in (9.42) refers to the ideal solution and the excess term g^E characterizes deviation from the ideal behaviour, where g_{AB} is a function of temperature and pressure. As we know, the molar Gibbs free energy of a thermodynamically stable phase must satisfy the condition of intrinsic stability:

$$\left(\frac{\partial^2 g}{\partial x_A^2}\right)_{T,p} > 0. \tag{9.45}$$

Substituting g given by expressions (9.42)–(9.44) into (9.45), we obtain the following condition:

$$\frac{RT}{x_A x_B} - 2g_{AB}(T, p) > 0. \tag{9.46}$$

In what follows we assume constant pressure, thus, g_{AB} depends only on temperature. A critical point is such a point in the $x_A T$ plane, at which a homogeneous liquid phase ceases to be stable and the mixture begins to separate into two liquid phases with different composition. It occurs when condition (9.45) breaks down, i.e., when

$$\left(\frac{\partial^2 g}{\partial x_A^2}\right)_{T,p} = 0. \tag{9.47}$$

If at a given temperature inequality (9.46) is satisfied for the composition that minimizes the first term, i.e., for $x_A = 1/2$, then it is also satisfied for any x_A. Therefore, the temperature at which condition (9.46) breaks down first, i.e., the critical temperature, must satisfy the equation

$$4RT - 2g_{AB}(T) = 0, \tag{9.48}$$

and the critical composition (for the simple solution model) is equal to $x_{cr} = 1/2$. Note that Eq. (9.48) can have a solution only if $g_{AB}(T) > 0$, i.e., in the case of positive deviation from the ideal behaviour.

It is convenient to introduce the function

$$a(T) = \frac{g_{AB}(T)}{RT}. \tag{9.49}$$

Then two possibilities follow from (9.46) and (9.48):

1. Complete miscibility of components if $a(T) < 2$,
2. Miscibility gap if $a(T) > 2$,

and a critical temperature is a solution of the equation $a(T) = 2$. Whether it is an upper or lower critical solution temperature depends on the behaviour of the function $a(T)$. If $a(T)$ decreases monotonically with increasing temperature then complete miscibility occurs in a high temperature region and a miscibility gap exists in a low temperature region, as shown in Fig. 9.7. However, if $a(T)$ increases monotonically with temperature, then we have the situation presented in Fig. 9.8.

A closed miscibility curve (Fig. 9.9) occurs when the function $a(T)$ has one maximum with $a_{max} > 2$. Then the equation $a(T) = 2$ has two solutions: $T = T_{cr}^{l}$ and $T = T_{cr}^{u}$, where $T_{cr}^{l} < T_{cr}^{u}$. Since $a(T) < 2$ for $T > T_{cr}^{u}$ or $T < T_{cr}^{l}$, these temperature ranges correspond to complete miscibility of components, whereas in the range $T_{cr}^{l} < T < T_{cr}^{u}$, in which $a(T) > 2$, a miscibility gap exists.

Finally, two separate miscibility gaps (Fig. 9.10) occur when the function $a(T)$ has one minimum with $a_{min} < 2$. As in the previous case, the equation $a(T) = 2$ has two solutions, but now $T_{cr}^{u} < T_{cr}^{l}$. Then $a(T) < 2$ in the range $T_{cr}^{u} < T < T_{cr}^{l}$, which corresponds to complete miscibility of components, and $a(T) > 2$, if $T > T_{cr}^{l}$ (upper miscibility gap) or $T < T_{cr}^{u}$ (lower miscibility gap).

Note that the derivative of $a(T)$ with respect to temperature is related to the excess molar enthalpy. From the Gibbs–Helmholtz relation (see (8.29)), we obtain

$$\frac{h^E}{RT^2} = -\left(\frac{\partial g^E/RT}{\partial T}\right)_{p,x_A} = -\left(\frac{\partial a}{\partial T}\right)_p x_A x_B. \tag{9.50}$$

Thus, the occurrence of a miscibility gap with an upper critical point is related to $h^E > 0$, since $a(T)$ is then a decreasing function of T, whereas a miscibility gap with a lower critical point occurs when $h^E < 0$. If both an upper and lower critical points occur then h^E changes the sign because $a(T)$ has a minimum or maximum between the upper and lower critical solution temperature.

Determination of the Miscibility Curve If the phase α, of the composition x_A^{α}, is in equilibrium with the phase β, of the composition x_A^{β}, then the molar fractions must satisfy the equations:

$$\mu_A(x_A^{\alpha}) = \mu_A(x_A^{\beta}), \tag{9.51}$$

$$\mu_B(x_A^{\alpha}) = \mu_B(x_A^{\beta}), \tag{9.52}$$

where the dependence on T and p has been suppressed for simplicity. Using the relation

$$g'(x_A) = \mu_A - \mu_B \tag{9.53}$$

(see (7.16)), we can also express the equations for x_A^α and x_A^β in terms of the molar Gibbs free energy, i.e.,

$$g'(x_A^\alpha) = g'(x_A^\beta), \tag{9.54}$$

where g' denotes the derivative with respect to x_A. The second equation is derived from the relation $g = x_A\mu_A + x_B\mu_B$ as follows:

$$\begin{aligned} g(x_A^\alpha) - g(x_A^\beta) &= x_A^\alpha\mu_A^\alpha + x_B^\alpha\mu_B^\alpha - x_A^\beta\mu_A^\beta - x_B^\beta\mu_B^\beta \\ &= (x_A^\alpha - x_A^\beta)(\mu_A^\alpha - \mu_B^\alpha), \end{aligned} \tag{9.55}$$

where we have used the equalities: $\mu_A^\alpha = \mu_A^\beta$, $\mu_B^\alpha = \mu_B^\beta$, $x_B^\alpha = 1 - x_A^\alpha$ and $x_B^\beta = 1 - x_A^\beta$. Substituting (9.53) into (9.55), we obtain

$$g(x_A^\alpha) - g(x_A^\beta) = (x_A^\alpha - x_A^\beta)g'(x_A^\alpha). \tag{9.56}$$

The geometric interpretation of Eqs. (9.54) and (9.56) is simple. The compositions x_A^α and x_A^β are determined by the double tangent line to the curve $g = g(x_A)$, which is shown in Fig. 9.11. The figure presents a schematic graph of the function $g(x_A)$, which in the case of a simple solution is given by expressions (9.42)–(9.44). If the temperature is such that the components are completely miscible then the function $g(x_A)$ has the shape shown in Fig. 9.11a, i.e., it is everywhere convex, as required by stability condition (9.45). At the critical temperature, $g''(x_{cr}) = 0$, and in the temperature range in which a miscibility gap occurs, $g(x_A)$ has the shape shown in Fig. 9.11b. Note that in the last figure the function $g(x_A)$ contains an unphysical part for which $g'' < 0$. This is because we use a model function, which is only an approximation to the true molar Gibbs free energy.[2]

The model function $g(x_A)$ can be corrected to become physically acceptable for all compositions, including the two-phase region $x_A^\beta < x_A < x_A^\alpha$. Since the Gibbs free energy is an extensive quantity, in the two-phase region we have

$$G = n^\alpha g(x_A^\alpha) + n^\beta g(x_A^\beta), \tag{9.57}$$

where n^α and n^β are the mole numbers of the phases α and β, respectively. They satisfy the lever rule (cf. (8.23))

$$n^\alpha(x_A - x_A^\alpha) = n^\beta(x_A^\beta - x_A) \tag{9.58}$$

[2] A similar problem appeared in Sect. 6.7 in relation to the van der Waals equation of state. The unphysical part of the van der Waals isotherm corresponds to negative isothermal compressibility, i.e., it does not satisfy the condition of mechanical stability.

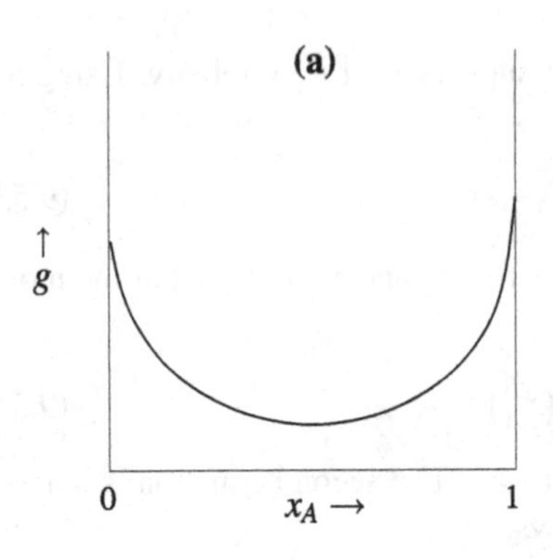

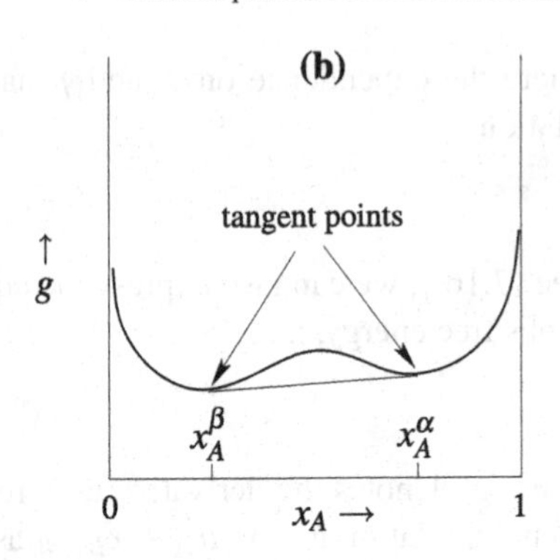

Fig. 9.11 Schematic graph of the molar Gibbs free energy vs. composition, for the simple solution model, at two temperatures. In (**a**) the components are completely miscible, and in (**b**) a miscibility gap exists in the range of composition $x_A^\beta < x_A < x_A^\alpha$. The compositions of the coexisting phases correspond to the tangent points of the curve $g = g(x_A)$ with the double tangent line. In a real system, $g(x_A)$ is a linear function of x_A in the two-phase region. To mimic this in our model system, we form a *convex envelop* of the model function $g(x_A)$, as shown in (**b**)

and the condition $n^\alpha + n^\beta = n$, where n is the total mole number of the mixture. From the lever rule, we determine the ratio

$$\frac{n^\alpha}{n} = \frac{x_A - x_A^\beta}{x_A^\alpha - x_A^\beta}. \tag{9.59}$$

Dividing G by n and using (9.57) and (9.59), we obtain the function $g(x_A)$ for the two-phase region:

$$g(x_A) = g(x_A^\beta) + \left[g(x_A^\alpha) - g(x_A^\beta)\right]\frac{x_A - x_A^\beta}{x_A^\alpha - x_A^\beta}. \tag{9.60}$$

It is a linear function which satisfies conditions (9.54) and (9.56). In Fig. 9.11b, it is represented by the double tangent to the model function $g(x_A)$ at the compositions x_A^α and x_A^β, for which the phases α and β coexist. Replacing the part of the model function between x_A^α and x_A^β with the linear function, we form a *convex envelop* of $g(x_A)$. This construction has a similar meaning to the Maxwell construction for the van der Waals isotherms (see Sect. 6.7.1).

9.3 Liquid–Vapour Equilibrium in Presence of Miscibility Gap

In the previous section, we focused our attention on the liquid–liquid equilibrium in solutions with a miscibility gap. Here we take into account also the liquid–vapour equilibrium. There are two possible situations: (1) the system has an upper critical point, as in Fig. 9.7 or Fig. 9.9, thus, only a single liquid phase exists before the solution reaches the bubble point isotherm, (2) the system does not possess an upper

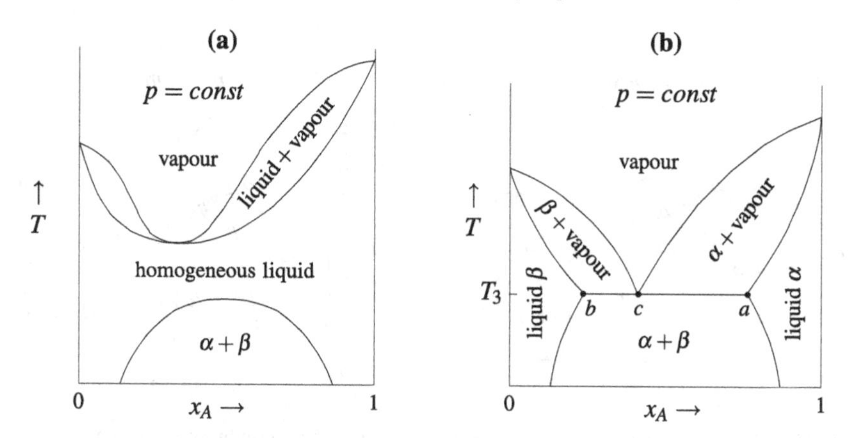

Fig. 9.12 (**a**) Phase diagram of a positive azeotrope with a miscibility gap in the liquid phase and the upper critical point. (**b**) Phase diagram of a heteroazeotrope. The azeotropic point occurs inside the miscibility gap. The *horizontal* segment at $T = T_3$ represents coexistence of three phases: the liquids α and β and the gaseous phase, whose compositions are represented by the points a, b and c, respectively

critical point because the liquid–liquid two-phase region extends right to the liquid–vapour one.

First, we consider case (1). We have shown (see Sect. 9.2.2), that a miscibility gap can exist only if the deviation from the ideal solution behaviour is positive. For this reason, above the temperature T_{cr}^{u} a positive azeotrope often forms, i.e., with a minimum on the bubble point isobar (Fig. 9.6a), which is also related to positive deviation from the ideal behaviour. This case is shown in Fig. 9.12a. In case (2), both liquids begin to boil before the mixture becomes completely miscible. The most common mixture of this type is the *heteroazeotrope*, whose phase diagram is shown in Fig. 9.12b. Three single-phase regions exist here: the liquids α and β and the vapour, which are separated by two-phase regions. Below the temperature T_3 a miscibility gap exists. Above T_3, the vapour can coexist with either the liquid α or β, depending on the mixture composition. At $T = T_3$, three phases coexist: the liquids α and β and the vapour. Their compositions are given by the points a, b and c, respectively, provided that $b < x_A < a$. Thus, the segment ba represents the three-phase region. The temperature T_3 is the lowest temperature at which the gaseous phase can exist. A mixture whose composition is equal to c separates into two liquid phases at a temperature lower than T_3, and at T_3 it boils like an ordinary azeotrope, i.e., the overall liquid composition is equal to the vapour composition. For this composition only, both liquids change simultaneously into vapour at constant temperature. For a composition in the miscibility gap but different from c, the liquids begin to boil at T_3 and the temperature does not change until one of them disappears. Then only one of the liquids remains in equilibrium with the gaseous phase.

Another phase diagram with a miscibility gap is shown in Fig. 9.13a. This type of a mixture is called the *heterozeotropic mixture*. As in the case of an ordinary

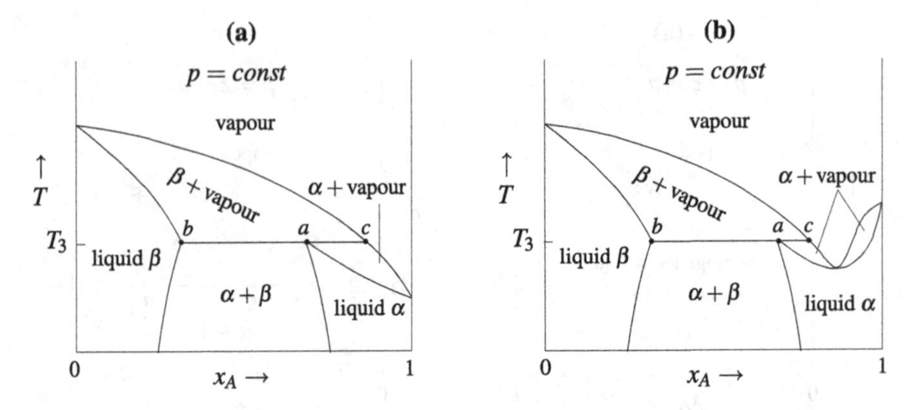

Fig. 9.13 (**a**) Phase diagram of a heterozeotropic mixture. The miscibility gap extends right to the boiling point isobar of a zeotropic mixture, i.e., the liquid and vapour compositions are always different. (**b**) Phase diagram of a homoazeotrope. The miscibility gap extends right to the boiling point isobar of a mixture with the azeotropic point in the region of the liquid α. In *both figures*, T_3 is the temperature of coexistence of the liquids α and β and the vapour, whose compositions correspond to the points a, b and c, respectively.

zeotropic mixture, the boiling point and dew point isobars do not meet at any point. At the temperature T_3, three phases can coexist: the liquids α and β and the vapour, whose compositions correspond to the points a, b and c, respectively. T_3 lies between the boiling points of pure substances. For an overall composition of the mixture between the points b and a, both liquids boil at constant temperature T_3 until the phase α disappears from the system. Then only the liquid β and vapour remains in equilibrium. If the overall mixture composition is between the points a and c, only the liquid α begins to boil and it boils until the boiling point reaches the value T_3, at which the liquid β also appears. Then the liquid–vapour transition proceeds at $T = T_3$ until the liquid α disappears from the system and only the liquid β remains in equilibrium with the vapour.

Figure 9.13b shows another system with a miscibility gap without an upper critical point. The parts of the bubble point and dew point isobars that bound the liquid α-vapour two-phase region have the shape characteristic of a positive azeotrope. This type of a mixture is called the *homoazeotrope*. The temperature of the azeotropic point lies below the temperature T_3 at which the liquid phases can simultaneously coexist with the vapour.

9.4 Liquid–Solid Equilibrium and Solid Solutions

In Sect. 8.2.3, we presented the phase diagram of a simple eutectic (Fig. 8.7). It was then assumed that the liquid solution is ideal and that components are completely immiscible in the solid phase. If, however, components can mix, completely or partially, also in the solid phase then the mixture is called a *solid solution*. The influence of pressure on phase equilibrium of condensed phases is much smaller than

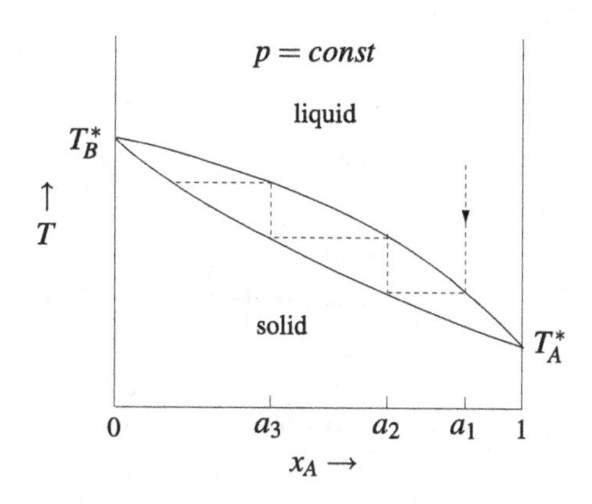

Fig. 9.14 Solid solution with monotonic dependence of the melting point on composition; T_A^* and T_B^* denote the melting points of pure substances. The freezing point curve (*upper line*) and the melting point curve (*lower line*) bound the liquid–solid two-phase region. A few steps in the process of fractional crystallization (*dashed line*) are shown

in the case of liquid–vapour equilibrium. Therefore, we can assume, for simplicity, that the pressure is equal to the standard value p^0. The types of phase diagram for solid solutions to a great extend correspond to those discussed above in the case of liquid–vapour and liquid–liquid equilibrium. Here we present only some of them.

If components can mix in any proportion both in the liquid and solid phase then the simplest phase diagram, shown in Fig. 9.14, is very similar to the phase diagram of a zeotropic mixture at constant pressure (cf. Fig. 9.4). The melting point of a solid solution is a monotonic function of composition. The upper line in Fig. 9.14 is called the *freezing point curve* and the lower line is called the *melting point curve*.[3] In between, the liquid–solid two-phase region extends. If at a given temperature the overall composition of a mixture is in the two-phase region then the liquid and solid phases are in equilibrium. The compositions of the coexisting phases are determined by the intersection of the $T = const$ line with the melting point and freezing point curves. The solid phase is richer in the component with a higher melting point (component B). This property is used in the process of *fractional crystallization*, in which components of a solid solution can be separated from each other (Fig. 9.14). When a liquid with the initial composition a_1 freezes a solid phase with the composition $a_2 < a_1$ separates from it. Then the solid is removed from the system and changed into liquid again. The liquid has the same composition a_2 and when it freezes a solid with the composition $a_3 < a_2$ separates from it. By repeating this process, we can obtain an almost pure solid phase of the substance B. Phase diagrams of many metallic alloys, e.g., copper–nickel, cobalt–nickel or gold–silver, are of the type shown in Fig. 9.14.

Some solid solutions exhibit a minimum on the melting point curve, e.g., copper–manganese, copper–gold, cobalt–manganese. Their phase diagrams are very similar in shape to the phase diagram of a positive azeotrope (cf. Fig. 9.6a). The counterpart of a negative azeotrope (Fig. 9.6b), i.e., with a maximum on the melting point curve, is not very common. In the case of large positive deviation from the ideal behaviour, a miscibility gap in the solid phase region appears. Such a situation is shown in

[3] In the literature, also the terms *liquidus* and *solidus* are used, respectively.

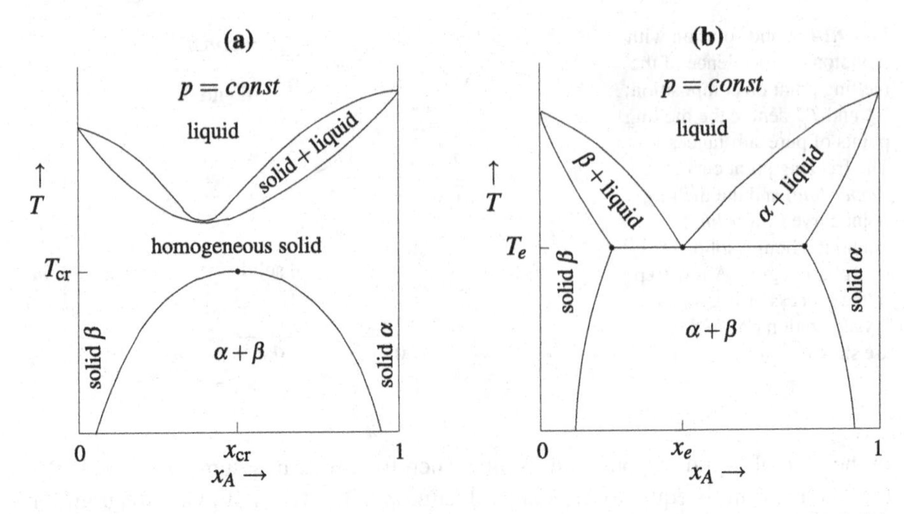

Fig. 9.15 (a) Solid solution with a miscibility gap below the critical temperature T_{cr}. (b) Solid solution of the eutectic type. Unlike the simple eutectic, the solid phases are not pure substances. A liquid with the eutectic composition x_e freezes at constant temperature T_e, and the solid phases α and β separate from the solution

Fig. 9.15a. In the high temperature region, the liquid phase is in equilibrium with a homogeneous solid phase and the melting point curve has a minimum. The homogeneous solid phase is stable above the critical temperature T_{cr}. For $T < T_{cr}$, the solid phases α and β can coexist. At low temperatures, the solid phases become almost pure crystals.

A solid solution that forms an *eutectic system* is shown in Fig. 9.15b. In this case, the phase diagram resembles that of the heteroazeotrope (cf. Fig. 9.12b). Unlike the simple eutectic (Fig. 8.7), the solid phases are not pure crystals of A and B, but solid solutions. At the eutectic temperature T_e, the solid phases α and β and a liquid phase, of the composition x_e, can coexist. T_e is the lowest temperature at which the solution can exist as a liquid. A liquid with the eutectic composition freezes like a pure substance at constant temperature equal to T_e. Many alloys form eutectic solutions, e.g., aluminium–copper, copper–zinc, antimony–lead.

9.5 Exercises

9.1 At the temperature of 298 K, the vapour pressure above pure liquids A and B amounts to 0.031 bar and 0.029 bar, respectively. The vapour pressure above the solution $A+B$, of the composition $x_A = 0.2$, amounts to $p = 0.041$ bar, and $y_A = 0.44$ is the vapour composition. Calculate the activity coefficients of both components in the solution, assuming that vapour is an ideal gas. The behaviour of which component is closer to the predictions of Raoult's law?

9.2 Liquids A and B form an azeotrope, whose composition at the azeotropic point amounts to $x_A = x_a = 0.70$, at the temperature of 298 K. The ratio of vapour pressures above pure liquids at $T = 298$ K amounts to $p_A^*/p_B^* = 2.5$. Assuming the simple solution model, determine the coefficient g_{AB}. Is it a positive or negative azeotrope? Assume that the vapour above the solution is an ideal gas and g_{AB} does not depend on pressure.

9.3 At a given temperature T, the coefficient $g_{AB}(T) = 0.9RT$. Does the simple solution characterized by this value of g_{AB} form an azeotrope if the ratio of the vapour pressures above pure components amounts to $p_A^*/p_B^* = 3$ at the temperature T?

9.4 At atmospheric pressure, two liquids form a simple solution for which

$$g_{AB}(T)/RT = 5(T - T_0)\ \mathrm{K}^{-1} - 2(T - T_0)^2\ \mathrm{K}^{-2},$$

where T_0 is a reference temperature. Does a miscibility gap occur in this mixture, and if so, what is the shape of the miscibility curve, how many critical points does it have and what are the values of the critical temperatures?

9.5 Liquids A and B form an azeotrope. Assuming that at a given temperature T we know the vapour pressures of pure components, $p_A^* \neq p_B^*$, and the composition at the azeotropic point, $x_A = x_a$, calculate the Henry constant of the solute B in the solvent A at the temperature T. Assume the simple solution model with the coefficient g_{AB} independent of pressure and that vapour is an ideal gas.

9.6 Apply Eqs. (9.20) and (9.21) for the bubble point and dew point isotherms to an ideal solution of A and B. Derive the expressions for the change in the partial molar volumes, Δv_A and Δv_B, at the liquid–vapour transition, that follow from these equations.

Part III
Chemical Thermodynamics

Chapter 10
Systems with Chemical Reactions

10.1 Condition of Chemical Equilibrium

During a chemical reaction substances forming a mixture undergo an internal change. Each reaction proceeds according to a certain equation, for instance,

$$3H_2 + N_2 \rightleftharpoons 2NH_3. \tag{10.1}$$

The substances whose amounts decrease during the reaction are called *reactants* and the substances whose amounts increase are called *products*. Reaction (10.1) can proceed from the left to the right or in the reverse direction, depending on the external conditions and initial amounts of the components. The direction from the left to the right is regarded as positive and the reverse direction is regarded as negative. In the example above, during the reaction in the positive direction the amounts 3ξ of H_2 and 1ξ of N_2 disappear from the system and the amount 2ξ of NH_3 is produced, where ξ is a common factor, whose physical dimension is the mole, which determines the progress of the reaction. It is often convenient to write all substances on the right-hand side of the reaction equation, i.e.,

$$0 \rightleftharpoons \nu_{H_2}H_2 + \nu_{N_2}N_2 + \nu_{NH_3}NH_3, \tag{10.2}$$

where $\nu_{H_2} = -3$, $\nu_{N_2} = -1$ and $\nu_{NH_3} = 2$. In general, the equation of a reaction which can proceed in either direction can be written in the following form:

$$\sum_{i=1}^{C'} |\nu_i| A_i \rightleftharpoons \sum_{i=C'+1}^{C} \nu_i A_i, \tag{10.3}$$

where A_i denotes the ith component: a reactant, if $1 \le i \le C'$, or a product, if $C' + 1 \le i \le C$. The coefficients ν_i are called the *stoichiometric coefficients*. The convention is used that $\nu_i < 0$ for reactants and $\nu_i > 0$ for products. If $\nu_i = 0$ for a certain value of i it means that the component A_i does not participate in the given

R. Hołyst, A. Poniewierski, *Thermodynamics for Chemists, Physicists and Engineers*,

DOI 10.1007/978-94-007-2999-5_10,

reaction. Writing all components on the right-hand side of the reaction equation and using the convention for the sign of stoichiometric coefficients, we obtain

$$0 \rightleftharpoons \sum_{i=1}^{C} \nu_i A_i. \tag{10.4}$$

During the reaction, the amounts of individual components change according to the following formula:

$$n_i(\xi) = n_i(0) + \nu_i \xi, \tag{10.5}$$

where $i = 1, \ldots, C$ and $n_i(0)$ is the value of n_i at $\xi = 0$, hence

$$dn_i = \nu_i d\xi. \tag{10.6}$$

The parameter ξ is called the *extent of reaction*. If the reaction proceeds from the left to the right then $\xi > 0$ and if it proceeds in the reverse direction then $\xi < 0$. First we consider a reaction which proceeds in the positive direction. Since $\nu_i < 0$ for reactants, there exists a maximum value $\xi = \xi_{max}$ that $n_i(\xi_{max}) = 0$ for one of the reactants. Obviously all n_i must be positive or zero, thus, the reaction stops when ξ reaches the maximum value. Similarly, a reaction which proceeds in the negative direction stops when ξ reaches a minimum value ξ_{min}. Therefore, the following inequality holds:

$$\xi_{min} \le \xi \le \xi_{max}, \tag{10.7}$$

where ξ_{min} and ξ_{max} depend on the initial amounts of the components, $n_i(0)$.

We consider a multi-component system in which a reaction occurs at constant temperature and pressure, which is a common situation in practice. Therefore, the Gibbs free energy is to be used to study the system. Then we consider a virtual process in which the mole numbers n_i change according to formula (10.5). Suppose also that we can control the reaction, so that at any moment the state of the system is an equilibrium state with some constraints. For such states, we can treat the Gibbs free energy as a function of ξ, using to the relation

$$G(T, p, \xi) = G\big(T, p, n_1(\xi), \ldots, n_C(\xi)\big). \tag{10.8}$$

An infinitesimal change in G in the reaction amounts to

$$dG = \sum_{i=1}^{C} \left(\frac{\partial G}{\partial n_i}\right)_{T,p,n_{j\neq i}} dn_i = \sum_{i=1}^{C} \mu_i \nu_i d\xi, \tag{10.9}$$

where μ_i is the chemical potential of the ith component. The quantity

$$A = -\left(\frac{\partial G}{\partial \xi}\right)_{T,p} = -\sum_{i=1}^{C} \nu_i \mu_i \tag{10.10}$$

is called the *affinity of reaction*. If G is a decreasing function of ξ $(A > 0)$ then the reaction which proceeds in the positive direction is a spontaneous process, whereas if G increases with ξ $(A < 0)$ then the reverse reaction proceeds spontaneously.

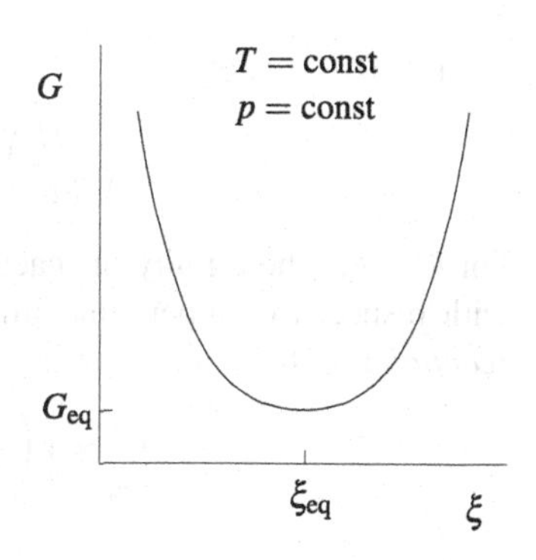

Fig. 10.1 Gibbs free energy as a function of the extent of reaction. The values G_{eq} and ξ_{eq} correspond to the state of chemical equilibrium. The reaction proceeds spontaneously in one or the other direction when the affinity of reaction $A \neq 0$

The *condition of chemical equilibrium* corresponds to the minimum of the function $G(T, p, \xi)$ at constant T and p, which is shown in Fig. 10.1. Using relation (10.10), we obtain the following form of the chemical equilibrium condition:

$$A(T, p, \xi) = -\sum_{i=1}^{C} \nu_i \mu_i = 0. \tag{10.11}$$

We denote by ξ_{eq} the value of ξ which satisfies Eq. (10.11). When the system is in chemical equilibrium a change in n_i caused by the reaction that proceeds in the positive direction is counterbalanced by a change in n_i caused by the reverse reaction, hence, the resultant change in n_i is equal to zero. Condition (10.11) is analogous to the conditions of thermal equilibrium (equality of temperatures), mechanical equilibrium (equality of pressures) or equilibrium with respect to matter flow (equality of chemical potentials). Note that the system reaches the state of chemical equilibrium or not, depending on the initial amounts of components (see (10.7)). If $\xi_{eq} > \xi_{max}$ or $\xi_{eq} < \xi_{min}$, then the system reaches the state of the lowest possible value of G and the reaction stops because one of the reactants has been used up.

10.1.1 Enthalpy of Reaction

In *endothermic reactions* the system absorbs heat, whereas in *exothermic reactions* it gives off heat to the surroundings. If a reaction proceeding in the positive direction is endothermic then the reverse reaction is exothermic and vice versa. A system in chemical equilibrium neither absorbs nor gives off heat. To observe a thermal effect of the reaction, the system must deviate slightly from the state of chemical equilibrium. We consider an infinitesimal deviation from chemical equilibrium at constant T and p. The heat absorbed or given off by the system at constant pressure is equal to the change in its enthalpy. Using the identity

$$H = G + TS = G - T\left(\frac{\partial G}{\partial T}\right)_{p,n_i}, \tag{10.12}$$

and (10.10), we get

$$\left(\frac{\partial H}{\partial \xi}\right)_{T,p} = -A + T\left(\frac{\partial A}{\partial T}\right)_{p,\xi}. \tag{10.13}$$

For $\xi = \xi_{\text{eq}}$, the affinity of reaction vanishes ($A = 0$), whereas the derivative of A with respect to temperature differs from zero, in general, hence, the *enthalpy of reaction* is defined as follows:

$$\left(\frac{\partial H}{\partial \xi}\right)_{\text{eq}} = \left[\left(\frac{\partial H}{\partial \xi}\right)_{T,p}\right]_{\xi=\xi_{\text{eq}}} = T\left(\frac{\partial A}{\partial T}\right)_{p,\xi=\xi_{\text{eq}}}. \tag{10.14}$$

10.2 Effect of External Perturbation on Chemical Equilibrium

The state of chemical equilibrium is determined from the condition

$$A(T, p, \xi) = 0, \tag{10.15}$$

which is an equation for ξ at constant T and p. The solution of (10.15) is denoted by $\xi_{\text{eq}}(T, p)$. Now we want to investigate how ξ_{eq} changes when the external conditions, i.e., the temperature or pressure, vary. To do this, we use the identity

$$A\big(T, p, \xi_{\text{eq}}(T, p)\big) = 0. \tag{10.16}$$

10.2.1 Effect of Temperature

First, we differentiate (10.16) with respect to temperature at constant p:

$$\left(\frac{\partial A}{\partial T}\right)_{p,\xi_{\text{eq}}} + \left(\frac{\partial A}{\partial \xi_{\text{eq}}}\right)_{T,p}\left(\frac{\partial \xi_{\text{eq}}}{\partial T}\right)_p = 0. \tag{10.17}$$

Since G has a minimum at ξ_{eq}, it follows from (10.10) that

$$-\left(\frac{\partial A}{\partial \xi_{\text{eq}}}\right)_{T,p} = \left[\left(\frac{\partial^2 G}{\partial \xi^2}\right)_{T,p}\right]_{\xi=\xi_{\text{eq}}} = G_{\xi\xi}^{\text{eq}} > 0. \tag{10.18}$$

Using (10.14), (10.17) and (10.18), we find that at $p = const$,

$$\left(\frac{\partial \xi_{\text{eq}}}{\partial T}\right)_p = \left(T G_{\xi\xi}^{\text{eq}}\right)^{-1}\left(\frac{\partial H}{\partial \xi}\right)_{\text{eq}}. \tag{10.19}$$

From relation (10.19), we conclude that the chemical equilibrium shifts in the direction in which the enthalpy of the system increases. In the case of an endothermic reaction, we have $(\partial H/\partial \xi)_{\text{eq}} > 0$, thus, ξ_{eq} increases with temperature, i.e., the chemical equilibrium shifts in the direction of products. In the case of an exothermic reaction, $(\partial H/\partial \xi)_{\text{eq}} < 0$, hence, ξ_{eq} decreases with temperature and the chemical equilibrium shits in the direction of reactants.

Corollary 10.1 *An increase in the temperature of the system at constant pressure shifts the position of chemical equilibrium in the direction in which the enthalpy of the system increases.*

To better understand the behaviour of the system, we calculate the total change in its enthalpy caused by a change in temperature. We treat the enthalpy as a function of T, p and ξ, and substitute $\xi = \xi_{eq}(T, p)$. For $p = const$, we have:

$$\mathrm{d}H = C_p \mathrm{d}T + \left(\frac{\partial H}{\partial \xi}\right)_{eq} \mathrm{d}\xi_{eq}, \tag{10.20}$$

where C_p is the heat capacity at constant pressure. The first term in (10.20) is related to the transfer of heat in the state of chemical equilibrium and the second term is related to the shift of chemical equilibrium. Substituting

$$\mathrm{d}\xi_{eq} = \left(T G_{\xi\xi}^{eq}\right)^{-1} \left(\frac{\partial H}{\partial \xi}\right)_{eq} \mathrm{d}T \tag{10.21}$$

(see (10.19)) into (10.20), we get

$$\mathrm{d}H = \left[C_p + \left(T G_{\xi\xi}^{eq}\right)^{-1} \left(\frac{\partial H}{\partial \xi}\right)_{eq}^{2}\right] \mathrm{d}T, \tag{10.22}$$

which means that the shift of chemical equilibrium increases the total heat capacity of the system. Since a system with large heat capacity changes its temperature less than a system with small heat capacity, for the same amounts of heat absorbed, the shift of chemical equilibrium reduces the increase in the temperature of the system. In that way, the effect of external perturbation on the system is reduced.

10.2.2 Effect of Pressure

Because of the relation

$$V = \left(\frac{\partial G}{\partial p}\right)_{T,n_i}, \tag{10.23}$$

the volume of the system, V, can be treated as a function of temperature, pressure and the extent of reaction. Differentiating V with respect to ξ at constant T and p, we get

$$\left(\frac{\partial V}{\partial \xi}\right)_{eq} = \left[\left(\frac{\partial V}{\partial \xi}\right)_{T,p}\right]_{\xi=\xi_{eq}} = -\left(\frac{\partial A}{\partial p}\right)_{T,\xi=\xi_{eq}}. \tag{10.24}$$

Differentiation of (10.16) with respect to pressure at constant T gives

$$\left(\frac{\partial A}{\partial p}\right)_{T,\xi_{eq}} + \left(\frac{\partial A}{\partial \xi_{eq}}\right)_{T,p} \left(\frac{\partial \xi_{eq}}{\partial p}\right)_{T} = 0, \tag{10.25}$$

hence, using (10.10), (10.24) and (10.25), we get

$$\left(\frac{\partial \xi_{eq}}{\partial p}\right)_T = -\left(G_{\xi\xi}^{eq}\right)^{-1}\left(\frac{\partial V}{\partial \xi}\right)_{eq}. \tag{10.26}$$

From relation (10.26), it follows that the position of chemical equilibrium shifts in the direction in which the volume of the system decreases. If in a given reaction the volume of the system decreases, i.e., $(\partial V/\partial \xi)_{eq} < 0$, the chemical equilibrium shifts towards products, and if the volume increases, the chemical equilibrium shifts towards reactants.

Corollary 10.2 *An increase in the pressure of the system at constant temperature shifts the position of chemical equilibrium in the direction in which the volume of the system decreases.*

Analogously to the case of enthalpy, we calculate the total change in the volume of the system caused by a change in pressure. In chemical equilibrium, $V = V(T, p, \xi_{eq})$ depends only on T and p. An infinitesimal change in V caused by a change in pressure at $T = const$ amounts to

$$\mathrm{d}V = -V\kappa_T \mathrm{d}p + \left(\frac{\partial V}{\partial \xi}\right)_{eq} \mathrm{d}\xi_{eq}, \tag{10.27}$$

where κ_T is the isothermal compressibility. Using (10.26), we obtain

$$\mathrm{d}V = -V\left[\kappa_T + \left(VG_{\xi\xi}^{eq}\right)^{-1}\left(\frac{\partial V}{\partial \xi}\right)_{eq}^2\right]\mathrm{d}p, \tag{10.28}$$

which means that the shift of chemical equilibrium increases the total isothermal compressibility of the system. Since for the same relative change in volume, the pressure changes less in a system with large compressibility than in a system with small compressibility, the shift of chemical equilibrium reduces the increase in the pressure of the system. In that way, the effect of external perturbation on the system is reduced.

10.2.3 Le Chatelier–Braun Principle

A concise summary of Corollaries 10.1, 10.2 and relations (10.22), (10.28) is given by the *Le Chatelier–Braun principle* formulated below.

Corollary 10.3 *If a system in chemical equilibrium experiences external perturbation of the equilibrium state, then the equilibrium shifts to minimize the effect of the perturbation.*

We illustrate the Le Chatelier–Braun principle by the reaction

$$3H_2 + N_2 \rightleftharpoons 2NH_3. \tag{10.29}$$

Example 10.1 The synthesis of ammonia is an exothermic reaction. An increase in the temperature of the surroundings causes flow of heat into the system. If there was no chemical reaction in the system, the flow of heat would increase its temperature by ΔT. However, the external perturbation launches the reverse endothermic reaction of ammonia decomposition, which absorbs a part of the heat supplied to the system. As a result, the increase in the temperature is smaller than ΔT. Eventually, a new equilibrium state with a smaller concentration of ammonia settles down at a higher temperature.

Example 10.2 During the synthesis of ammonia the volume of the system decreases because 4 mol of hydrogen and nitrogen change into 2 mol of ammonia. Suppose we perturb the equilibrium state, increasing the external pressure, which reduces the volume occupied by the gases and increases the pressure in the system. To counteract the increase in the pressure, the synthesis reaction is launched, which reduces the total mole number. As a result, the pressure increases less than it would do if there was no reaction in the system. Eventually, a new equilibrium state with a higher concentration of ammonia settles down at a higher pressure.

10.3 Law of Mass Action for Ideal Gases

We consider a mixture of ideal gases in which a chemical reaction occurs. In this case (see Sect. 7.4.2), we have

$$\mu_i = \mu_i^0(T) + RT \ln \frac{p_i}{p^0}, \tag{10.30}$$

where p_i, μ_i^0 and p^0 denote the partial pressure and standard chemical potential of the ith component, and the standard pressure, respectively. Therefore, the affinity of reaction adopts the following simple form:

$$A = -\sum_{i=1}^{C} \nu_i \mu_i^0(T) - RT \ln \prod_{i=1}^{C} \left(\frac{p_i}{p^0}\right)^{\nu_i}. \tag{10.31}$$

The first sum in (10.31) is called the *standard Gibbs free energy of reaction*, for which we use the symbol $\Delta_r G^0$, i.e.,

$$\Delta_r G^0 = \sum_{i=1}^{C} \nu_i \mu_i^0. \tag{10.32}$$

It represents a hypothetical process in which unmixed reactants react, giving separated products, and both the reactants and products are in their standard states. It is also convenient to introduce the *standard equilibrium constant*

$$K^0 = \exp\left(-\frac{\Delta_r G^0}{RT}\right). \tag{10.33}$$

This definition of K^0 is general, i.e., it does not refer to ideal gases only.

The condition of chemical equilibrium, $A = 0$, applied to (10.31) leads to a relation called the *law of mass action*:

$$\prod_{i=1}^{C} \left(\frac{p_i}{p^0} \right)^{\nu_i} = K^0. \tag{10.34}$$

K^0 is only a function of temperature, which follows from its definition. The law of mass action can also be expressed in the form of a relation between the molar fractions. Substituting $p_i = px_i$ into (10.34), we obtain

$$\prod_{i=1}^{C} x_i^{\nu_i} = K_x, \tag{10.35}$$

where K_x is called the *equilibrium constant*, and the index x means that the law of mass action is expressed in terms of the molar fractions. Comparing (10.34) with (10.35), we find that

$$K_x = \left(\frac{p}{p^0} \right)^{-\Delta n} K^0(T), \tag{10.36}$$

where

$$\Delta n = \sum_{i=1}^{C} \nu_i. \tag{10.37}$$

Thus, K_x is a function of both temperature and pressure. We recall that the stoichiometric coefficients are negative for reactants and positive for products, which means that on the left-hand side of the mass-action law, in the form (10.34) or (10.35), we have a certain quotient. For instance, (10.35) can be expressed as

$$\frac{\prod_i x_i^{\nu_i}}{\prod_j x_j^{|\nu_j|}} = K_x, \tag{10.38}$$

where the indices i and j number the products and reactants, respectively.

Example 10.3 We illustrate the law of mass action by reaction (10.29). From (10.38), we get

$$\frac{x_{NH_3}^2}{x_{H_2}^3 x_{N_2}} = K_x. \tag{10.39}$$

Since $x_{NH_3} = 1 - x_{H_2} - x_{N_2}$, (10.39) is a relation between the molar fractions of hydrogen and nitrogen in chemical equilibrium, for the given T and p.

10.3.1 Effect of Temperature on the Equilibrium Constant

From definition (10.33) of the standard equilibrium constant K^0 and the Gibbs-Helmholtz relation (see (8.29)), one derives the *van 't Hoff equation*:

$$\frac{\mathrm{d}\ln K^0(T)}{\mathrm{d}T} = -\frac{1}{R}\frac{\mathrm{d}}{\mathrm{d}T}\left(\frac{\Delta_r G^0}{T}\right) = \frac{\Delta_r H^0}{RT^2}, \tag{10.40}$$

where $\Delta_r H^0$ denotes the *standard enthalpy of reaction*. From definition (10.32), we get

$$\Delta_r H^0 = \sum_{i=1}^{C} \nu_i h_i^0(T), \tag{10.41}$$

where h_i^0 is the molar enthalpy of the pure substance i in its standard state. Relation (10.40) can be used to determine $\Delta_r H^0$ from the dependence of the equilibrium constant on temperature, which can be obtained from the law of mass action.

Now we calculate the enthalpy of reaction defined by (10.14). Using (10.32) and (10.33), we express A as follows:

$$A(T, p, \xi) = RT\ln K^0(T) - RT\sum_{i=1}^{C} \nu_i\left(\ln\frac{p}{p^0} + \ln x_i\right). \tag{10.42}$$

The molar fractions depend only on ξ because $n_i = n_i(0) + \nu_i\xi$, therefore,

$$T\left(\frac{\partial A}{\partial T}\right)_{p,\xi} = A + RT^2\frac{d\ln K^0(T)}{dT}. \tag{10.43}$$

As we know $A = 0$ in chemical equilibrium, thus, using (10.14) and the van 't Hoff equation, we get

$$\left(\frac{\partial H}{\partial \xi}\right)_{\mathrm{eq}} = \Delta_r H^0. \tag{10.44}$$

Relation (10.44) follows from the fact that the enthalpy of mixing vanishes for ideal gases (see Sect. 7.4.3), therefore, the total thermal effect of reaction comes only from the standard enthalpy of reaction.

10.3.2 Effect of Pressure on the Equilibrium Constant

The constant K^0 depends only on temperature. Therefore, to determine the effect of pressure on the position of chemical equilibrium for a mixture of ideal gases, we differentiate $\ln K_x$ with respect to pressure, using (10.36), hence

$$\left(\frac{\partial \ln K_x}{\partial p}\right)_T = \frac{-\Delta n}{p} = \frac{-\Delta V}{RT}. \tag{10.45}$$

We have used here the ideal gas equation of state: $pV = nRT$, hence

$$p\Delta V = RT\Delta n, \tag{10.46}$$

where ΔV is the total change in the volume of the mixture for the reaction in the positive direction. For instance, in the synthesis of ammonia (see (10.29)), $\Delta n = -3 - 1 + 2 = -2$. Thus, the value of K_x increases with pressure, which means that the equilibrium shifts in the direction of higher concentration of NH_3 (see (10.39)).

10.4 Thermochemistry

Chemical reactions are usually irreversible processes, i.e., they proceed in one direction only. The equation of a irreversible chemical reaction has the following general form:

$$0 \rightarrow \sum_{i=1}^{C} \nu_i A_i, \tag{10.47}$$

where ν_i are negative for reactants and positive for products. If the initial and final states are equilibrium states then all state functions and their changes in the reaction, are well defined quantities. *Thermochemistry* studies thermal effects of chemical reactions.

10.4.1 Hess' Law

If the reaction occurs at constant pressure then the heat absorbed or given off by the system is equal to the change in its enthalpy, i.e.,

$$Q = \Delta H. \tag{10.48}$$

Since enthalpy is a state function, the standard enthalpy of reaction $\Delta_r H^0$ (see (10.41)) depends only on the initial state (unmixed reactants) and the final state (separated products) of the system. In other words, it does not depend on the process that brings the system from the initial to final state. Making use of this fact, we can calculate $\Delta_r H^0$ for reactions whose thermal effect is difficult to measure. To achieve this, a given reaction should be expressed as a sum of reactions for which $\Delta_r H^0$ is known.

Corollary 10.4 *The standard enthalpy of a given reaction is equal to the sum of standard enthalpies of reactions into which the given reaction can be decomposed.*

For historical reasons, the above conclusion is called *Hess' law*, although it is only a consequence of the fact that enthalpy is a state function.

In general, $\Delta_r H^0$ is a function of temperature but often its value at the temperature of 298.15 K (25 °C) is given. An example of application of Hess' law is given below. For convenience, we mark the phase in which a given compound exists at the standard pressure $p^0 = 1$ bar and temperature of 25 °C, i.e., the gaseous (g), liquid (l) or solid (s) phase.

Example 10.4 The reaction

$$C_2H_4(g) + H_2(g) \rightarrow C_2H_6(g), \tag{10.49}$$

in which three gases participate, can be decomposed into the following three reactions:

$$\begin{aligned} C_2H_4(g) + 3O_2(g) &\rightarrow 2CO_2(g) + 2H_2O(l), \\ H_2(g) + \frac{1}{2}O_2(g) &\rightarrow H_2O(l), \\ C_2H_6(g) + \frac{7}{2}O_2(g) &\rightarrow 2CO_2(g) + 3H_2O(l), \end{aligned}$$

with $\Delta_r H^0$ equal to, respectively:

$$-1411.3\ \mathrm{kJ\,mol^{-1}}, \quad -285.8\ \mathrm{kJ\,mol^{-1}}, \quad -1559.8\ \mathrm{kJ\,mol^{-1}}.$$

Using the same rules as in the case of algebraic equations, we add the first reaction to the second one and then subtract the third reaction, to obtain reaction (10.49). Then we do the same with the standard enthalpy of reaction, i.e., we subtract the third enthalpy from the sum of the first and second one, which gives $\Delta_r H^0 = -137.3\ \mathrm{kJ\,mol^{-1}}$ for reaction (10.49).

10.4.2 Standard Enthalpy of Formation

If all reactants of a given reaction are elements then the standard enthalpy of reaction is called the *standard enthalpy of formation* and is denoted by $\Delta_f H^0$.

Example 10.5 The reaction of formation of methane from the elements,

$$C(s) + 2H_2(g) \rightarrow CH_4(g), \tag{10.50}$$

can be decomposed into the following three reactions:

$$\begin{aligned} CH_4(g) + 2O_2(g) &\rightarrow CO_2(g) + 2H_2O(l), \\ C(s) + O_2(g) &\rightarrow CO_2(g), \\ 2H_2(g) + O_2(g) &\rightarrow 2H_2O(l), \end{aligned}$$

for which $\Delta_r H^0$ amounts to, respectively:

$$-890.4\ \mathrm{kJ\,mol^{-1}}, \quad -393.5\ \mathrm{kJ\,mol^{-1}}, \quad -571.6\ \mathrm{kJ\,mol^{-1}}.$$

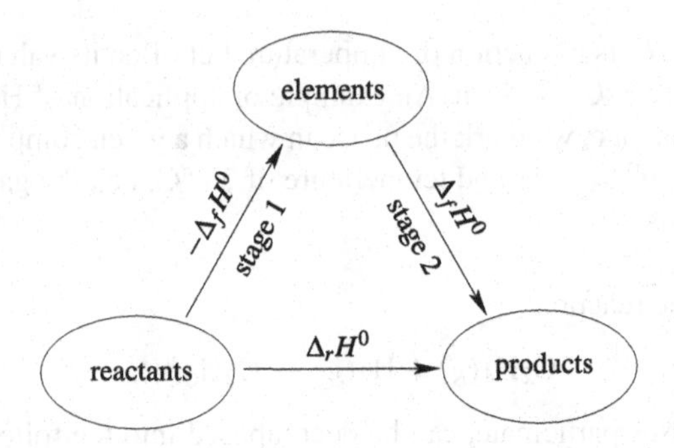

Fig. 10.2 Application of Hess' law to the calculation of the standard enthalpy of reaction $\Delta_r H^0$. At the first stage, the reactants are decomposed into the elements. At the second stage, the products of the original reaction are synthesized from the elements obtained at the first stage. $\Delta_f H^0$ denotes the standard enthalpy of formation

Subtracting the first reaction from the sum of the second and third one, we obtain reaction (10.50), hence, the standard enthalpy of formation amounts to $\Delta_f H^0 = -74.7\,\mathrm{kJ\,mol^{-1}}$.

The standard enthalpy of formation is of great importance because each reaction can be split, at least mentally, into two stages. At the first stage, each reactant is decomposed into the elements. At the second stage, the products of the original reaction are synthesized from the elements obtained at the first stage. Figure 10.2 is a visual presentation of this idea. Note that the first stage is the reverse of the synthesis of a given reactant from the elements, hence, the corresponding enthalpy of reaction for that reactant is equal to $-\Delta_f H^0$. Therefore, $\Delta_r H^0$ for a given reaction can be formally presented as follows:

$$\Delta_r H^0 = \sum_{\text{products}} \nu_i \Delta_f H_i^0 - \sum_{\text{reactants}} |\nu_i| \Delta_f H_i^0, \tag{10.51}$$

where $\Delta_f H_i^0$ denotes the enthalpy of formation of the ith compound participating in the reaction.

Example 10.6 We want to calculate $\Delta_f H^0$ for the combustion of liquid benzene, i.e.,

$$C_6H_6(l) + \frac{15}{2} O_2(g) \rightarrow 6CO_2(g) + 3H_2O(l). \tag{10.52}$$

According to (10.51) we have

$$\Delta_r H^0 = 6\Delta_f H^0_{CO_2} + 3\Delta_f H^0_{H_2O} - \Delta_f H^0_{C_6H_6} - \frac{15}{2}\Delta_f H^0_{O_2}.$$

Substituting the value of $\Delta_f H^0$ for each compound, we obtain

$$\begin{aligned}\Delta_r H^0 &= \left[6(-393.5) + 3(-285.8) - 49.0 - \frac{15}{2} \times 0\right] \mathrm{kJ\,mol^{-1}} \\ &= -3267.4\,\mathrm{kJ\,mol^{-1}}.\end{aligned}$$

Note that $\Delta_f H^0_{O_2} = 0$ because the standard enthalpy of formation refers to the ground states of elements, i.e., to their most stable form at a given temperature and the pressure of 1 bar. Therefore, the standard enthalpy of formation of an element is zero by definition. For example, at the temperature of 25 °C and pressure of 1 bar, the most stable form of oxygen is the molecule O_2, hence $\Delta_f H^0_{O_2} = 0$.

10.4.3 Kirchhoff Equation

The temperature 25 °C, for which the standard enthalpy of reaction is usually given, has been selected arbitrarily. Obviously chemical reactions can occur at very different temperatures. Therefore, a problem arises how to obtain $\Delta_r H^0$ as a function of temperature if we know its value at 25 °C or at another reference temperature. We can do it if we know the molar heat capacity at constant pressure for each compound participating in a given reaction. Differentiating (10.41) with respect to temperature, we get the relation

$$\Delta_r c_p^0 = \left(\frac{\partial \Delta_r H^0}{\partial T} \right)_p = \sum_{i=1}^{C} \nu_i c_{p,i}^0, \tag{10.53}$$

called the *Kirchhoff equation*, where

$$c_{p,i}^0 = \left(\frac{\partial h_i^0}{\partial T} \right)_p \tag{10.54}$$

denotes the molar heat capacity at constant pressure of the ith compound in its standard state. Then we integrate (10.53) from T_1 to T_2, to get

$$\Delta_r H^0(T_2) - \Delta_r H^0(T_1) = \int_{T_1}^{T_2} \Delta_r c_p^0(T) \mathrm{d}T. \tag{10.55}$$

If $\Delta_r c_p^0$ is approximately constant in the range of temperature considered, then

$$\Delta_r H^0(T_2) = \Delta_r H^0(T_1) + \Delta_r c_p^0(T_1)(T_2 - T_1). \tag{10.56}$$

If, however, the dependence of $\Delta_r c_p^0$ on temperature in the interval $T_1 < T < T_2$, cannot be neglected, then the integral needs to be calculated. For instance, in the case of linear dependence on temperature, i.e.,

$$\Delta_r c_p^0(T) = a + bT, \tag{10.57}$$

we obtain

$$\Delta_r H^0(T_2) = \Delta_r H^0(T_1) + \overline{\Delta_r c_p^0}(T_2 - T_1), \tag{10.58}$$

where

$$\overline{\Delta_r c_p^0} = \frac{1}{2}\left[\Delta_r c_p^0(T_1) + \Delta_r c_p^0(T_2)\right] \tag{10.59}$$

denotes the average value.

Example 10.7 At the temperature $T_1 = 25$ °C, the standard enthalpy of formation of water vapour amounts to $\Delta_f H^0(T_1) = -241.83\ \mathrm{kJ\,mol^{-1}}$ and we want to calculate $\Delta_f H^0(T_2)$ at $T_2 = 100$ °C. To do it, we apply (10.58). In the range of temperature considered, the average value of c_p^0 for individual reactants amounts to (in units of $\mathrm{J\,K^{-1}\,mol^{-1}}$): 28.95 for $H_2(g)$, 29.46 for $O_2(g)$ and 33.60 for $H_2O(g)$. The reaction proceeds according to the equation

$$H_2(g) + \frac{1}{2}O_2(g) \rightarrow H_2O(g), \tag{10.60}$$

hence (see (10.53)),

$$\overline{\Delta_r c_p^0} = \overline{c_{p,H_2O}^0} - \overline{c_{p,H_2}^0} - \frac{1}{2}\overline{c_{p,O_2}^0} = -10.08\ \mathrm{J\,K^{-1}\,mol^{-1}},$$

and after substitution into (10.58) we get

$$\Delta_f H^0(T_2) = -242.59\ \mathrm{kJ\,mol^{-1}}.$$

10.5 Phase Rule for Chemical Systems

Several reactions can occur simultaneously in a system, and each of them proceeds according to a certain equation of reaction such as Eq. (10.4). To each reaction a different set of stoichiometric coefficients corresponds and some of the coefficients can be equal to zero. For R reactions, there are $r = 1, \ldots, R$ equations of the form

$$0 \rightleftharpoons \sum_{i=1}^{C} \nu_i^{(r)} A_i. \tag{10.61}$$

An infinitesimal change in the mole numbers n_i associated with the rth reaction has the same form as in (10.6), i.e.,

$$\mathrm{d}n_i^{(r)} = \nu_i^{(r)}\mathrm{d}\xi^{(r)}, \tag{10.62}$$

where $\xi^{(r)}$ denotes the extent of that reaction. Summing up over all reactions, we obtain

$$\mathrm{d}n_i = \sum_{r=1}^{R} \nu_i^{(r)}\mathrm{d}\xi^{(r)}, \tag{10.63}$$

hence

$$\mathrm{d}G = \sum_{i=1}^{C} \mu_i \mathrm{d}n_i = \sum_{r=1}^{R}\left(\sum_{i=1}^{C} \nu_i^{(r)}\mu_i\right)\mathrm{d}\xi^{(r)}. \tag{10.64}$$

The state of chemical equilibrium corresponds to the minimum of G over the parameters $\xi^{(1)}, \ldots, \xi^{(r)}$ at constant T and p. At the minimum, all coefficients at the differentials $d\xi^{(r)}$ in Eq. (10.64) must vanish, which leads to R equations for the chemical potentials:

$$\sum_{i=1}^{C} \nu_i^{(r)} \mu_i = 0. \tag{10.65}$$

Now we consider the general case of a C-component system in which P phases coexist and R chemical reactions occur simultaneously. The system is in chemical equilibrium, thus, conditions (10.65) must be satisfied. To find the number of degrees of freedom f for the system (see Sect. 7.7), we have to subtract from the total number of intensive variables the number of independent equations which these variables must satisfy in thermodynamic equilibrium. The intensive variables are: the chemical potentials of all components in all phases and the temperature and pressure, i.e., $CP + 2$ variables altogether. The independent equations that must be satisfied are: P Gibbs–Duhem equations (see (7.143)), $P - 1$ equalities of the chemical potentials for each component (see (7.144)), which gives $C(P - 1)$ equations, and R equations (10.65) for the chemical equilibrium. Therefore, the total number of independent equations is equal to $P + CP - C + R$, hence, the number of degrees of freedom amounts to

$$f = C + 2 - P - R. \tag{10.66}$$

This is the *phase rule for a chemical system.* If no chemical reaction occurs in the system ($R = 0$) we simply recover (7.145).

Example 10.8 In a two-component system in which two phases coexist and one chemical reaction occurs, $f = 2 + 2 - 2 - 1 = 1$. The temperature or pressure can be chosen as an independent variable.

Note that if a certain reaction proceeds to the completion, i.e., until one of the components is exhausted, then both R and C decrease by one, whereas the number of degrees of freedom does not change. An important conclusion follows from this observation. To study equilibrium states of a given chemical system, there is no need to consider all potential reactions that can occur if some of them proceed to the completion and some components practically disappear from the system.

Hitherto we have not discriminated between a *compound* and *component.* However, in the case of a system with chemical reactions for which conditions of chemical equilibrium (10.65) are satisfied, such a discrimination can be necessary (see Chap. 11). A *compound* is a particle in the chemical sense, e.g., an atom, molecule, ion, or a collection of such particles, whereas a *component* of a mixture is a compound whose amount can be varied independently of amounts of other compounds. For instance, when hydrogen and nitrogen are mixed, we obtain three compounds: H_2, N_2 and NH_3, because of the reaction

$$3H_2 + N_2 \rightleftharpoons 2NH_3.$$

However, only the amounts of two of them can be varied independently, therefore, we have two components. For instance, if we know the amounts of hydrogen and nitrogen, then the amount of ammonia follows from the condition of chemical equilibrium. Then C in Eq. (10.66) denotes the number of all compounds, whereas the number of components is equal to $C - R$. After this modification of meaning of a component, the phase rule adopts form (7.145) whether chemical reactions occur in a given system or not.

10.6 Exercises

10.1 The condition of chemical equilibrium, $(\partial G/\partial \xi)_{T,p} = 0$, corresponds to a minimum of the Gibbs free energy, thus, the condition of stability, $(\partial^2 G/\partial \xi^2)_{T,p} > 0$, must be satisfied for $\xi = \xi_{\mathrm{eq}}$. Prove that in the case of ideal gases the stability condition is always satisfied.

10.2 Derive a relation between the change in the enthalpy, ΔH, and the change in the internal energy, ΔU, for a reaction at constant temperature, if all reactants and products are ideal gases.

10.3 Find the change in the equilibrium constant K_x for the following reactions:

$$\mathrm{CO}(g) + \mathrm{Cl_2}(g) \rightleftharpoons \mathrm{COCl}(g) + \mathrm{Cl}(g),$$
$$2\mathrm{SO_2}(g) + \mathrm{O_2}(g) \rightleftharpoons 2\mathrm{SO_3}(g),$$

assuming a twofold increase in pressure at constant temperature.

10.4 The reaction

$$3A(g) + B(g) \rightleftharpoons C(g) + 2D(g)$$

takes place at constant pressure and temperature in a vessel occupied initially by 2 mol of A, $\frac{1}{3}$ mol of B, 1 mol of C and $\frac{1}{2}$ mol of D. Determine the minimum and maximum value of the extent of reaction. Assuming that all reactants and products are ideal gases and the state of chemical equilibrium corresponds to $\xi = \xi_{\mathrm{eq}} = \frac{1}{4}$ mol, calculate the reaction constant K_x.

10.5 The reaction

$$A(g) + B(g) \rightleftharpoons 2C(g),$$

where A, B and C are ideal gases, occurs at constant temperature and at the pressure of 1 bar. Calculate the molar fractions of all gases in the state of chemical equilibrium if the standard equilibrium constant of the reaction $K^0 = 1$. Consider two sets of the initial values of molar fractions: (a) $x_A(0) = 0.5$, $x_B(0) = 0.5$, $x_C(0) = 0$ and (b) $x_A(0) = 0.25$, $x_B(0) = 0.75$, $x_C(0) = 0$.

10.6 Phosphorus pentachloride undergoes a decomposition into phosphorus trichloride and chlorine above the temperature of 200 °C, i.e.,

$$\mathrm{PCl_5}(g) \rightleftharpoons \mathrm{PCl_3}(g) + \mathrm{Cl_2}(g).$$

Before the reaction started the sample of PCl_5 had a mass of 1.9 g. The reaction proceeds at a temperature of 320 °C and pressure of 0.314 bar. When the system reaches the equilibrium state it occupies the volume $V_{eq} = 2.4$ L. Calculate the standard equilibrium constant K^0 and the percentage of decomposed PCl_5 molecules, assuming that all compounds participating in the reaction are ideal gases.

10.7 The decomposition reaction

$$A(g) \rightleftharpoons B(g) + C(g)$$

is endothermic. At the temperature T_0 and pressure p_0, the system has reached an equilibrium state in which the molar fraction of the component A is equal to $x_A(T_0, p_0)$. What is the direction of the shift of chemical equilibrium if: (a) the temperature increases from T_0 to T_1 at constant pressure $p = p_0$, (b) the pressure increases from p_0 to p_1 at constant temperature $T = T_0$?

10.8 For the given reaction, we know the standard equilibrium constant K^0 and the standard enthalpy of reaction, $\Delta_r H^0$, at the temperature T_0. What is the value of K^0 at the temperature $T_1 > T_0$ if T_1 does not differ too much from T_0?

10.9 The standard enthalpy of reaction at the temperature $T_0 = 298$ K is equal to $\Delta_r H^0 = 32\ \mathrm{kJ\,mol^{-1}}$. Estimate the value of the standard equilibrium constant K^0 at the temperature $T_1 = 310$ K, with respect to $K^0(T_0)$.

10.10 The decomposition reaction

$$2A(g) \rightleftharpoons B(g) + C(g)$$

proceeds at constant pressure and temperature. At the temperature $T_1 = 300$ K, 40 % of molecules of the substance A are decomposed, whereas at the temperature $T_2 = 315$ K the percentage of decomposed molecules increases to 42 %. What is the value of the standard enthalpy of reaction $\Delta_r H^0$? Assume that all compounds are ideal gases, and $\Delta_r H^0$ does not depend on temperature.

10.11 The decomposition reaction of the substance A is characterized by the equation

$$A(g) \rightleftharpoons \nu_B B(g) + \nu_C C(g).$$

The degree of dissociation for the decomposition reaction is defined as follows. We denote by n_A the mole number of the substance A present in the system in chemical equilibrium, and by n_d the mole number of that part of A which is decomposed into B and C. Then the degree of dissociation $\alpha = n_d/n_i$, where $n_i = n_A + n_d$ is the initial mole number of A present in the system before the reaction started. Find the relation between α and the equilibrium constant K_x if A, B and C are ideal gases.

10.12 Calculate the standard enthalpy of the reaction

$$C_2H_4(g) + H_2(g) \rightarrow C_2H_6(g),$$

using the reactions:

$$C_2H_4(g) + 3O_2 \rightarrow 2CO_2(g) + 2H_2O(l),$$

$$H_2(g) + \frac{1}{2}O_2(g) \rightarrow H_2O(l),$$

$$C_2H_6(g) + \frac{7}{2}O_2(g) \rightarrow 2CO_2(g) + 3H_2O(l),$$

for which the standard enthalpy of reaction amounts to -1411.3 kJ mol^{-1}, -285.8 kJ mol^{-1} and -1559.8 kJ mol^{-1}, respectively.

10.13 Calculate the standard enthalpy of formation of 1 mol of N_2O_5, using the data for the following reactions:

$$2NO(g) + O_2(g) \rightarrow 2NO_2(g),$$

$$4NO_2(g) + O_2(g) \rightarrow 2N_2O_5(g),$$

$$N_2(g) + O_2(g) \rightarrow 2NO(g).$$

The standard enthalpy for these reactions amounts to -114.1 kJ mol^{-1}, -110.2 kJ mol^{-1} and 180.5 kJ mol^{-1}, respectively.

10.14 The reaction of propane combustion has the following form:

$$C_3H_8(g) + 5O_2(g) \rightarrow 3CO_2(g) + 4H_2O(l).$$

The standard enthalpy of formation amounts to, respectively: $\Delta_f H^0_{C_3H_8} = -103.7$ kJ mol^{-1}, $\Delta_f H^0_{CO_2} = -393.5$ kJ mol^{-1}, $\Delta_f H^0_{H_2O} = -285.8$ kJ mol^{-1}. Assuming that 4 mol of water are formed in the reaction and the gases participating in it are ideal, calculate the standard enthalpy of reaction and the change in the volume and internal energy of the system. The temperature of the system before and after the reaction amounts to 25 °C.

10.15 Calculate the change in the internal energy and enthalpy for the synthesis of ammonia:

$$3H_2(g) + N_2(g) \rightarrow 2NH_3(g).$$

Assume that all gases participating in this reaction are ideal gases in their standard states. Calculate the work done in the reaction. The temperature at the beginning and at the end of the reaction amounts to 298 K, and the enthalpy of ammonia formation $\Delta_f H^0_{NH_3} = -46.1$ kJ mol^{-1}.

10.16 Calculate the heat of formation of methane from the elements:

$$C(s) + 2H_2(g) \rightarrow CH_4(g),$$

at the temperature $T = 298$ K. Consider two cases: (1) $p = const = 1$ bar, (2) $V = const$. Make use of the following reactions of formation:

$$H_2(g) + \frac{1}{2}O_2(g) \to H_2O(l),$$
$$C(s) + O_2(g) \to CO_2(g),$$

for which $\Delta_f H^0_{H_2O} = -285.8\ \mathrm{kJ\,mol^{-1}}$, $\Delta_f H^0_{CO_2} = -393.5\ \mathrm{kJ\,mol^{-1}}$, and the reaction of methane combustion

$$CH_4(g) + 2O_2(g) \to CO_2(g) + 2H_2O(l),$$

with $\Delta_r H^0 = -890.3\ \mathrm{kJ\,mol^{-1}}$. Assume that hydrogen and methane are ideal gases.

10.17 The standard enthalpy of the reaction

$$2A(g) + B(g) \to 3C(g),$$

amounts to $\Delta_r H^0 = -20\ \mathrm{kJ\,mol^{-1}}$ at the temperature of 298 K . The substances A and B are composed of linear molecules, whereas the molecules of the substance C have 6 degrees of freedom per molecule. Assuming that A, B and C are ideal gases, calculate $\Delta_r H^0$ at a temperature of 340 K.

10.18 The reaction

$$A(l) \rightleftharpoons 2B(l) + D(g)$$

occurs in a three-component system in thermodynamic equilibrium. Find the number of degrees of freedom of the system. Consider two cases: (1) A and B mix in any proportion, (2) A and B form two liquid phases.

10.19 A three-component system, in which the following reaction takes place:

$$A(l) \rightleftharpoons B(l) + C(g),$$

is in a condition of liquid–vapour equilibrium at the temperature T, and the liquids A and B form an ideal solution. Assuming that C is an ideal gas, and the presence of C in the solution can be neglected, express the pressure above the solution as a function of the partial pressure p_C. Assume also that the standard equilibrium constant $K^0(T)$, as well as the pressures $p_A^*(T)$ and $p_B^*(T)$ for the liquid–vapour coexistence in pure substances, are known.

10.20 Assuming that hydrogen and water vapour are ideal gases, write the condition of chemical equilibrium for the reaction

$$CuO(s) + H_2(g) \rightleftharpoons Cu(s) + H_2O(g).$$

Chapter 11
Electrochemical Systems

11.1 Electrolyte Solutions

11.1.1 Dissociation

In this chapter, we turn our attention to substances whose molecules *dissociate* in a solution into free ions. Such substances are called *electrolytes*. An ion with an elementary charge of the electron or its multiple is called the *anion*, and an ion with an elementary charge of the proton or its multiple is called the *cation*. For example, in an aqueous solution of common salt the following reaction occurs:

$$\mathrm{NaCl} \rightleftharpoons \mathrm{Na}^{+} + \mathrm{Cl}^{-}. \tag{11.1}$$

In this reaction, water is a solvent and common salt is a solute. Therefore, four kinds of molecules (compounds) are present in the solution: NaCl, the cation Na^+, the anion Cl^- and H_2O.[1] There are only two components, however, water and common salt, because of the conditions of chemical equilibrium and *electric neutrality*. In the case of reaction (11.1), for instance, the number of the cations Na^+ must be equal to the number of the anions Cl^-.

In what follows, we consider liquid solutions of electrolytes. An electrolyte dissolved in a solvent, such as water or methanol, dissociates according to the equation:

$$\mathrm{X}_{\nu_+}\mathrm{Y}_{\nu_-} \rightleftharpoons \nu_+\mathrm{X}^{z_+ +} + \nu_-\mathrm{Y}^{|z_-|-}, \tag{11.2}$$

where z_+ and z_- denote the charges of the cation and anion, respectively, in units of the elementary charge e. Each molecule of the electrolyte introduces ν_+ of the cations and ν_- of the anions into the solution. For instance, in the case of reaction (11.1), we have $\nu_+ = \nu_- = 1$, $z_+ = 1$ and $z_- = -1$. The equality

$$\nu_+ z_+ + \nu_- z_- = 0 \tag{11.3}$$

---

[1] For the time being, we ignore the fact that a vary small part of water molecules dissociate into the ions H^+ and OH^-.

R. Hołyst, A. Poniewierski, *Thermodynamics for Chemists, Physicists and Engineers*,
DOI 10.1007/978-94-007-2999-5_11,

expresses the condition of electric neutrality for the molecule $X_{\nu_+}Y_{\nu_-}$.

The Gibbs free energy of a two-component mixture has the following form (see (7.17)):

$$G = n_A\mu_A + n_B\mu_B, \tag{11.4}$$

where n_A, n_B denote the mole numbers and μ_A, μ_B denote the chemical potentials of the solvent A and electrolyte B, respectively. On the other hand, taking into account all compounds present in the solution, we obtain

$$G = n_A\mu_A + n_u\mu_u + n_+\mu_+ + n_-\mu_-, \tag{11.5}$$

where the indices u, $+$ and $-$ refer to the undissociated molecules of the electrolyte, to the cations and to the anions, respectively. For reaction (11.2), the numbers n_B, n_u, n_+ and n_- satisfy the following relations:

$$n_u = (1-\alpha)n_B, \tag{11.6}$$

$$n_+ = \alpha\nu_+ n_B, \tag{11.7}$$

$$n_- = \alpha\nu_- n_B, \tag{11.8}$$

where the parameter $0 \le \alpha \le 1$ is called the *degree of dissociation* of the electrolyte. It means that in the total amount of the electrolyte n_B, the amount αn_B is dissociated and the amount $(1-\alpha)n_B$ remains undissociated.

From the condition of chemical equilibrium for reaction (11.2),

$$\mu_u = \nu_+\mu_+ + \nu_-\mu_-, \tag{11.9}$$

and from relations (11.6)–(11.8), it follows that

$$n_u\mu_u + n_+\mu_+ + n_-\mu_- = n_B\mu_u. \tag{11.10}$$

Substituting (11.10) into (11.5) and comparing with (11.4), we obtain

$$\mu_B = \mu_u, \tag{11.11}$$

hence

$$\mu_B = \nu_+\mu_+ + \nu_-\mu_-. \tag{11.12}$$

Thus, we have related the chemical potential of the electrolyte (component B) to the chemical potentials of ions present in the solution due to the dissociation of the electrolyte.

11.1.2 Chemical Potential of the Electrolyte

We concentrate here on the chemical potential of the electrolyte, μ_B, since the chemical potential of the solvent follows from the Gibbs–Duhem equation (see (7.23)), i.e.,

$$(1-x_B)\left(\frac{\partial \mu_A}{\partial x_B}\right)_{T,p} = -x_B\left(\frac{\partial \mu_B}{\partial x_B}\right)_{T,p}, \tag{11.13}$$

where x_B is the molar fraction of the electrolyte. In what follows, we express the composition of the solution in terms of the molality of the electrolyte, m. From the definition of the molality (see Definition 7.4), we have

$$m = \frac{n_B}{M_A n_A} = \frac{x_B}{M_A(1 - x_B)}, \tag{11.14}$$

where M_A is the molar mass of the solvent. Multiplying both sides of (11.13) by dx_B/dm and making use of (11.14), we obtain

$$\left(\frac{\partial \mu_A}{\partial m}\right)_{T,p} = -M_A m \left(\frac{\partial \mu_B}{\partial m}\right)_{T,p}. \tag{11.15}$$

Note that the molality of the cations, m_+, anions, m_-, and the undissociated molecules, m_u, are related to m in the same way as in the case of the mole numbers (see (11.6)–(11.8)), i.e.,

$$m_u = (1 - \alpha)m, \tag{11.16}$$

$$m_+ = \alpha \nu_+ m, \tag{11.17}$$

$$m_- = \alpha \nu_- m, \tag{11.18}$$

The chemical potentials of the cations, anions and undissociated molecules are expressed in the following form:

$$\mu_+ = \mu_+^0 + RT \ln \frac{m_+}{m^0} + RT \ln \gamma_+, \tag{11.19}$$

$$\mu_- = \mu_-^0 + RT \ln \frac{m_-}{m^0} + RT \ln \gamma_-, \tag{11.20}$$

$$\mu_u = \mu_u^0 + RT \ln \frac{m_u}{m^0} + RT \ln \gamma_u, \tag{11.21}$$

where γ_+, γ_- and γ_u denote the corresponding activity coefficients (see (7.131) and (7.132)), which tend to unity in the limit of infinite dilution. Due to the presence of electrostatic interaction, the positive and negative ions are always close to each other, and any macroscopic fragment of the solution is electrically neutral. Therefore, it is not possible to determine experimentally the properties of a solution of anions or cations and the activity coefficients γ_+ and γ_-, in particular. Nevertheless, it is possible to determine a combination of these coefficients, using relation (11.12). Substituting (11.19) and (11.20) into (11.12), we obtain

$$\mu_B = \mu_B^0 + RT \ln\left[\left(\frac{m_+}{m^0}\right)^{\nu_+} \left(\frac{m_-}{m^0}\right)^{\nu_-}\right] + RT \ln\left(\gamma_+^{\nu_+} \gamma_-^{\nu_-}\right), \tag{11.22}$$

where

$$\mu_B^0 = \nu_+ \mu_+^0 + \nu_- \mu_-^0 \tag{11.23}$$

is the standard chemical potential of the electrolyte. The first two terms in (11.22) correspond to the ideal dilute solution, i.e., to the limit $m \to 0$. The last term characterizes deviation from the ideal behaviour. We will see that it is peculiar to electrostatic interaction that relatively large deviation from the ideal behaviour occur even in dilute solutions.

It is convenient to introduce the *average activity coefficient of ions*, $\gamma_\pm$, defined as a geometrical average of $\gamma_+^{\nu_+}$ and $\gamma_-^{\nu_-}$, i.e.,

$$\gamma_\pm^\nu = \gamma_+^{\nu_+}\gamma_-^{\nu_-}, \tag{11.24}$$

where $\nu = \nu_+ + \nu_-$. In other words, we arbitrarily attribute to the cations and anions the same contribution to the deviation from the ideal dilute solution, assuming that

$$\mu_+ = \mu_+^0 + RT\ln\frac{m_+}{m^0} + RT\ln\gamma_\pm, \tag{11.25}$$

$$\mu_- = \mu_-^0 + RT\ln\frac{m_-}{m^0} + RT\ln\gamma_\pm. \tag{11.26}$$

Substituting (11.25) and (11.26) into (11.12), we recover relation (11.22). We can simplify (11.22), using relations (11.17) and (11.18), hence

$$\left(\frac{m_+}{m^0}\right)^{\nu_+}\left(\frac{m_-}{m^0}\right)^{\nu_-} = \left(\frac{\alpha m}{m^0}\right)^{\nu}\nu_\pm^\nu, \tag{11.27}$$

where we have introduced the average $\nu_\pm$:

$$\nu_\pm^\nu = \nu_+^{\nu_+}\nu_-^{\nu_-}. \tag{11.28}$$

Finally, substituting (11.24) and (11.28) into (11.22), we obtain

$$\mu_B = \mu_B^0 + RT\nu\ln\left(\frac{\gamma\nu_\pm m}{m^0}\right), \tag{11.29}$$

where $\gamma = \alpha\gamma_\pm$. When $m \to 0$ the solution becomes an ideal dilute solution in which all molecules of the electrolyte are dissociated. Therefore, the degree of dissociation $\alpha = 1$, which means that $\gamma \to 1$ when $m \to 0$. Thus, the coefficient γ measures the deviation of a real solution from a completely dissociated ideal dilute solution.

11.1.3 Debye–Hückel Limiting Law

We have already mentioned that solutions of electrolytes exhibit considerable deviations from the ideal dilute solution even at strong dilution. The electrostatic interaction between ions is responsible for such behaviour. As we know from elementary electrostatics, the Coulomb potential of a single charge decays inversely proportionally to the distance r from the charge. By comparison, the potential energy of interaction of two electrically neutral atoms decays much faster, i.e., as r^{-6}. Therefore, it can be assumed that for a sufficiently dilute solution, the deviation from the ideal behaviour is only due to the electrostatic interaction between ions. This assumption allows one to calculate, in an approximate way, the contribution of the electrostatic interaction to the chemical potential of ions. The matter is a little complicated because each ion is surrounded by a cloud of ions with the opposite charge, called the *counter ions*, which changes the effective electrostatic potential of a single ion. This effect of *screening* of electric charges leads to the *screened Coulomb potential* which decays with the distance as $r^{-1}e^{-r/r_D}$, where r_D is called the *Debye screening length*. It means that the screened Coulomb potential is much smaller than the unscreened one for $r \gg r_D$. On the basis of the above model of the electrostatic interaction between ions in a solution, and with some additional assumptions, it is possible to derive an expression for the activity coefficient of ions, $\gamma_\pm$. The derivation is beyond the scope of this book, however. Therefore, we present only the final result. According to the *Debye–Hückel limiting law* for an electrolyte with two kinds of ions (see (11.2)),

$$\ln \gamma_\pm = -|z_+ z_-| a' \sqrt{I}, \tag{11.30}$$

where the quantity

$$I = \frac{z_+^2 m_+ + z_-^2 m_-}{2m^0} \tag{11.31}$$

is called the *ionic strength*, and the dimensionless parameter a' depends only on the properties of the pure solvent. Relation (11.30) holds when the ionic strength of the solution tends to zero. From (11.31) we infer, that the dependence of I on the ion charges is stronger than on the molality of the electrolyte. In application, the Debye–Hückel limiting law is often expressed in terms of the decimal logarithm, i.e.,

$$\log \gamma_\pm = -|z_+ z_-| a \sqrt{I}, \tag{11.32}$$

where $a = a'/\ln 10$. The coefficient a is defined by the following formula:

$$a = \frac{F^3}{4\pi N_A \ln 10} \sqrt{\frac{\rho m^0}{2(\varepsilon RT)^3}}, \tag{11.33}$$

where ρ is the density of the solvent, ε denotes its electric permeability, N_A is the Avogadro constant, and

$$F = eN_A = 9{,}6485 \times 10^4 \text{ C mol}^{-1} \tag{11.34}$$

is called the *Faraday constant*. For aqueous solutions, $a = 0.509$ at the temperature of 25 °C.

Substituting (11.17) and (11.18) into (11.30), we obtain

$$\ln \gamma_\pm = -|z_+ z_-| a' I(m^0) \sqrt{\frac{m}{m^0}}, \tag{11.35}$$

where $I(m^0) = (z_+^2 \nu_+ + z_-^2 \nu_-)/2$ and we have assumed $\alpha = 1$, since the limit of infinite dilution is considered. From (11.35), it follows that $\gamma_\pm \to 1$ when $m \to 0$, and substituting $\gamma_\pm = 1 - |\Delta\gamma_\pm|$, we get

$$|\Delta\gamma_\pm| \approx |z_+ z_-| a' I(m^0) \sqrt{\frac{m}{m^0}}. \tag{11.36}$$

We recall that in the case of a non-electrolyte solution, it is assumed that the activity coefficient of the solute can be expanded in the Taylor series around $m_B = 0$ (or $x_B = 0$). Therefore, $\gamma_B = 1 + \Delta\gamma_B$ with the correction term proportional to m_B. As we can see, the presence of ions in a solution causes that the activity coefficient tends to unity more slowly, when $m \to 0$, than in the case of non-electrolyte solutions. Such behaviour means simply that the approximation of the ideal dilute solution has a narrower range of applicability in the case of electrolyte solutions than in the case of non-electrolyte solutions.

11.2 Aqueous Solutions of Acids and Bases

11.2.1 Brønsted–Lowry Theory of Acids and Bases

According to the *Brønsted–Lowry* theory, molecules or ions capable of donating a proton (ion H^+) are called *acid*, and those capable of accepting a proton are called *base*. In other words, acid is a proton *donor* and base is a proton *acceptor*. For example, H_2O, H_3O^+, H_2SO_4, HCl, NH_3 are acids, and OH^-, H_2O, HSO_4^-, SO_4^{2-}, Cl^- are bases. A concept of a *conjugate acid–base pair* is introduced. The acid BH^+ formed when the base B accepts a proton is called the conjugate acid of B, and B is called the conjugate base of BH^+. The conjugate acid of the given base has always a charge larger by one unit of the positive charge e than the charge of the base. The absolute charges of the species considered are not relevant, however. A compound behaves as an acid if it reacts with a base which forms a covalent bond with a proton. A reaction in which a proton transfer occurs[2] can be expressed as follows:

$$\mathrm{HA} + \mathrm{B} \rightleftharpoons \mathrm{A}^- + \mathrm{BH}^+, \tag{11.37}$$

where HA, A^- and BH^+, B are the conjugate acid–base pairs. For instance, in the reaction proceeding from the left to the right, the acid HA loses a proton, to form the conjugate base A^-, and the base B accepts a proton, to form the conjugate acid BH^+.

Water is an example of an *amphoteric* compound, which means that it behaves both as an acid and base. For instance, the reaction of proton transfer can involve two water molecules, i.e.,

$$2\mathrm{H_2O}(l) \rightleftharpoons \mathrm{H_3O^+}(aq) + \mathrm{OH^-}(aq). \tag{11.38}$$

One of the water molecules behaves as a base because it accepts H^+, to form the *hydronium ion* H_3O^+. The other water molecule behaves as an acid because it loses H^+, to form the *hydroxide ion* OH^-. The reaction (11.38) is called *autodissociation of water*. Another example of amphoteric nature of water is the reaction with acetic acid:

$$\mathrm{CH_3OOH} + \mathrm{H_2O} \rightleftharpoons \mathrm{CH_3COO^-} + \mathrm{H_3O^+}, \tag{11.39}$$

[2]The reaction of proton transfer is also called *protolysis*, however, this term is no longer recommended because of its misleading similarity to *hydrolysis* or *photolysis*.

in which water behaves as a base, and the reaction with ammonia:

$$NH_3 + H_2O \rightleftharpoons NH_4^+ + OH^-, \tag{11.40}$$

in which water behaves as an acid. Reactions of proton transfer in aqueous solution of the acid HA or base B have the following general form:

$$HA(aq) + H_2O(l) \rightleftharpoons H_3O^+(aq) + A^-(aq), \tag{11.41}$$

$$B(aq) + H_2O(l) \rightleftharpoons BH^+(aq) + OH^-(aq). \tag{11.42}$$

11.2.2 pH of a Solution

The parameter *pH* is used to determine the acidity or basicity of a solution. It is defined as

$$pH = -\log a_{H_3O^+}, \tag{11.43}$$

where $a_{H_3O^+}$ denotes the activity of hydronium ions in the solution, and log stands for the decimal logarithm. Often $a_{H_3O^+}$ is referred to as the activity of hydrogen ions and denoted by a_{H^+}. In the case of dilute solutions, the activity can be replaced by the molality m or molar concentration c, provided that the corresponding activity coefficient can be approximated by 1. It is convenient to introduce a dimensionless molar concentration defined as the ratio c/c^0, where $c^0 = 1\ \mathrm{mol\,L^{-1}}$ is the standard molar concentration. We use the symbol $[A]$ for the dimensionless molar concentration of the compound A, hence, its activity amounts to $a_A = \gamma_A[A]$. For strong dilution, the activity coefficient $\gamma_A \approx 1$ and $a_A \approx [A]$. In this way we recover the school definition of pH, i.e.,

$$pH = -\log[H_3O^+], \tag{11.44}$$

which overlaps with (11.43) in the range of small molar concentration.

The molar concentration of hydronium ions can be related to the standard equilibrium constant of autodissociation of water (11.38):

$$K^0 = \frac{(a_{H_3O^+})(a_{OH^-})}{(a_{H_2O})^2}. \tag{11.45}$$

Since only a very small percentage of water molecules dissociate into ions, the activity of undissociated water is practically equal to the activity of pure water, $a_{H_2O} \approx 1$, hence

$$K^0 \approx K_w = (a_{H_3O^+})(a_{OH^-}), \tag{11.46}$$

where K_w is called the *ion product of water*; at the temperature of 25 °C,

$$K_w = 1{,}008 \times 10^{-14}. \tag{11.47}$$

Replacing the activity with molar concentration in (11.46), we get

$$K_w = [H_3O^+][OH^-]. \tag{11.48}$$

In pure water, $[H_3O^+] = [OH^-]$. From (11.47) and (11.48), we obtain

$$[H_3O^+] = [OH^-] \approx 10^{-7}, \tag{11.49}$$

which means that at 25 °C, the molar concentration of each ion in pure water amounts to 10^{-7} mol L^{-1}, hence

$$\text{pH} = 7. \tag{11.50}$$

In the case of dilute aqueous solutions of acids or bases, the molar ion concentration of H_3O^+ changes, whereas the ion product of water does not. The reason is that K_w is approximately equal to the equilibrium constant K^0 for the autodissociation of water which does not depend on other reactions in a given solution. According to Eqs. (11.41) and (11.42), the number of H_3O^+ ions increases in aqueous solutions of acids, whereas in aqueous solutions of bases, the number of ions OH^- increases. Higher molar concentration of $[H_3O^+]$ in aqueous solutions of acids than in pure water means that pH < 7, whereas for aqueous solutions of bases pH > 7.

11.2.3 Dissociation Constant

The dissociation reaction of the acid HA can be expressed as follows:

$$\text{HA} \rightleftharpoons \text{H}^+ + \text{A}^-. \tag{11.51}$$

The equilibrium constant of this reaction is a measure of the acid strength. In aqueous solution, the dissociation of the acid proceeds according to Eq. (11.41). The H^+ ions combine with water molecules to form hydronium ions. To write the chemical equilibrium condition (see (10.11)) for reaction (11.41), we express chemical potentials in terms of activities, i.e.,

$$\mu_i = \mu_i^0 + RT \ln a_i, \tag{11.52}$$

where i numbers the compounds. Then, using the definition of the standard equilibrium constant K^0 (see (10.33)), we obtain the equilibrium condition in the following form:

$$K^0 = \frac{(a_{A^-})(a_{H_3O^+})}{a_{HA} a_{H_2O}}. \tag{11.53}$$

In what follows we assume that the solution is sufficiently dilute, to replace the activities with the molar concentrations, hence

$$K^0 = \frac{[A^-][H_3O^+]}{[HA][H_2O]}. \tag{11.54}$$

Moreover, it can be assumed that in dilute solution the molar concentration of water is practically constant and amounts to about 55 mol L^{-1}. Therefore, the product $K^0[H_2O]$ is also constant, and the equilibrium condition for acid dissociation (11.51) adopts the following form:

$$K_a = \frac{[A^-][H_3O^+]}{[HA]}, \tag{11.55}$$

where the equilibrium constant K_a is called the *acid dissociation constant*. Large values of K_a correspond to strong acids, i.e., high concentration of ions in proportion to the concentration of undissociated molecules. Strong acids have $K_a > 1$, and weak acids have $K_a < 1$. For instance, hydrochloric acid (HCl) has $K_a \approx 1.6 \times 10^6$ (strong acid), and for acetic acid (CH_3COOH), $K_a \approx 1.7 \times 10^{-5}$ (weak acid). Note that the range of K_a amounts to many orders of magnitude, therefore, it is more convenient to use logarithmic scale, i.e., to define the parameter

$$\mathrm{p}K_a = -\log K_a. \tag{11.56}$$

The lower value of $\mathrm{p}K_a$, the stronger the acid. The molar concentration of hydronium ions in pure water is very small, therefore, it can be assumed that almost all hydronium ions in the solution comes from the acid dissociation. Hence, the molar concentration of hydronium ions in the solution is practically equal to the molar concentration of A^- ions, i.e.,

$$[H_3O^+] = [A^-]. \tag{11.57}$$

If we substitute (11.57) into (11.55) then take decimal logarithm of both sides and make use of the definition of pH (cf. (11.44)), we obtain

$$\mathrm{pH} = -\frac{1}{2}\log\big(K_a[\mathrm{HA}]\big) = \frac{1}{2}\mathrm{p}K_a - \frac{1}{2}\log[\mathrm{HA}]. \tag{11.58}$$

In the case of base dissociation in aqueous solution (see (11.42)) we use a similar reasoning as for acid dissociation. The condition of chemical equilibrium has the following form:

$$K_b = \frac{[\mathrm{BH}^+][\mathrm{OH}^-]}{[\mathrm{B}]}, \tag{11.59}$$

where the equilibrium constant K_b is called the *base dissociation constant*. Since BH^+ and B form a conjugate acid–base pair, thus, the constant K_b is related to the constant K_a for the acid BH^+. Substituting $[OH^-]$ obtained from the ion product of water (see (11.48)) into (11.59), we get

$$K_b = \frac{[\mathrm{BH}^+]K_w}{[\mathrm{B}][\mathrm{H_3O}^+]} = \frac{K_w}{K_a}. \tag{11.60}$$

From (11.60), it follows that a strong acid (large K_a) corresponds to a weak base (small K_b) and vice versa.

11.3 Electrochemical Cells

11.3.1 Daniell Cell

In electrochemical cells chemical reactions are used to perform work by means of electric current. An example of such a cell is the *Daniell cell* shown in Fig. 11.1. It

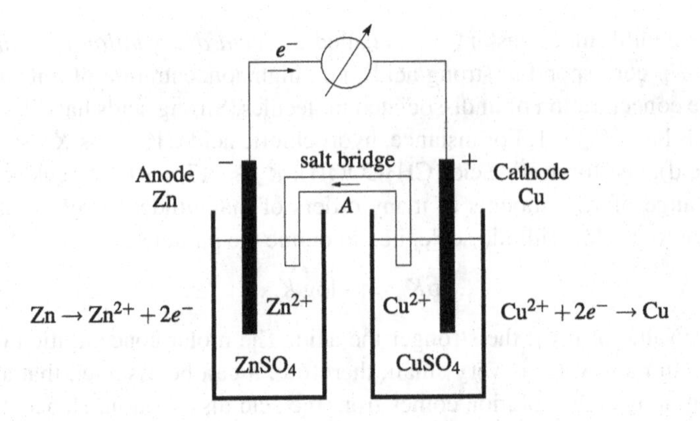

Fig. 11.1 Daniell cell consists of two half-cells: a zinc electrode immersed in aqueous solution of zinc sulfate and a copper electrode immersed in aqueous solution of copper sulfate. At the zinc electrode (*anode*), zinc is oxidized, and at the copper electrode (*cathode*), copper is reduced. Electrons (e^-) flow from the anode to the cathode. The solutions in the two half-cells are connected with a salt bridge, which allows the anions (A) to move between the solutions but prevents the solutions from mixing

consists of two electrodes: a zinc (Zn) electrode and a copper (Cu) electrode, immersed in aqueous solutions of their salts, which are electrolytes, i.e., in a solution of zinc sulfate ($ZnSO_4$) and a solution of copper sulfate ($CuSO_4$), respectively. A single electrode and the electrolyte solution, in which the electrode is immersed, form a *half-cell.* If electric current does not flow between the electrodes, then each half-cell is in chemical equilibrium. In the electrolyte solutions, the following dissociation reactions occur:

$$\mathrm{ZnSO_4} \rightleftharpoons \mathrm{Zn^{2+}} + \mathrm{SO_4^{2-}}, \tag{11.61}$$

$$\mathrm{CuSO_4} \rightleftharpoons \mathrm{Cu^{2+}} + \mathrm{SO_4^{2-}}. \tag{11.62}$$

At the electrodes, either *reduction* (the gain of electrons by a molecule, atom or ion) or *oxidation* (the loss of electrons by a molecule, atom or ion) occurs, which are called the *redox* (reduction–oxidation) reactions, i.e.,

$$\mathrm{Zn^{2+}}(aq) + 2e^- \rightleftharpoons \mathrm{Zn}(s), \tag{11.63}$$

$$\mathrm{Cu^{2+}}(aq) + 2e^- \rightleftharpoons \mathrm{Cu}(s), \tag{11.64}$$

where (s) refers to the solid phase. When the zinc electrode is immersed in the solution, atoms of the electrode have a tendency to leave electrons on the electrode and pass into the solution in the form of positive zinc ions. In the case of the copper electrode, positive copper ions in the solution exhibit a tendency to attach electrons from the electrode and deposit on it.

When the electrodes are connected with a wire, chemical equilibrium in the half-cells is disturbed. Electrons begin to flow from the zinc electrode, where they are in excess, to the copper electrode. Then the reaction at the zinc electrode proceeds in the direction of oxidation, and the reaction at the copper electrode proceeds in the direction of reduction, i.e.,

$$Zn(s) \rightarrow Zn^{2+}(aq) + 2e^-, \tag{11.65}$$

$$Cu^{2+}(aq) + 2e^- \rightarrow Cu(s). \tag{11.66}$$

Simultaneously with a flow of electrons, a flow of anions SO_4^{2-} between the half-cells must be ensured. When zinc atoms are oxidized an excess of cations Zn^{2+} appears in the solution. It must be counterbalanced by anions coming from the second half-cell, in which the number of cations Cu^{2+} decreases due to reduction of copper atoms. Therefore, anions flow in the direction opposite to the flow of electrons. The flow of anions is ensured by the *salt bridge*, which also prevents the solutions from mixing. Instead of the salt bridge, a porous membrane permeable to anions only can be used. The electrode at which reduction occurs is called the *cathode* because it attracts cations. The electrode at which oxidation occurs is called the *anode* because anions flow towards it. From half-cell reactions (11.66) and (11.65), we obtain the total reaction for the Daniell cell:

$$Cu^{2+}(aq) + Zn(s) \rightarrow Cu(s) + Zn^{2+}(aq). \tag{11.67}$$

11.3.2 Galvanic and Electrolytic Cells

A *galvanic cell* converts chemical energy into electrical energy. Chemical reactions at the electrodes of a galvanic cell occur spontaneously when the electrodes are connected with a conductor of electric current. An *electrolytic cell* converts electrical energy into chemical energy. Chemical reactions do not occur spontaneously when the electrodes are connected with a conductor. A chemical reaction in the cell starts only when an external source of electricity is connected to the cell. Electrolytic cells are used to store electrical energy in the form of chemical energy. They are also used for decomposition of compounds by means of electric current in a process called *electrolysis*. An example of electrolysis is decomposition of water into hydrogen and oxygen.

In what follows, we concentrate on galvanic cells. An example of a galvanic cell is the Daniell cell discussed above. A galvanic cell in its simplest form is similar to the Daniell cell, i.e., each half-cell consists of a metallic electrode immersed in a solution of a salt of that metal, and the two solutions are connected with a salt bridge. The reactions taking place at the electrodes X and Y can be symbolically expressed as follows:

$$X^{n+} + ne^- \rightleftharpoons X, \tag{11.68}$$

$$Y^{m+} + me^- \rightleftharpoons Y. \tag{11.69}$$

Multiplying both sides of (11.69) by $\nu = n/m$ and then subtracting from (11.68), we get the total reaction for the galvanic cell:

$$X^{n+} + \nu Y \rightleftharpoons X + \nu Y^{m+}, \tag{11.70}$$

We assume that the electrode X is the cathode and the electrode Y is the anode. When the electrodes are connected with an electric current conductor the reaction

$$X^{n+} + \nu Y \rightarrow X + \nu Y^{m+}, \tag{11.71}$$

proceeds, in which reduction occurs at the electrode X, and oxidation occurs at the electrode Y. It should be emphasized that transfer of electrons from Y to X does not take place due to a direct reaction but through a conductor connecting the electrodes. In a galvanic cell, the anode is the negative electrode and the cathode is the positive electrode. In the case of an electrolytic cell, it is the other way round, i.e., the anode is the positive polarity contact. However, the anode is always the electrode at which oxidation takes place.

For diagrams representing galvanic cells, the convention is used that the cathode is placed on the right-hand side and the anode is placed on the left-hand side. The difference between the potential of the right electrode and the potential of the left electrode of a galvanic cell is the *electric potential difference* E. Usually the limiting value of E for zero current flowing through the cell is assumed.[3] Then chemical equilibrium in the half-cells is not disturbed. In practice, the measurement of E should be performed with a voltmeter of very large internal resistance.

11.4 Reversible Cell

A chemical reaction in which ions take part can be controlled by means of an external electric field. It means that a reaction which would proceed spontaneously in the absence of electric field, can be stopped or reversed if an appropriate electric potential difference is applied. A reversible process must be quasi-static, i.e., its rate must tend to zero. Since the rate of chemical reactions in a galvanic cell can be controlled, we assume that all processes taking place in the cell are reversible. Then it is said to be a *reversible cell*, i.e., working in a reversible way.

11.4.1 Work of Chemical Reaction

We consider a closed system at constant temperature T and constant pressure p. An infinitesimal change in the Gibbs free energy, $G = U - TS + pV$, in a isothermal–isobaric process amounts to

$$\mathrm{d}G = \mathrm{d}U - T\mathrm{d}S + p\mathrm{d}V. \tag{11.72}$$

In a reversible process, $T\mathrm{d}S = đQ$, hence, $\mathrm{d}G = đW + p\mathrm{d}V$. As we know, $-p\mathrm{d}V$ is the mechanical work performed on the system whose volume changes by $\mathrm{d}V$. The

[3] E is sometimes called the *electromotive force*, although this name is no longer recommended, since the potential difference is not a force.

system can also perform other kinds of work. Here we are interested in the work done by the system by means of an electric current, due to chemical reactions in the system. In general,

$$đW = -p\mathrm{d}V + đW', \tag{11.73}$$

where $đW'$ denotes other kinds of work. From (11.73) and (11.72), it follows that at constant T and p,

$$\mathrm{d}G = đW' \tag{11.74}$$

(see (5.33)). Below, we make use of Eq. (11.74), to derive a relation between reactions taking place in the half-cells and the electric potential difference of a galvanic cell, E, for zero electric current.

11.4.2 Nernst Equation

We assume a reversible cell, thus, (11.74) is satisfied at constant T and p. To obtain the limit of zero electric current, the voltage $\Phi = -E$ is to be applied to the cell from an external source of electricity. The infinitesimal work done on the cell by the external source during transfer of a positive electric charge $\mathrm{d}q$ from the left half-cell to the right half-cell amounts to

$$đW' = \Phi \mathrm{d}q = -E\mathrm{d}q. \tag{11.75}$$

The charge $\mathrm{d}q$ can be related to the extent of the cell reaction, $\mathrm{d}\xi$. For reaction (11.71), a change in ξ by 1 mol is equivalent to the flow of n mol of the positive elementary charge e through the cell, hence

$$\mathrm{d}q = N_A n e \mathrm{d}\xi = nF\mathrm{d}\xi, \tag{11.76}$$

where F is the Faraday constant (see (11.34)). Using (11.74), (11.75) and (11.76), we get

$$-nEF\mathrm{d}\xi = \mathrm{d}G = -A\mathrm{d}\xi, \tag{11.77}$$

where A is the affinity of reaction (see (10.10)), hence

$$E = \frac{A}{nF}. \tag{11.78}$$

In this way, we have related the electric potential difference of the cell, E, to the affinity of the cell reaction in a reversible cell. Using reaction equation (11.70) and relation (10.10), we obtain

$$A = -(\mu_{\mathrm{X}} - \mu_{\mathrm{X^{n+}}}) + \nu(\mu_{\mathrm{Y}} - \mu_{\mathrm{Y^{m+}}}). \tag{11.79}$$

It is convenient to express the chemical potentials in term of the activities, i.e.,

$$\mu_i = \mu_i^0 + RT \ln a_i, \tag{11.80}$$

where i numbers the compounds participating in reaction (11.70). Substituting (11.80) into (11.79), we obtain the *Nernst equation*:

$$E = E^0 - \frac{RT}{nF} \ln Q_r, \tag{11.81}$$

where E^0 is called the *standard cell potential*, and

$$Q_r = \frac{a_X}{a_{X^{n+}}} \left(\frac{a_{Y^{m+}}}{a_Y} \right)^{\nu} \tag{11.82}$$

is the reaction quotient. It is easy to verify that E^0 is related to the standard Gibbs free energy of reaction (see (10.32)), i.e.,

$$E^0 = -\frac{\Delta_r G^0}{nF}. \tag{11.83}$$

Note that relation (11.83) can also be expressed in terms of the standard equilibrium constant K^0 (see (10.33)):

$$E^0 = \frac{RT}{nF} \ln K^0. \tag{11.84}$$

Thus, from the Nernst equation and relation (11.84), an important conclusion follows that measuring the cell potential E as a function of the electrolyte concentration, we can determine E^0, and hence, the standard equilibrium constant K^0. If X and Y are pure substances, as in the case of the Daniell cell for instance, then $a_X = 1$ and $a_Y = 1$, for $p = p^0$. On the other hand, the ions activity is equal to its molality for dilute solutions.

Substituting (11.84) to (11.81), we get

$$E = \frac{RT}{nF} \ln \frac{K^0}{Q_r}. \tag{11.85}$$

If initially the cell reaction proceeds spontaneously, chemical equilibrium is reached after some time, in which $E = 0$. Then the cell cannot perform any more work, because the Gibbs free energy has reached the minimum value. This situation corresponds to the relation

$$Q_r = K^0, \tag{11.86}$$

which expresses the law of mass action (cf. (10.35)) with activities in place of molar fractions.

Example 11.1 For the Daniell cell (see (11.67)), we have X = Cu, Y = Zn, $n = m = 2$, $\nu = 1$, hence, the Nernst equation adopts the following form:

$$E = E^0 - \frac{RT}{2F} \ln \frac{a_{Cu}(a_{Zn^{2+}})}{(a_{Cu^{2+}})a_{Zn}}. \tag{11.87}$$

11.4.3 Half-Cell Potential

Substituting (11.82) to (11.81), we get

$$E = E_R - E_L, \tag{11.88}$$

where the potentials

$$E_R = E_R^0 - \frac{RT}{nF} \ln \frac{a_X}{a_{X^{n+}}}, \tag{11.89}$$

$$E_L = E_L^0 - \frac{RT}{mF} \ln \frac{a_Y}{a_{Y^{m+}}}, \tag{11.90}$$

and the standard potentials

$$E_R^0 = -\frac{(\Delta_r G^0)_R}{nF}, \tag{11.91}$$

$$E_L^0 = -\frac{(\Delta_r G^0)_L}{mF}. \tag{11.92}$$

correspond to the right (R) and left (L) half-cell reaction, respectively. This means that the cell potential can be expressed as a difference between the potentials of the right and left half-cell.

Discussing the Daniell cell, we mentioned that the half-cell reactions are *redox* (reduction–oxidation) reactions. Therefore, it is convenient to treat X and Y as the reduced forms (denoted Red), and X^{n+} and Y^{m+} as the oxidized forms (denoted Ox). To change the oxidized form of a substance into the reduced form, the charge ze^- needs to be delivered, which we express shortly as

$$\text{Ox} + ze^- \rightleftharpoons \text{Red}. \tag{11.93}$$

It follows from (11.89) and (11.90) that the half-cell potential, called also the *redox potential*, can be expressed in the following general form:

$$E_{\text{Red}} = E_{\text{Red}}^0 - \frac{RT}{zF} \ln \frac{a_{\text{Red}}}{a_{\text{Ox}}}. \tag{11.94}$$

Relation (11.94) is an alternative form of the Nernst equation, which refers to the half-cell.

It should be added that there exist half-cells, in which a metallic electrode only provides electrons for a redox reaction, but it does not take part in the reaction. For instance, if the solution contains iron ions Fe^{2+} and Fe^{3+}, then in the presence of a platinum electrode, the following reaction occurs:

$$\text{Fe}^{3+} + e^- \rightleftharpoons \text{Fe}^{2+}, \tag{11.95}$$

$$E_{\text{Red}} = E_{\text{Red}}^0 - \frac{RT}{F} \ln \frac{a_{\text{Fe}^{2+}}}{a_{\text{Fe}^{3+}}}. \tag{11.96}$$

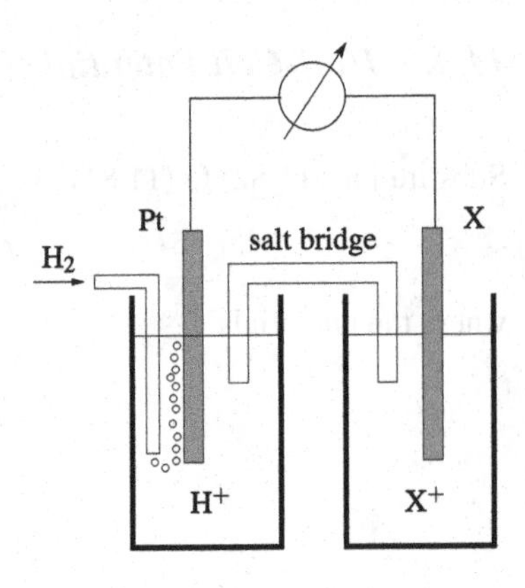

Fig. 11.2 Potential of the electrode X in the right half-cell is measured with respect to the *standard hydrogen electrode*. The platinum electrode (Pt) serves only as a catalyzer for the reaction $H^+ + e^- \rightleftharpoons \frac{1}{2}H_2$ taking place in the left half-cell, in which H^+ ions in the solution are in equilibrium with gaseous hydrogen at the standard pressure p^0

11.4.4 Standard Hydrogen Electrode

If the standard potentials of two half-cells are known we can determine the standard potential of the cell composed of them, using the relation

$$E^0 = E_R^0 - E_L^0. \tag{11.97}$$

However, the potential of a half-cell can be determined only in reference to the potential of another half-cell. Therefore, it is convenient to select a particular half-cell and treat it as a reference electrode, with respect to which the standard potentials of other half-cells are determined. Without loss of generality, we can assume that the potential of the reference electrode is equal to zero by definition. In practice, the *standard hydrogen electrode* is used as the reference electrode. It is a half-cell containing H^+ ions (protons) in the solution. In the presence of platinum, which serves as a catalyzer, the reaction

$$H^+ + e^- \rightleftharpoons \frac{1}{2}H_2. \tag{11.98}$$

takes place in the solution. Hydrogen dissolved in the solution is in equilibrium with gaseous hydrogen at the standard pressure p^0. The concentration of ions H^+ in the solution also corresponds to the standard conditions, i.e., their activity $a_{H^+} = 1$.

A half-cell whose potential we want to determine is placed by convention on the right-hand side, and the standard hydrogen electrode is placed on the left-hand side, as shown in Fig. 11.2. The potential difference E of the cell built in this way is by definition the potential of the given half-cell. Measuring E as a function of the electrolyte molality for very dilute solutions, it is possible to determine the standard potential of the cell, E^0, which is assumed to be the standard potential of the given half-cell. The standard potential of different metals, ordered from the lowest to the highest value, form the *electrochemical series* (Table 11.1). The sign of E^0 provides information about the direction of spontaneous reaction when all compounds are in their standard states. If for a certain metal X we have $E^0 > 0$, then X is said to be more *electropositive* than hydrogen (e.g. Cu). This means that reduction occurs on the electrode X, i.e., metal X deposits on the electrode. If $E^0 < 0$, then X is more

Table 11.1 Electrochemical series at the temperature of 25 °C. For the hydrogen electrode, $E^0 = 0$ for any temperature by definition

Electrode	Electrode reaction	E^0 [V]
Li^+/Li	$Li^+ + e^- \rightleftharpoons Li$	−3.05
K^+/K	$K^+ + e^- \rightleftharpoons K$	−2.93
Na^+/Na	$Na^+ + e^- \rightleftharpoons Na$	−2.71
Mg^{2+}/Mg	$Mg^{2+} + 2e^- \rightleftharpoons Mg$	−2.37
Al^{3+}/Al	$Al^{3+} + 3e^- \rightleftharpoons Al$	−1.66
Mn^{2+}/Mg	$Mn^{2+} + 2e^- \rightleftharpoons Mn$	−1.18
Zn^{2+}/Zn	$Zn^{2+} + 2e^- \rightleftharpoons Zn$	−0.76
Fe^{2+}/Fe	$Fe^{2+} + 2e^- \rightleftharpoons Fe$	−0.44
Ni^{2+}/Ni	$Ni^{2+} + 2e^- \rightleftharpoons Ni$	−0.25
Sn^{2+}/Sn	$Sn^{2+} + 2e^- \rightleftharpoons Sn$	−0.14
Pb^{2+}/Pb	$Pb^{2+} + 2e^- \rightleftharpoons Pb$	−0.13
$H^+/\frac{1}{2}H_2$	$H^+ + e^- \rightleftharpoons \frac{1}{2}H_2$	0
Cu^{2+}/Cu	$Cu^{2+} + 2e^- \rightleftharpoons Cu$	+0.34
Ag^+/Ag	$Ag^+ + e^- \rightleftharpoons Ag$	+0.80
Pt^{2+}/Pt	$Pt^{2+} + 2e^- \rightleftharpoons Pt$	+1.19
Au^{3+}/Au	$Au^{3+} + 3e^- \rightleftharpoons Au$	+1.50

electronegative than hydrogen (e.g. Zn). Then reduction occurs on the platinum (Pt) electrode, i.e., hydrogen is given off by the solution.

Example 11.2 To determine the standard potential of the Daniell cell, we find in Table 11.1 the value of E^0 for copper (+0.34 V) and subtract from it the value of E^0 for zinc (−0.76 V), which gives +1, 10 V.

11.5 Exercises

11.1 Write the condition of chemical equilibrium for the following reactions:

$$NH_3(g) + H_2O(l) \rightleftharpoons NH_4^+(aq) + OH^-(aq),$$
$$BaSO_4(s) \rightleftharpoons Ba^{2+}(aq) + SO_4^{2-}(aq).$$

In the first reaction, ammonia dissolves in water, which causes a weak electrolyte to form. The second reaction concerns a saturated aqueous solution of slightly soluble salt. Assume that slightly soluble salt $BaSO_4(s)$ in aqueous solution is completely dissociated.

11.2 Generalize the concept of solubility product (see Exercise 11.1), i.e., consider a saturated aqueous solution of slightly soluble salt A_xB_y which dissociates according to the equation

$$A_xB_y(s) \rightleftharpoons xA(aq) + yB(aq),$$

where A and B denote the positive and negative ions, respectively. Apply the expression derived, to calculate the molar concentration of ions in a saturated aqueous solution of salt Ag_2CO_3, assuming that the ion activities can be replaced by their molar concentrations, and the solubility constant at the temperature of 25 °C amounts to $K_s = 6.2 \times 10^{-12}$.

11.3 The standard potentials of the zinc and copper electrodes at 298 K amount to −0.763 V and +0.339 V, respectively. Calculate the equilibrium constant for the reaction $Cu^{2+} + Zn \rightleftharpoons Cu + Zn^{2+}$.

11.4 In melted common salt (NaCl), cations Na^+ and anions Cl^- can move freely, therefore, the liquid conducts electric current (ion conductivity), despite the fact that crystalline salt is an insulator. At the cathode, connected to the negative terminal of the battery, the reaction

$$Na^+ + e^- \to Na,$$

takes place, and at the anode, connected to the positive terminal, the reaction

$$2Cl^- \to Cl_2 + 2e^-$$

takes place. The process is called *electrolysis*. Assume that the intensity of electric current flowing through the salt amounts to 10 A. How long does it take to obtain 46 g of metallic sodium at the cathode and how much gaseous chlorine is produced at the anode during this time?

11.5 A chemical reaction in a reversible cell generates the potential difference $E = 1.015$ V at the temperature of 0 °C and at atmospheric pressure. From the measurement of E as a function of temperature, the derivative $(\partial E/\partial T)_p = -4.02 \times 10^{-4}$ V K^{-1} was determined at 0 °C. Using these data, calculate the enthalpy of the cell reaction, assuming that during oxidation each metal atom passing into solution leaves 2 electrons. What part of the heat produced in the reaction cannot be used to perform work?

11.6 The hydrogen electrode is composed of a platinum electrode immersed in a solution of hydrogen ions which are in equilibrium with gaseous hydrogen. Platinum does not take part in the reaction but serves as an electric contact and a catalyzer for the reaction

$$H^+(aq) + e^- \rightleftharpoons \frac{1}{2}H_2(g).$$

Find a relation between the potential of the hydrogen half-cell, the pressure of gaseous hydrogen and the activity of hydrogen ions in the solution. Calculate the change in the potential of the hydrogen half-cell at the temperature 25 °C if: (1) the pressure of hydrogen has decreased 9 times at constant activity of hydrogen ions, (2) the activity of hydrogen ions has increased 3 times at constant pressure.

11.7 Calculate the potential of a half-cell composed of the zinc electrode in equilibrium with zinc ions in solution, at the temperature of 25 °C and for the activity of zinc ions $a_{\mathrm{Zn}^{2+}} = 0.1$, assuming the value of the standard potential $E^0_{\mathrm{Zn}} = -0.763$ V.

11.8 A galvanic cell is composed of the hydrogen half-cell and another half-cell, whose electrode is made of pure silver (Ag) in contact with solid silver chloride (AgCl). Both electrodes are immersed in the same electrolyte, hydrochloric acid (HCl). The oxidation takes place at the hydrogen electrode, i.e.,

$$\frac{1}{2}\mathrm{H}_2(g) \rightleftharpoons \mathrm{H}^+(aq) + e^-,$$

and the reduction takes place at the Ag–AgCl electrode:

$$\mathrm{AgCl}(s) + e^- \rightleftharpoons \mathrm{Ag}(s) + \mathrm{Cl}^-(aq).$$

Write the total cell reaction and derive an expression for the potential difference E. What quantities does E depend on? Can this cell be used to determine the average activity coefficient of the ions, $\gamma_\pm$, for the electrolyte?

Solutions

Exercises of Chapter 2

2.1 We calculate the total mass of air M. The mass is equal to the volume multiplied by the density. It is proper to notice the units used here: 1 g cm^{-3} and $1\ \text{L} = 1000\,\text{cm}^3$, whose product gives the unit of mass, i.e., 1 kg. We get

$$\text{M} = \text{V}\rho = 22 \times 1000\ \text{cm}^3 \times 10^{-3}\ \text{g cm}^{-3} = 22\ \text{g} = 0.022\ \text{kg}.$$

We neglect the fact that air is composed mainly of N_2 (78 %) and O_2 (21 %). Then we use the expression for the kinetic energy. Assuming that all molecules have an average speed of about $v = 300\ \text{m s}^{-1}$, which corresponds roughly to the speed of sound in air, we obtain the kinetic energy:

$$E = \frac{Mv^2}{2} = 22\ \text{g} \times (300)^2\ \text{m}^2\,\text{s}^{-2}/2 = 1980\ \text{J}/2 \approx 1\ \text{kJ}.$$

It is proper to notice here the way we calculate numerical values. We can represent each number in the following form: $a10^b$, where a is of the order 1, and b is an integer. If we have large numbers (b is positive) or small numbers (b is negative), then this way of representation of numbers save us a lot of time during calculation and also improves its quality, because multiplication of two numbers can be written as $a10^b \times c10^d = ac \times 10^{b+d}$. In the equation above, we have $0.022 \times 90000 = 2.2 \times 10^{-2} \times 9 \times 10^4 = 9 \times 2.2 \times 10^{-2+4} = 1980$. As we can see, when a and c are small, of the order 1, then this way of calculation is quick. The second remark concerns rounding off. Sometimes we need very accurate numbers, but most often we want only to find out the order of magnitude of a given quantity and for this reason, we write $1980\ \text{J} \approx 2\ \text{kJ}$. The value of 2 kJ is also easier to remember than 1980 J, and the rounding-off error amounts to 1 % only.

2.2 We use the formula for the potential energy of a body in the gravitational field of the earth, calculated with respect to the earth surface (at $h = 0$). In thermodynamics, we do not know the internal energy of a body. We can only calculate changes in its internal energy with respect to some reference states, as in the case of this exercise.

R. Hołyst, A. Poniewierski, *Thermodynamics for Chemists, Physicists and Engineers*,
DOI 10.1007/978-94-007-2999-5, © Springer Science+Business Media Dordrecht 2012

Here the initial state corresponds to the energy E_i at $h = 0$, and the final state corresponds to the energy E_f at $h = 5$ km $= 5000$ m $= 5 \times 10^3$ m. The mass of one mole of water (H_2O) $m = 18$ g. The change in its potential energy, ΔE, amounts to

$$E_f - E_i = mgh = 18 \times 10^{-3}\ \text{kg} \times 9.81\ \text{m}\,\text{s}^{-2} \times 5 \times 10^3\ \text{m} = 882.9\ \text{J} \approx 0.9\ \text{kJ}.$$

Why is the difference in the potential energy ΔE independent of the way we reach the height $h = 5$ km? We often hear about climbers who have found a new and more difficult route to the top of a known mountain, K2 for instance. However, no matter which way they choose and how much effort it requires to get to the top, each time they do it, their potential energy increases exactly by the same amount. We say that the potential energy is a function of the height above sea level, and not a function of the path along which we move, to reach a given height. The essence of thermodynamics consists in investigation of quantities, e.g., the internal energy, which depend only on the state of a system and not on the way the given state is reached. We note by chance that the kinetic energy of one mole of air is comparable with the change in its gravitational energy when it is lifted a few kilometers above sea level. This means that in a typical chemical experiment, the potential energy can be neglected.

2.3 Heat of evaporation amounts to 40 kJ mol^{-1}. To evaporate 1 mol of water, 40 kJ of heat is to be supplied to the system. We denote this heat by Δh, because it is related to a state function called the *enthalpy*, H (see Chap. 3). The molar mass of water is $M = 18$ g mol^{-1}. Since the mass of water amounts to $m = 9$ g, the total heat needed to evaporate this mass of water amounts to

$$Q = \frac{m}{M}\Delta h = \frac{9\ \text{g}}{18\ \text{g}\,\text{mol}^{-1}} \times 40\ \text{kJ}\,\text{mol}^{-1} = 20\ \text{kJ}.$$

Molecules in liquid water are close to one another (density of liquid water is 1 g cm^{-3}) and they strongly interact (forming hydrogen bonds). The density of water vapour, which escapes from the kettle, for instance, amounts to 10^{-3} g m^{-3}, thus, it is 1000 times smaller than the density of liquid water. It means, that water molecules in the vapour are 10 times farther away from one another than in liquid water (think how to determine this ratio from the ratio of the densities of liquid water and water vapour). 1 mol of water contains $N_A = 6.022 \times 10^{23}$ molecules, and each H_2O molecule has 4 neighbours (oxygen forms two hydrogen bonds and each hydrogen atom in the molecule takes part in one hydrogen bond with another water molecule). Measuring the heat of evaporation (40 kJ mol^{-1}), we can estimate the energy of interaction of water molecules (see Fig. 3.2), i.e., 40 kJ$/(4N_A) = 3.2 \times 10^{-20}$ J $= 0.1$ eV (we divide by 4, to take into account the number of neighbours of a molecule). We notice by chance that the energy of intermolecular interactions is much larger than the gravitational energy or the kinetic energy of molecules in water vapour (see the previous two exercises).

2.4 The mass of 1 mol of argon amounts to $M = 40$ g. We have $\Delta h = 6$ kJ mol^{-1}, hence, to evaporate the mass $m = 40$ g of argon, we need

$$Q = \frac{m}{M}\Delta h = \frac{40\ \text{g}}{40\ \text{g}\,\text{mol}^{-1}} \times 6\ \text{kJ}\,\text{mol}^{-1} = 6\ \text{kJ}.$$

Now we want to find the energy of interaction between argon atoms which follows from this value of Q. We calculate it in a similar way as in the previous exercise, i.e., $6\ \mathrm{kJ}/12N_\mathrm{A} = 0.17 \times 10^{-20}\ \mathrm{J} \approx 0.005\ \mathrm{eV}$ (we divide by 12, to take into account the number of neighbours of a single argon atom). The number of the closest neighbours of one atom, i.e., the atoms that are sufficiently close to the given atom to interact with it, follows from the local structure of a substance (arrangement of atoms or molecules in space). Argon at low temperature forms a crystalline structure in which each argon atom is surrounded by 12 closest neighbours. Water molecules interact with one another almost 20 times stronger than argon atoms. The interactions between argon atoms are weak because they result from the interaction of induced electric dipoles.

2.5 This exercise allows us to estimate how much the energy of molecular interactions differs from the energy contained in chemical bonds that binds atoms in a molecule. The heat of combustion amounts to $\Delta h = 400\ \mathrm{kJ\,mol^{-1}}$. The molar mass of carbon $M = 12\ \mathrm{g\,mol^{-1}}$, and the mass used for the combustion $m = 12$ g, hence, the amount of heat given off during the combustion amounts to

$$Q = \frac{m}{M}\Delta h = \frac{12\ \mathrm{g}}{12\ \mathrm{g\,mol^{-1}}} \times 400\ \mathrm{kJ\,mol^{-1}} = 400\ \mathrm{kJ}.$$

The reaction of combustion has the following form:

$$\mathrm{C} + \mathrm{O_2} \rightarrow \mathrm{CO_2}.$$

In this reaction, the double bond between two oxygen atoms is broken and two double bonds between two oxygen atoms and one carbon atom are formed. Due to the reconstruction of the chemical bonds, 400 kJ of heat per mole of carbon is given off. By comparison with the previous source of energy, we can see that the energy hidden in chemical bonds, i.e., in the electronic structure of molecules and interactions between negative electrons and positive nuclei of atoms, is tens of times larger than the energy of intermolecular interactions in a liquid (cf. the previous exercises).

2.6 Protons and neutrons are bound in the atom nuclei through interactions called the *nuclear* or *strong* interactions. The energy of nuclear interactions is released inside the sun at a temperature of millions degrees centigrade. Hans Bethe (1906–2005), a great physicist and a Nobel prize winner, was the first to propose these interactions as a source of the solar energy. A cycle of nuclear reactions inside the sun can be presented in short in the form of one nuclear reaction:

$$4\,{}^{1}\mathrm{H} + 2e \rightarrow {}^{4}\mathrm{He} + 2\nu + 6\gamma.$$

Four hydrogen nuclei (protons) and two electrons form one helium nuclei, which is composed of two protons and two neutrons, and some energy is released in the form of two neutrinos (ν) and six photons in the range of gamma-rays. The energy released in this nuclear reaction amounts to about 26 MeV, which gives 26 MeV$/4 =$ 6.5 MeV per each hydrogen atom used in the reaction. Thus, in the reaction of 1 mol of protons (${}^{1}\mathrm{H}$), the energy

$$\Delta U = 6.022 \times 10^{23} \times 6.5 \times 10^{6} \times 1.6 \times 10^{-19}\ \mathrm{J} = 6.26 \times 10^{11}\ \mathrm{kJ}$$

is released. Due to the nuclear synthesis, 1.5×10^9 times more heat is released than during carbon combustion. Construction of a device for a controllable thermonuclear synthesis would give the mankind practically inexhaustible source of energy and allow to achieve a great leap forward.

2.7 We make use of the Einstein formula. The speed of light in vacuum $c \approx 300\,000 \text{ km s}^{-1}$. Annihilation changes the whole mass into the energy of photons. The mass to be changed into pure energy amounts to $m = 12$ g. We calculate the energy contained in 12 g of carbon:

$$E = mc^2 = 12 \times 10^{-3} \text{ kg} \times 9 \times 10^{16} \text{ m}^2 \text{s}^{-2} = 1.08 \times 10^{12} \text{ kJ mol}^{-1}.$$

Thus, the annihilation of 1 mol of carbon provides almost 10^{10} times more energy than its combustion. Possibility of getting such a process under control would give the mankind the greatest access to energy resources.

2.8 We begin with writing down the parameters of the initial and final equilibrium states. In the initial state, 1 mol of water vapour, of the mass $m = 18$ g and density $\rho_i = 10^{-3}$ g cm^{-3}, occupies the volume $V_i = m/\rho_i = 18 \times 10^3$ cm^3. In the final state, liquid water has the same mass, and its density $\rho_f = 1$ g cm^{-3}, hence, the final volume $V_f = m/\rho_f = 18$ cm^3. Thus, the change in the volume amounts to

$$\Delta V = V_f - V_i = 18 \text{ cm}^3 - 18000 \text{ cm}^3 = -17982 \text{ cm}^3.$$

The change is negative, since the volume in the final state is smaller than the volume in the initial state.

2.9 In the initial state, we have separate compounds N_2 and H_2, and their mole numbers are $n_{N_2} = 1$ and $n_{H_2} = 3$. The total mole number in the initial state is

$$n_i = n_{H_2} + n_{N_2} = 1 + 3 = 4.$$

In the final state, we have only NH_3, and its mole number is $n_{NH_3} = 2$, hence, the final mole number is

$$n_f = n_{NH_3} = 2.$$

Note that in equilibrium, we would have a mixture of three compounds in the final state, but here we ignore this fact and assume that all reactants have been used up in the reaction, giving a pure product. The total mole number in the system has changed by

$$\Delta n = n_f - n_i = 2 - 4 = -2.$$

The total mole number has decreased by 2. The mole number of NH_3 has increased by 2, the mole number of H_2 has decreased by 3, and the mole number of N_2 has decreased by 1.

2.10 In the initial state, we have $V_{1i} = 100$ L, $n_{1i} = 3 + 4 + 1 = 8$ mol in vessel 1, and $V_{2i} = 100$ L, $n_{2i} = 5 + 2 + 1 = 8$ in vessel 2. In the final state, we have $V_f = V_{1i} + V_{2i} = 200$ L and $n_f = n_{1i} + n_{2i} = 16$ mol in the fused vessels.

2.11 The internal energy, volume and mole number are extensive parameters, thus, in the final state, after the fusion of the vessels, we have the following values of the state parameters: $U_f = 4U$, $V_f = 4V$, $n_f = 4n$ (cf. Fig. 2.4).

2.12 The heat given off by a man amounts to $Q = 2\,000$ kcal $= 8368$ kJ. A day and night have $t = 24 \times 60 \times 60$ s. Dividing a daily consumption of energy by time, we get the power, i.e., $Q/t = 96.85\ \mathrm{J\,s^{-1}} = 96.85$ W. It means that a man consumes approximately the same amount of energy as a 100 W bulb. Lavoisier compared the heat of carbon combustion with the heat given off by living beings and with the amounts of carbon dioxide and water produced. His study became a basis for establishing of a detailed energetic balance of the human organism.

2.13 Water evaporates too quickly and its density is too small (1 $\mathrm{g\,cm^{-3}}$), to be a good working substance in the barometer. It is easy to calculate how high the barometer would have to be if we used flaxseed oil (density $\rho_{\mathrm{oil}} = 0.94\ \mathrm{g\,cm^{-3}}$) instead of mercury, to measure atmospheric pressure. The density of mercury $\rho_{\mathrm{Hg}} = 13.6\ \mathrm{g\,cm^{-3}}$. First, we measure atmospheric pressure p, using flaxseed oil in the barometer, and then we measure p with the mercury barometer. Since the liquid density is different in each case, the height of the liquid column is also different. We calculate this difference, using the formula for the pressure measured with the barometer:

$$p = \rho_{\mathrm{oil}} g h_{\mathrm{oil}} = \rho_{\mathrm{Hg}} g h_{\mathrm{Hg}}.$$

Since $h_{\mathrm{Hg}} = 760$ mm at the pressure of 1 atm, we get

$$h_{\mathrm{oil}} = \frac{\rho_{\mathrm{Hg}}}{\rho_{\mathrm{oil}}} h_{\mathrm{Hg}} = \frac{13.6}{0.94} \times 760\ \mathrm{mm} = 10995.7\ \mathrm{mm} \approx 11\ \mathrm{m}.$$

As we can see, the oil barometer would have the height of 11 m and no doubt it would not be a very practical device.

2.14 The pressure in a car tyre amounts to about 2 atm, and a car weighing 1 ton stands on four wheels. In a mountain bicycle, we pump the wheels up to 4 atm. The pressure at the depth of 1000 km amounts to about 250 000 atm. As one can see, the question is not correctly formulated, as in order to say that something is small or large we have to compare it with other things. No quantity exists that could be said to be large or small *in itself*, as Immanuel Kant might have said.

Let us calculate the force acting on our bodies. First, we have to estimate the area of the body, A. We assume it to be equal to 2 $\mathrm{m^2}$. By definition, the force is the product of p and A, hence, $F = pA = 101325\ \mathrm{N\,m^{-2}} \times 2\ \mathrm{m^2} \approx 200000$ N. It means that air acts on the body with a force corresponding to the weight of 20 ton (1 kG $\approx$ 10 N). It is really a lot! We do not feel it because the pressure of our blood and other body fluids is exactly the same as atmospheric pressure, in accord with mechanical equilibrium, i.e., $p_{\mathrm{body}} = p$. If the external pressure would fall down to zero our bodies would burst due to the internal pressure, the blood would boil and partially evaporate (the boiling point decreases with lowering of the pressure) and the remaining blood and body fluids would freeze, because they would give the heat off to the vapour causing their temperature to fall below the freezing point.

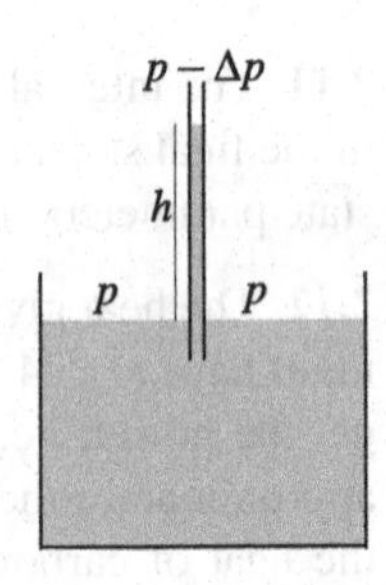

Fig. S2.1 To draw a liquid with a straw, we have to produce the underpressure Δp with the mouth

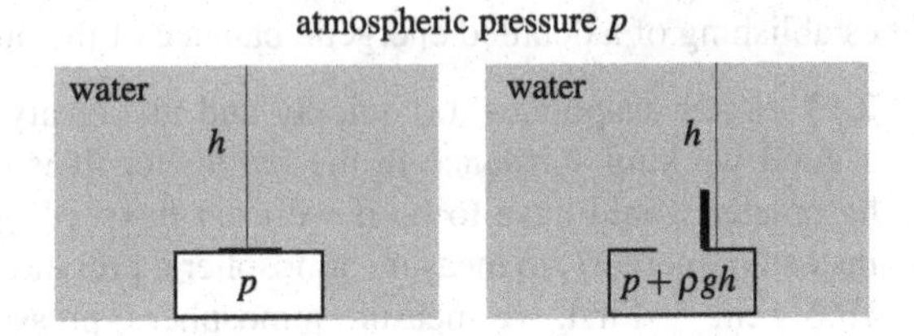

Fig. S2.2 To lift the lid, a sailor would have to lift a whole column of water above the lid. When the sluice-gate is filled with water the pressure above and below the lid is the and equal to $p+\rho gh$

2.15 The liquid must rise 20 cm above its surface in the glass, to reach the mouth. The liquid in the glass is at atmospheric pressure p, and the pressure above the liquid in the straw amounts to $p_0 < p$. From the balance of forces, it follows that to make the liquid (its density $\rho = 1\ \mathrm{g\,cm^{-3}}$) in the straw rise to the level h, the pressure difference $\Delta p = p - p_0$ must be equal to (see Fig. S2.1):

$$\Delta p = \rho gh = 1\ \mathrm{g\,cm^{-3}} \times 9.81\ \mathrm{m\,s^{-2}} \times 20\ \mathrm{cm} = 1962\ \mathrm{Pa}.$$

This underpressure we have to produce in the straw, to draw the liquid into the mouth. The pressure difference is about 50 times smaller than atmospheric pressure.

2.16 A column of water, of the height $h = 10$ m, presses on the lid (Fig. S2.2). It exerts the pressure

$$p_0 = \rho gh = 10^3\ \mathrm{kg\,m^{-3}} \times 9.81\ \mathrm{m\,s^{-2}} \times 10\ \mathrm{m} = 9.81 \times 10^4\ \mathrm{N\,m^{-2}}.$$

Since the lid area amounts to $A = 2\ \mathrm{m^2}$, the total force is equal to

$$F = p_0 A = 9.81 \times 10^4\ \mathrm{N\,m^{-2}} \times 2\ \mathrm{m^2} = 19.62 \times 10^4\ \mathrm{N}.$$

This is the force the sailor has to use to lift the lid (we neglect the weight of the lid). It is about 20 ton, which is not surprising, since lifting the lid, the sailor lifts at the same time a column of water 10 m in height. If the sluice-gate is filled with water, the pressure on both sides of the lid is the same and then the force needed to lift the lid is equal to its weight minus its buoyancy, which is about 10–20 kg. This is why the sluice-gate must be filled with water before the access door is opened.

2.17 We have to invert relation (2.18) between the Fahrenheit and Celsius scale:

$$t_F/°\mathrm{F} = \frac{9}{5} t_C/°\mathrm{C} + 32,$$

hence

$$t_C/^\circ\mathrm{C} = \frac{5}{9}(t_F/^\circ\mathrm{F} - 32).$$

The temperature of 0 °F corresponds to −17.8 °C (the lowest temperature of supercooled water obtained by Fahrenheit in his laboratory), 70 °F corresponds to 21.1 °C (room temperature), and 451 °F corresponds to 232.8 °C (the autoignition temperature of paper and the title of the famous sf novel by Ray Bradbury).

2.18 The Celsius and Kelvin scales are related to each other with the formula

$$T/\mathrm{K} = t_C/^\circ\mathrm{C} + 273.15,$$

hence, $t_C = -273.15$ °C corresponds to $T = 0$ K, the absolute zero temperature, unattainable experimentally.

2.19 We know that 0 K corresponds to −273.15 °C. Converting to the Fahrenheit scale, we get −459.67 °F.

2.20 From the expression

$$t_C/^\circ\mathrm{C} = \frac{5}{9}(t_F/^\circ\mathrm{F} - 32),$$

we get $t_C = 37.8$ °C. The body temperature of a healthy man amounts to 36.6 °C, which means that Fahrenheit calibrated his thermometer, measuring the temperature of a sick person. In general, it is not a good idea to use living organisms to calibrate thermometers. For instance, the temperature of the human body can vary by even a few degrees, depending on the condition of the organism. It means that if we calibrate the thermometer one day we can get a result which differs by a few degrees from the result obtained another day. Therefore, the calibration of thermometers is usually based on reproducible phenomena, which occur always at the same temperature, such as coexistence of the vapour, liquid and solid of a pure substance at its triple point. At the triple point of water (water vapour, liquid water and ice coexist), the temperature and pressure amount always to 273.16 K (0.01 °C) and 611 Pa (4.6 torr), respectively.

2.21 We take a body with the characteristics corresponding to the perfect blackbody and put it into a mixture of ice, liquid water and water vapour (the triple point of water). Then we measure the electromagnetic radiation emitted by the body and use the formula

$$U = \gamma V T^4,$$

where the internal energy U is proportional to the intensity of radiation. In this way, we determine the quantity $U_0 = U(T_0 = 273.16)$ K. To determine the given temperature T, we use the relation between T and U, hence

$$T = 273.16\ \mathrm{K}\left(\frac{U}{U_0}\right)^{1/4}.$$

2.22 The internal energy per mole for the system of the internal energy U and mole number n is given by

$$u = \frac{U}{n}.$$

When we join together m identical systems, each of the internal energy U and mole number n, we obtain a composite system of the internal energy mU and mole number mn, since the internal energy and amount of substance are extensive quantities. The internal energy per mole of the composite system amounts to

$$\frac{mU}{mn} = \frac{U}{n} = u,$$

thus, it does not depend on the size of the composite system. It has the same value for the system composed of m subsystems and for each of the subsystems.

2.23 The molar mass of water amounts to $M = 18\ \mathrm{g\,mol^{-1}}$. The volume $V_1 = 18\ \mathrm{cm^3}$ contains the mass

$$m_1 = V_1 \rho = 18\ \mathrm{g},$$

hence, the mole number of water amounts to

$$n_1 = \frac{m_1}{M} = 1\ \mathrm{mol}.$$

For the second vessel, we find $m_2 = 36$ g and $n_2 = 2$ mol. The molar density in both vessels amounts to

$$\rho_n = \frac{n_1}{V_1} = \frac{n_2}{V_2} = \frac{1}{18}\ \mathrm{mol\,cm^{-3}}.$$

The total volume after the fusion:

$$V = V_1 + V_2 = 54\ \mathrm{cm^3}$$

and the total mole number:

$$n = n_1 + n_2 = 3,$$

hence, for the molar density, we get

$$\rho_n = \frac{n}{V} = \frac{1}{18}\ \mathrm{mol\,cm^{-3}}.$$

The molar density is an intensive quantity, since it does not depend on the size of the system. The volume and amount of substance are extensive quantities. The mass density ρ is an intensive quantity, because

$$\rho = \frac{m}{V} = \frac{nM}{V} = M\rho_n.$$

2.24 We denote by v the volume occupied by 1 mol of water, i.e., its molar volume. It is obtained from the formula

$$v = \frac{M}{\rho} = 18\ \mathrm{cm^3\,mol^{-1}}.$$

To determine the volume per molecule, v_m, we divide v by N_A, which gives

$$v_m = \frac{v}{N_A} = \frac{18}{6.022 \times 10^{23}} \text{ cm}^3 \approx 3 \times 10^{-23} \text{ cm}^3.$$

We can express this volume, using $1 \text{ Å} = 10^{-8}$ cm as a unit of length, which corresponds roughly to the atom size. We obtain $v_m = 30 \text{ Å}^3$. We can estimate the linear size of the molecule, l_m, assuming that $v_m = l_m^3$, which gives $l_m \approx 3$ Å. In reality, the size of water molecule amounts to 2.76 Å. Liquids are very dense and the crowd of molecules resembles the crowds of people travelling in Tokyo by underground during the rush-hour, when people almost seat on one another. In a liquid, one molecule touches other molecules and altogether they fill up the space rather closely.

2.25 We express V for the ideal gas as a function of T, p and n:

$$V(T, p, n) = \frac{nRT}{p}.$$

According to the definition of an infinitesimal change of a state function, $\mathrm{d}V$ depends on $\mathrm{d}p$ and $\mathrm{d}T$ as follows:

$$\mathrm{d}V = V(T + \mathrm{d}T, p + \mathrm{d}p, n) - V(T, p, n) = \left(\frac{\partial V}{\partial T}\right)_{p,n} \mathrm{d}T + \left(\frac{\partial V}{\partial p}\right)_{T,n} \mathrm{d}p.$$

Since

$$\left(\frac{\partial V}{\partial T}\right)_{p,n} = \frac{nR}{p}, \qquad \left(\frac{\partial V}{\partial p}\right)_{T,n} = -\frac{nRT}{p^2},$$

we get

$$\mathrm{d}V = \frac{nR}{p}\mathrm{d}T - \frac{nRT}{p^2}\mathrm{d}p.$$

2.26 For a monatomic van der Waals gas, we have

$$U(T, V, n) = \frac{3}{2}nRT - \frac{an^2}{V}.$$

The increase in the internal energy at constant mole number amounts to

$$\mathrm{d}U = U(T + \mathrm{d}T, V + \mathrm{d}V, n) - U(T, V, n) = \left(\frac{\partial U}{\partial T}\right)_{V,n} \mathrm{d}T + \left(\frac{\partial U}{\partial V}\right)_{T,n} \mathrm{d}V.$$

Since

$$\left(\frac{\partial U}{\partial T}\right)_{V,n} = \frac{3}{2}nR, \qquad \left(\frac{\partial U}{\partial V}\right)_{T,n} = \frac{an^2}{V^2},$$

we get

$$\mathrm{d}U(T, V, n) = \frac{3}{2}nR\mathrm{d}T + \frac{an^2}{V^2}\mathrm{d}V.$$

2.27 The relation between the pressure and temperature of the photon gas is

$$p = \frac{1}{3}\gamma T^4,$$

hence, we get an infinitesimal increase in the pressure

$$\mathrm{d}p = p(T + \mathrm{d}T) - p(T) = \frac{\partial p}{\partial T}\mathrm{d}T.$$

Since

$$\frac{\partial p}{\partial T} = \frac{4}{3}\gamma T^3,$$

we find that

$$\mathrm{d}p = \frac{4}{3}\gamma T^3 \mathrm{d}T.$$

2.28 For $đ\omega = \omega_x \mathrm{d}x + \omega_y \mathrm{d}y$, we have to check if $\partial\omega_x/\partial y = \partial\omega_y/\partial x$. In case (1), we have

$$\frac{\partial\omega_x}{\partial y} = 6xy^2, \qquad \frac{\partial\omega_y}{\partial x} = 6xy^2,$$

thus, the equality of the derivatives is satisfied, which means that $đ\omega = \mathrm{d}f$, where f is a function of x and y. It is easy to verify that $f(x, y) = x^2y^3 + const$. In case (2), we have

$$\frac{\partial\omega_x}{\partial y} = 4xy^3, \qquad \frac{\partial\omega_y}{\partial x} = 2xy^2,$$

thus, the derivatives differ. No function f exists, whose differential $\mathrm{d}f$ would be equal to $đ\omega$.

2.29 We consider a differential form $đ\omega = \omega_x(x, y)\mathrm{d}x + \omega_y(x, y)\mathrm{d}y$, defined in the xy plane, hence

$$\int_i^f đ\omega = \int_i^f \left[\omega_x(x, y)\mathrm{d}x + \omega_y(x, y)\mathrm{d}y\right],$$

where the path linking the initial point with the final point is to be defined. Both paths of integration are shown in Fig. S2.3. In case (1), we have

$$\int_i^f đ\omega = \int_0^1 \omega_x(x, 0)\mathrm{d}x + \int_0^1 \omega_y(1, y)\mathrm{d}y,$$

since we integrate first along the x axis, at $y = 0$, and then along the y axis, at $x = 1$. In the previous exercise, we showed that $đ\omega = xy^4\mathrm{d}x + x^2y^2\mathrm{d}y$ is not a function differential, hence, we substitute $\omega_x = xy^4$ and $\omega_y = x^2y^2$. The first integral vanishes, since $\omega_x(x, 0) = 0$. In the second integral, $\omega_y(1, y) = y^2$, hence

$$\int_i^f đ\omega = \int_0^1 y^2\mathrm{d}y = \frac{1}{3}.$$

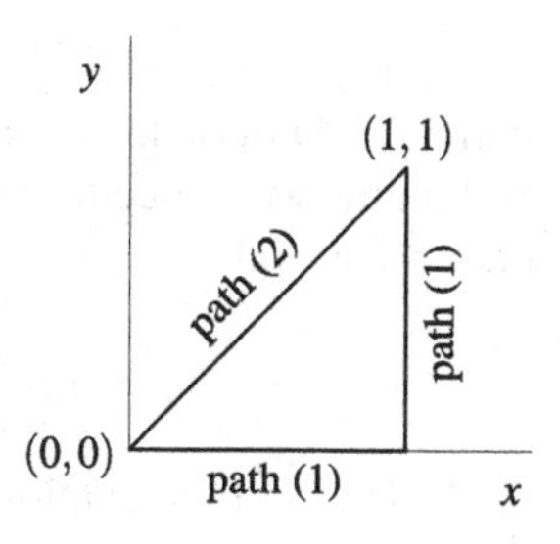

Fig. S2.3 Two paths linking the point (0, 0) with the point (1, 1)

In case (2), we have to substitute $y = x$ and $dy = dx$, hence

$$\int_i^f đ\omega = \int_0^1 \left[\omega_x(x,x) + \omega_y(x,x)\right]dx = \int_0^1 \left(x^5 + x^4\right)dx = \frac{11}{30}.$$

Thus, we have shown that the value of the integral depends on the path between the points i and f.

Now we integrate along the same paths the differential $df = 2xy^3dx + 3x^2y^2dy$, where $f(x, y) = x^2y^3 + const$. For path (1), we get

$$\int_i^f df = \int_0^1 3y^2dy = 1,$$

and for path(2), we get

$$\int_i^f df = \int_0^1 5x^4dx = 1.$$

In both cases, the value of the integral is the same, equal to $\Delta f = f(1,1) - f(0,0) = 1$.

Below we give a general method of integration of differential forms $đ\omega$. We restrict ourselves to two variables but it is easy to generalize the result. A path of integration is defined by a certain curve, which can be expressed in the parametric form: $\tau \mapsto (x(\tau), y(\tau))$, where the parameter τ changes from the initial value τ_i to the final value τ_f. The path may consist of several parts, as in case (1), but we assume for simplicity that the curve is smooth. If it is not, then we integrate along each smooth part of the curve and add the integrals. The parameter τ_i corresponds to the initial point (x_i, y_i), where $x_i = x(\tau_i)$, $y_i = y(\tau_i)$, and analogously for the final point (x_f, y_f). The integral of $đ\omega$ along the curve is defined as follows:

$$\int_i^f đ\omega = \int_{\tau_i}^{\tau_f} \left[\omega_x\big(x(\tau), y(\tau)\big)\frac{dx}{d\tau} + \omega_y\big(x(\tau), y(\tau)\big)\frac{dy}{d\tau}\right]d\tau.$$

Thus, we have reduced the problem to the integration of a function of one variable. Note that if $đ\omega = df$, then

$$\int_i^f df = \int_{\tau_i}^{\tau_f} \left(\frac{\partial f}{\partial x}\frac{dx}{d\tau} + \frac{\partial f}{\partial y}\frac{dy}{d\tau}\right)d\tau = \int_{\tau_i}^{\tau_f} \frac{df(x(\tau), y(\tau))}{d\tau}d\tau$$
$$= f\big(x(\tau_f), y(\tau_f)\big) - f\big(x(\tau_i), y(\tau_i)\big) = f(x_f, y_f) - f(x_i, y_i),$$

which means that the value of the integral depends only on the initial and final points, and not on the path linking these points.

Finally, we notice that if the curve can be presented as a set of points $(x, y(x))$, then $\tau = x$ and

$$\int_i^f đ\omega = \int_{x_i}^{x_f} \left[\omega_x\big(x, y(x)\big) + \omega_y\big(x, y(x)\big)\frac{\mathrm{d}y}{\mathrm{d}x} \right] \mathrm{d}x.$$

2.30 We use the equation of state of the ideal gas. In general, we should use an equation of state which describes gases better, e.g., the van der Waals equation of state. Then, however, we would have to know which gas it is, and find the parameters a and b for this gas in the tables. Since this information is not provided, we assume the equation of state independent of the gas, i.e., the ideal gas equation of state:

$$pV = nRT.$$

The initial state: $V_i = 120$ L, $T_i = (273.15 + 25)$ K $= 298.15$ K, $n_i = 5$ mol.
The final state: $V_f = 120$ L $= 120 \times 10^{-3}$ m^3, $T_f = 298.15$ K, $p_f = 0.5$ atm.

Substituting the parameters of the final state into the equation of state, we get the final value of the mole number n_f, i.e.,

$$n_f = \frac{p_f V_f}{RT_f} = \frac{0.5 \times 101325\ \mathrm{N\,m^{-2}} \times 120 \times 10^{-3}\ \mathrm{m^3}}{8.314\ \mathrm{J\,K^{-1}\,mol^{-1}} \times 298.15\ \mathrm{K}} = 2.453\ \mathrm{mol},$$

hence

$$\Delta n = n_f - n_i = 2.453 - 5 = -2.547\ \mathrm{mol}.$$

Δn is negative because the mole number has decreased; 2.547 mol of the gas has escaped from the vessel.

2.31 From the equation of state of the ideal gas, we get

$$p = \frac{nRT}{V} = \frac{1\ \mathrm{mol} \times 8.314\ \mathrm{J\,mol^{-1}\,K^{-1}} \times 298\ \mathrm{K}}{10^{-4}\ \mathrm{m^3}} \approx 2.48 \times 10^7\ \mathrm{Pa}.$$

The internal energy of a two-atomic gas is given by

$$U = \frac{5}{2}nRT = 2.5 \times 1\ \mathrm{mol} \times 8.314\ \mathrm{mol^{-1}\,K^{-1}} \times 298\ \mathrm{K} = 6193.9\ \mathrm{J}.$$

2.32 The van der Waals equation of state describes nitrogen better than the ideal gas equation of state. Performing calculations for both equations of state, we want to learn how good an approximation to a real gas the ideal gas is. From the van der Waals equation of state, we get

$$p = \frac{nRT}{V - nb} - \frac{an^2}{V^2} = \frac{8.314\ \mathrm{J\,K^{-1}} \times 298\ \mathrm{K}}{10^{-4}\ \mathrm{m^3} - 3.85 \times 10^{-5}\ \mathrm{m^3}} - \frac{0.1358\ \mathrm{J\,m^3}}{10^{-8}\ \mathrm{m^6}}$$
$$\approx 4.029 \times 10^7\ \mathrm{Pa} - 1.358 \times 10^7\ \mathrm{Pa} \approx 2.67 \times 10^7\ \mathrm{Pa}.$$

Thus, the pressure of the ideal gas differs by a few percent from the pressure obtained from the van der Waals equation of state. We notice, however, that this good agreement is fortuitous. It results from partial cancellation of two large terms in the van der Waals equation of state. Note that when the molar volume $v = V/n$ is small, i.e., comparable with the constant b, then the first term in the van der Waals equation is large and dominates over the second term.

To calculate the internal energy, we use the expression

$$U = \frac{5}{2}nRT - \frac{an^2}{V} = 6193.9\ \mathrm{J} - 1358\ \mathrm{J} \approx 4836\ \mathrm{J}. \tag{S2.1}$$

The first term is the same as for the ideal gas. This part of the internal energy comes from the kinetic energy of N_2 molecules. The second term comes from the potential energy of intermolecular attraction. At large densities and low temperatures, its absolute value becomes large compared to the first term, which leads to condensation of the gas.

2.33 When the gas density is small, as in this case, we obtain practically the same result for the pressure and internal energy, using the ideal gas and van der Waals equations of state. In the previous exercise, we assumed the volume $V = 0.1$ L, which corresponds to the molar density of 10 $\mathrm{mol\,L^{-1}}$. In the present case, the molar density amounts only to 0.001 $\mathrm{mol\,L^{-1}}$. We calculate the pressure and internal energy of the gas, using the van der Waals equation:

$$\begin{aligned} p &= \frac{nRT}{V - nb} - \frac{an^2}{V^2} = \frac{8.314\ \mathrm{J\,K^{-1}} \times 298\ \mathrm{K}}{1\ \mathrm{m^3} - 3.85 \times 10^{-5}\ \mathrm{m^3}} - \frac{0.1358\ \mathrm{J\,m^3}}{1\ \mathrm{m^6}} \\ &\approx 2477.67\ \mathrm{Pa} - 0.1358\ \mathrm{Pa} \approx 2477.53\ \mathrm{Pa}, \\ U &= \frac{5}{2}nRT - \frac{an^2}{V} = 6193.93\ \mathrm{J} - 0.1358\ \mathrm{J} \approx 6193.79\ \mathrm{J}. \end{aligned}$$

For the ideal gas, we get $p = 2477.57$ Pa and $U = 6193.93$ J. Thus, in the case of small molar densities, the corrections coming from intermolecular interactions are very small and can be neglected.

2.34 We use the expression for the internal energy of the photon gas: $U = \gamma V T^4$, where $\gamma = 7.56 \times 10^{-16}\ \mathrm{J\,m^{-3}\,K^{-4}}$. For $T = 298$ K, we get

$$U = 7.56 \times 10^{-16}\ \mathrm{J\,m^{-3}\,K^{-4}} \times 1\ \mathrm{m^3} \times 298^4\ \mathrm{K^4} \approx 6 \times 10^{-6}\ \mathrm{J},$$

and for $T = 400$ K, we get $U \approx 2 \times 10^{-5}$ J. The pressure is determined from the expression: $p = \gamma T^4/3 = U/(3V)$. For $T = 298$ K, we get $p \approx 2 \times 10^{-6}$ Pa, and for $T = 400$ K, $p \approx 6.5 \times 10^{-6}$ Pa.

Therefore, the internal energy of radiation contained in the photon gas is very small compared to the internal energy of the ideal gas, and the same concerns the pressure. The photon gas can be used to a fast transfer of energy (in lasers). We can imagine that the energy $E = 10^{-5}$ J is transferred during the time $t = 1$ fs $= 10^{-15}$ s (femtosecond lasers). It requires the power $P = E/t = 10^{10}$ W $= 10$ GW, which is the power generated by a big power station. For instance, the Bełchatów power station in Poland, generating the power of 4 GW, is the biggest conventional power station in Europe (it uses brown coal as fuel). The biggest water power stations in Brasil or USA generate the power of about 10 GW. Thus, to generate great power from a small amount of energy, the energy has to be delivered in a very short time.

2.35 For the ideal gas, we have $pV = nRT$ and $U = fnRT/2$, where f is the number of degrees of freedom per molecule ($f = 3$ for one atom), hence

$$p(U, V, n) = \frac{2U}{fV} = \frac{2mU}{fmV} = p(mU, mV, mn).$$

In the case of the van der Waals gas, we determine T from the expression for the internal energy, i.e.,

$$T = \frac{2}{fnR}\left(U + \frac{an^2}{V}\right),$$

and then substitute it into the expression for the pressure, hence

$$p = \frac{2U}{f(V - nb)} + \frac{2an^2}{fV(V - nb)} - \frac{an^2}{V^2}.$$

Using the last formula we verify that $p(mU, mV, mn) = p(U, V, n)$.

2.36 For the ideal gas $V(T, p, n) = nRT/p$, hence

$$-\frac{1}{V}\left(\frac{\partial V}{\partial p}\right)_{T,n} = \frac{nRT}{Vp^2} = \frac{1}{p}.$$

In the case of the van der Waals gas, to find the relation $V = V(T, p, n)$, the equation of state is to be written in the form of a third order equation for V, i.e.,

$$pV^2(V - nb) - nRTV^2 + an^2(V - nb) = 0,$$

and then the root of this equation is to be determined.[1] However, it is more convenient to calculate the derivative

$$\left(\frac{\partial p}{\partial V}\right)_{T,n} = -\frac{nRT}{(V - nb)^2} + \frac{2an^2}{V^3},$$

hence, we get

$$-\frac{1}{V}\left(\frac{\partial V}{\partial p}\right)_{T,n} = \left[\frac{nRTV}{(V - nb)^2} - \frac{2an^2}{V^2}\right]^{-1}.$$

In the case of the photon gas, the mole number n is not a thermodynamic variable, and from the equation of state $p = \gamma T^4/3$, it follows that $(\partial p/\partial V)_T = 0$, thus, $(\partial V/\partial p)_T$ is not well defined. However, we can express V as a function of U and p, i.e., $V = U/(3p)$, hence

$$-\frac{1}{V}\left(\frac{\partial V}{\partial p}\right)_U = \frac{1}{p}.$$

Thus, we obtain the same result as for the ideal gas, for which we differentiated at constant temperature. Note that from the equations of state for the ideal gas: $pV = nRT$ and $U = fnRT/2$, the relation $V = 2U/(fp)$ follows, hence

$$-\frac{1}{V}\left(\frac{\partial V}{\partial p}\right)_U = \frac{1}{p}.$$

---

[1] The equation has only one real root at sufficiently high temperatures.

2.37 For the ideal gas,

$$\frac{1}{V}\left(\frac{\partial V}{\partial T}\right)_{p,n} = \frac{nR}{pV} = \frac{1}{T}.$$

In the case of the photon gas, fixing pressure, we also fix temperature. Therefore, the above expression does not make sense. However, we can calculate

$$\frac{1}{V}\left(\frac{\partial V}{\partial U}\right)_{p} = \frac{1}{3Vp} = \frac{1}{U}.$$

2.38 For the ideal gas,

$$\left(\frac{\partial U}{\partial T}\right)_{V,n} = \frac{f}{2}nR,$$

and we obtain the same result for the van der Waals gas. In the case of the photon gas, we get

$$\left(\frac{\partial U}{\partial T}\right)_{V} = 4\gamma V T^3.$$

2.39 We make an assumption that the atmosphere is in thermodynamic equilibrium. This is not quite correct, since neither the temperature nor pressure of the atmosphere are constant, and also a macroscopic flow of air exists. Nevertheless, the assumption is good enough to estimate the amount of oxygen in the atmosphere. We calculate the volume V occupied by air. It is the volume of a spherical layer whose internal radius amounts to $R = 6500$ km (radius of the earth), and the external radius is equal to $R + h$, where $h = 10$ km. The temperature amounts to $T = 273.15 + 14$ K, and the pressure $p = 1$ atm. Since $h \ll R$, we get

$$V = \frac{4}{3}\pi(R+h)^3 - \frac{4}{3}\pi R^3 \approx 4\pi R^2 h \approx 5.3 \times 10^{18}\ \mathrm{m}^3.$$

The mole number is determined from the ideal gas equation of state:

$$n = \frac{pV}{RT} = \frac{101325\ \mathrm{Pa} \times 5.3 \times 10^{18}\ \mathrm{m}^3}{8.314\ \mathrm{J\,mol^{-1}\,K^{-1}} \times 287.15\ \mathrm{K}} \approx 2 \times 10^{20}\ \mathrm{mol}.$$

Since oxygen makes 21 % of the whole, the amount of oxygen in the atmosphere is equal to $n_{O_2} = 0.21n \approx 0.4 \times 10^{20}$ mol. The amount of 0.5×10^{16} mol of oxygen is used up yearly by living organisms, which means that even if the oxygen supply was not renewed it would disappear from the atmosphere only after $n_{O_2}/0.5 \times 10^{16} =$ 8000 years. Obviously, if the concentration of oxygen fell well below 21 % we would have serious problems with breathing. After a few hundred years, we would begin to feel the lack of oxygen in the atmosphere.

2.40 The data: $R = 700000$ km, $T = 6000$ K, $c = 3 \times 10^8$ m/s, $\gamma = 7.56 \times 10^{-16}\ \mathrm{J K^{-4} m^{-3}}$. Using the expression for u, we obtain the power per unit area, i.e.

$$I = \frac{uc}{4} = \frac{\gamma T^4 c}{4}.$$

Substituting the numbers, we get

$$I = \frac{7.56}{4} \times 10^{-16}\,\mathrm{J K^{-4} m^{-3}} \times 3 \times 10^{8}\,\mathrm{m s^{-1}} (6 \times 10^{3}\,\mathrm{K})^{4} \approx 7 \times 10^{7}\,\mathrm{W m^{-2}}.$$

The total power radiated from the sun surface, of the area $A = 4\pi R^2$, amounts to

$$P = AI = 4\pi (7 \times 10^{8}\,\mathrm{m})^{2} \times 7 \times 10^{7}\,\mathrm{W m^{-2}} \approx 4 \times 10^{26}\,\mathrm{W}.$$

It is an unimaginably great number. By comparison, the total energy consumption in 2008 amounted to 4.74×10^{20} J $\approx 132\,000$ TWh. With this level of energy consumption, the energy radiated by the sun during 1 second only would be enough for us for 840 000 years.

Exercises of Chapter 3

3.1 The process takes place at constant pressure p, hence, the heat absorbed by the system due to the change in the temperature from T_i to T_f amounts to

$$Q = nc_p(T_f - T_i) = nc_p \Delta T,$$

provided that the molar heat capacity at constant pressure, c_p, does not depend on temperature. Here we consider a monatomic ideal gas, for which $c_p = 5R/2$. The temperature difference follows from the equation of state $pV = nRT$:

$$\Delta T = \frac{p\Delta V}{nR},$$

hence

$$Q = \frac{5}{2} p\Delta V.$$

This result can also be derived directly from the first law of thermodynamics: $Q = \Delta U - W$. The change in the internal energy of the gas amounts to

$$\Delta U = \frac{3}{2} nR\Delta T = \frac{3}{2} p\Delta V,$$

and the work done on the gas in an isobaric process is equal to $W = -p\Delta V$, hence, $Q = (5/2)p\Delta V$. Substituting the data: $p = 1$ bar and $\Delta V = 30$ L, we get

$$Q = 2.5 \times 10^{5}\,\mathrm{Pa} \times 30 \times 10^{-3}\,\mathrm{m^3} = 7500\,\mathrm{J}.$$

3.2 In this case, the external pressure p_{ext} is fixed, but the gas pressure changes during the process from the initial value $p_i = 2$ atm to the final value $p_f = p_{\mathrm{ext}} = 1$ atm. The process is not quasi-static, thus, it is not reversible. However, if the system performs work on a reversible source of work, then we can calculate the work in the same way as in the previous exercise, i.e., $W = -p_{\mathrm{ext}}\Delta V$. The change in the volume is obtained from the equation of state $pV = nRT$ as follows:

$$\Delta V = V_f - V_i = nR\left(\frac{T_f}{p_{\mathrm{ext}}} - \frac{T_i}{p_i}\right).$$

Fig. S3.1 Volume of the vessel doubles at constant temperature and the pressure of the ideal gas decreases by a factor of two

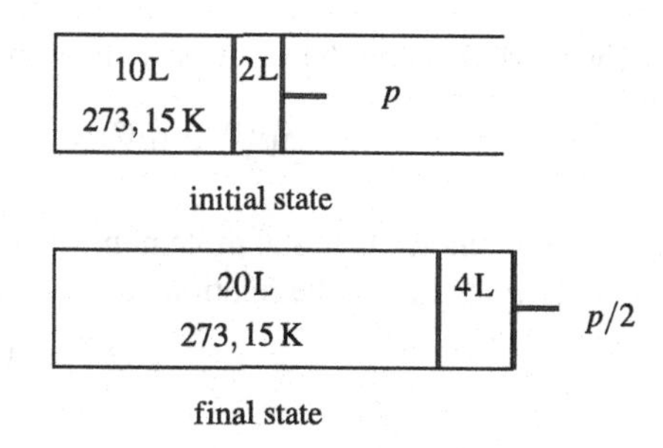

The change in the internal energy of the gas is equal to $\Delta U = (3/2)nR(T_f - T_i)$, hence

$$Q = \Delta U - W = nR\left[\frac{3}{2}(T_f - T_i) + \left(T_f - \frac{T_i p_{\text{ext}}}{p_i}\right)\right],$$

where $n = 5$, $T_i = 298.15$ K, $T_f = 293.15$ K. Substituting the data, we get $\Delta U = -311.8$ J and $W = -5989.2$ J, hence, $Q = 5677.4$ J. Thus, the system performs work ($W < 0$) and absorbs heat from the surroundings ($Q > 0$).

3.3 The system considered is shown in Fig. S3.1. Since the subsystems are in equilibrium, the gas pressure must be the same, hence

$$pV^{(1)} = n^{(1)}RT,$$
$$pV^{(2)} = n^{(2)}RT.$$

The temperature and mole numbers in the subsystems do not change, thus

$$p_i V_i^{(1)} = p_f V_f^{(1)},$$
$$p_i V_i^{(2)} = p_f V_f^{(2)},$$

hence

$$p_i V_i = p_f V_f,$$

where V_i and V_f denote the total volume of the system in the initial and final states. Thus, we have

$$\frac{V_f^{(1)}}{V_i^{(1)}} = \frac{V_f^{(2)}}{V_i^{(2)}} = \frac{p_i}{p_f} = \frac{V_f}{V_i} = 2.$$

The work done by each subsystem in the isothermal process amounts to

$$W^{(1)} = -n^{(1)}RT \ln \frac{V_f^{(1)}}{V_i^{(1)}} = -n^{(1)}RT \ln 2,$$

$$W^{(2)} = -n^{(2)}RT \ln \frac{V_f^{(2)}}{V_i^{(2)}} = -n^{(2)}RT \ln 2.$$

The work done by the whole system amounts to

$$W = W^{(1)} + W^{(2)} = -nRT\ln 2 = -nRT\ln\frac{V_f}{V_i},$$

where n denotes the total mole number of the gas. We know that $n^{(1)} = 10$, and $n^{(2)}$ is determined from the relation

$$\frac{n^{(2)}}{n^{(1)}} = \frac{V_i^{(2)}}{V_i^{(1)}} = \frac{1}{5},$$

hence $n^{(2)} = 2$ mol. Substituting the data, we get

$$W^{(1)} = -15741\ \text{J}, \qquad W^{(2)} = -3148\ \text{J}, \qquad W = -18889\ \text{J}.$$

The internal energy of the subsystems and whole system does not change, since $T = const$ and we consider the ideal gas, thus

$$Q = \Delta U - W = -W = 18889\ \text{J}.$$

3.4 Due to the condition of mechanical equilibrium, the gas pressure is the same in each subsystem. The temperature is also the same, since the system is in thermal equilibrium with the surroundings, hence

$$p_i V_i^{(j)} = n^{(j)} R T_i,$$
$$p_f V_f^{(j)} = n^{(j)} R T_f,$$

where $j = 1, 2, 3$ numbers the subsystems. Summing up over all subsystems, we get

$$p_i V = nRT_i,$$
$$p_f V = nRT_f,$$

where n denotes the total number of moles of the gas in the system, thus

$$\frac{T_i}{p_i} = \frac{T_f}{p_f},$$

hence it follows that

$$V_f^{(j)} = n^{(j)} R\frac{T_f}{p_f} = n^{(j)} R\frac{T_i}{p_i} = V_i^{(j)}.$$

The gas in the subsystems does not perform any work because the volume does not change. The mole number for each subsystem is determined from the equation of state

$$n^{(j)} = \frac{p_i V_i^{(j)}}{RT_i}.$$

The change in the internal energy of each subsystem amounts to

$$\Delta U^{(j)} = \frac{3}{2} n^{(j)} R(T_f - T_i) = \frac{3}{2} p_i V_i^{(j)}\left(\frac{T_f}{T_i} - 1\right).$$

Since the system performs no work, the heat Q absorbed by the system is equal to the change in its internal energy, i.e.

$$Q = \Delta U = \sum_{j=1}^{3} \Delta U^{(j)} = \frac{3}{2} p_i V \left(\frac{T_f}{T_i} - 1 \right).$$

Therefore, it does not matter if we heat up subsystems, which are thermally insulated from one another but remain in thermal contact with the surroundings, or if we heat up the whole system of the volume $V = V^{(1)} + V^{(2)} + V^{(3)}$.

3.5 We solve this problem in a similar way as in Exercise 3.3. The process is isothermal and the pressure is the same in each subsystem. From the isotherm equation $pV = const$, it follows that

$$p_f V_f^{(j)} = p_i V_i^{(j)},$$

for each subsystem j. Since $V_f^{(1)} / V_i^{(1)} = 4$, we have

$$\frac{V_f^{(j)}}{V_i^{(j)}} = \frac{p_i}{p_f} = 4$$

for all subsystems. The work done in a reversible isothermal process by the jth subsystem amounts to

$$W^j = -n^{(j)} RT \ln \frac{V_f^{(j)}}{V_i^{(j)}} = -n^{(j)} RT \ln 4.$$

To calculate the work done by the whole system, we sum up over all subsystems, hence

$$W = \sum_{i=1}^{3} W^{(i)} = -nRT \ln 4,$$

where n is the total mole number of the gas in the system. Since we know the initial volume of each subsystem, hence, also the total volume ($V_i = 16$ L), the initial pressure ($p_i = 1$ atm) and the temperature $T = 273.15$ K, we can determine n from the equation of state

$$n = \frac{p_i V_i}{RT}.$$

Substituting it into the expression for the work, we get

$$W = -p_i V_i \ln 4 = -101325 \text{ Pa} \times 16 \times 10^{-3} \text{ m}^3 \ln 4 \approx -2247 \text{ J}.$$

We notice that actually the temperature is not needed in this exercise. The internal energy does not change in the isothermal process, thus, the heat supplied to the system $Q = -W$.

3.6 The gas expanding to the vacuum performs no work ($W = 0$), since the external pressure $p_{\text{ext}} = 0$. The process is adiabatic, thus, by definition no heat is transferred between the system and surroundings ($Q = 0$), hence, the internal energy of the system does not change either ($\Delta U = Q + W = 0$). Since the internal energy in the initial state, U_i, is the same as in the final state, U_f, we have

$$U_f = \frac{3}{2} nRT_f = \frac{3}{2} nRT_i - \frac{an^2}{V_i} = U_i,$$

where we have used the assumption that $V_f = \infty$. From the above equation, we determine the change in the temperature

$$\Delta T = T_f - T_i = -\frac{2an}{3RV_i}.$$

Why does the temperature decrease when the van der Waals gas expands, even though it performs no work? Attractive intermolecular interactions cause molecules to slow down when they move away. Since the speed of molecules decreases, the gas temperature, which is proportional to the average kinetic energy of molecules, also decreases.

3.7 The work done against the constant external pressure p_{ext} amounts to $W = -p_{\text{ext}} \Delta V$, hence, the change in the internal energy in an adiabatic process is $\Delta U = W = -p_{\text{ext}} \Delta V$. For 1 mol of the ideal gas, $\Delta U = 3R\Delta T/2$, hence, also the work $W = 3R\Delta T/2$. Comparing both expressions for W, we determine the change in the volume:

$$\Delta V = -\frac{3R\Delta T}{2p_{\text{ext}}}.$$

3.8 Since we consider a reversible adiabatic process, we can make use of the adiabat equation: $TV^{2/3} = const$, hence

$$\frac{V_f}{V_i} = \left(\frac{T_i}{T_f}\right)^{3/2} = 4^{3/2} = 8,$$

for $T_f = T_i/4$. In this process, the volume has increased 8 times. The work done by the gas is equal to the change in its internal energy, hence

$$W = \Delta U = \frac{3}{2} nR(T_f - T_i) = -\frac{9}{8} nRT_i.$$

3.9 Substituting the temperature determined from the ideal gas equation of state, $T = pV/(nR)$, to the adiabat equation expressed in the variables T and V (see the previous exercise), we obtain the adiabat equation in the variables p and V, i.e.,

$$pV^{5/3} = const,$$

hence

$$\frac{p_f}{p_i} = \left(\frac{V_i}{V_f}\right)^{5/3} = \left(\frac{1}{4}\right)^{5/3} \approx 0.1.$$

For the initial pressure $p_i = 1$ bar, we get the final pressure $p_f \approx 0.1$ bar.

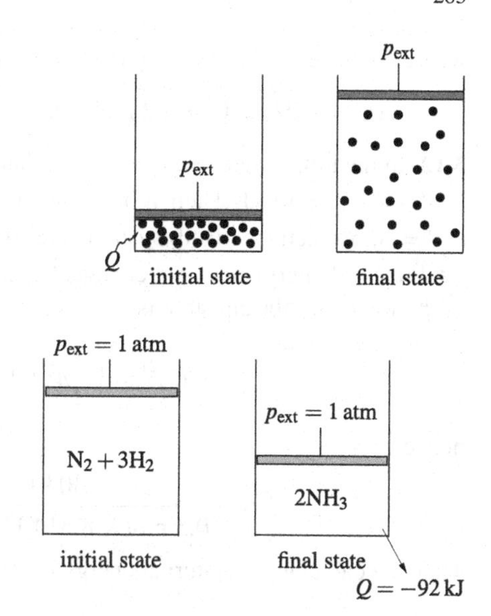

Fig. S3.2 Evaporation of water at the external pressure p_{ext}. In the initial state, water is a liquid, and in the final state, we have water vapour. Q is the heat supplied to change liquid water into vapour

Fig. S3.3 Synthesis of ammonia occurs at the pressure $p_{ext} = 1$ atm. As a result of the reaction, the heat $Q = -92$ kJ is given off to the surroundings and the final volume of the system is half of the initial volume

3.10 The process of water evaporation at constant pressure is shown schematically in Fig. S3.2. In the initial state, 1 mol of liquid water occupies the volume $V_i = 18$ cm^3 and fills up the whole available space of the vessel. The pressure is constant and equals $p_{ext} = 1$ atm. In the final state, only water vapour is present, which occupies the volume $V_f = 30.6$ L. A change of 1 mol of liquid water into vapour at the pressure of 1 atm requires supply of the heat $Q = 40670$ J. Moreover, when liquid water evaporates it performs the work W against the external pressure $p_{ext} = 1$ atm, where

$$W = -p_{ext}(V_f - V_i).$$

The change in the internal energy of the system amounts to

$$\Delta U = Q + W = Q - p_{ext}(V_f - V_i).$$

Substituting the data, we get

$$\Delta U = 40670\text{ J} - 101325\text{ Pa} \times \left(30.6 \times 10^{-3} - 18 \times 10^{-6}\right)\text{ m}^3 \approx 37569\text{ J}.$$

Thus, the work done by the system is much smaller then the heat supplied.

3.11 The process considered is shown schematically in Fig. S3.3. The heat given off in this reaction, $Q = -92$ kJ, comes from the energy of chemical bonds. $Q < 0$ since the system gives off the heat. In the energetic balance, we have to take into account also the work done by (or on) the system due to the change in its volume. All substances are treated as ideal gases. In the initial state, there are 3 mol of H_2 and 1 mol of N_2, which gives the total mole number $n_i = 4$ mol. In the final state, there are $n_f = 2$ mol of ammonia (NH_3). The temperature and pressure have the same values in the initial and final states: $T = 298$ K and $p_{ext} = 1$ atm. The change in the internal energy

$$\Delta U = Q + W = Q - p_{ext}(V_f - V_i) = Q - RT(n_f - n_i),$$

where we have used the ideal gas equation of state. Substituting the data, we get

$$\Delta U = -92 \times 10^3\ \mathrm{J} + 2\ \mathrm{mol} \times 8.314\ \mathrm{J\,K^{-1}\,mol^{-1}} \times 298\ \mathrm{K} \approx -87\ \mathrm{kJ}.$$

3.12 In the initial state, we have two isolated systems with the total internal energy $U_i^{(1)} + U_i^{(2)} = 30$ kJ. Their mole numbers are constant and equal to $n^{(1)} = 2$ and $n^{(2)} = 3$, respectively. Both systems are isolated from the surroundings, thus, their total internal energy does not change, hence, $U_f^{(1)} + U_f^{(2)} = 30$ kJ. The temperature of the final equilibrium state is T_f. We determine T_f from the equation

$$U_f^{(1)} + U_f^{(2)} = \left(\frac{3}{2}n^{(1)} + \frac{5}{2}n^{(2)}\right) RT_f = 30\ \mathrm{kJ},$$

hence

$$T_f = \frac{30\ \mathrm{kJ}}{10.5\ \mathrm{mol} \times 8.314\ \mathrm{J\,K^{-1}\,mol^{-1}}} \approx 343.65\ \mathrm{K}.$$

Then we calculate the internal energy of each system:

$$U_f^{(1)} = \frac{3}{2}n^{(1)} RT_f \approx 8571\ \mathrm{J},$$

$$U_f^{(2)} = \frac{5}{2}n^{(2)} RT_f \approx 21429\ \mathrm{J}.$$

3.13 The total internal energy in the final state is the same as in the initial state, hence

$$U_i = \gamma V\left(T^{(1)4} + T^{(2)4}\right) = U_f = 2\gamma V T_f^4,$$

and for the final temperature, we get

$$T_f = \left(\frac{T^{(1)4} + T^{(2)4}}{2}\right)^{1/4}.$$

3.14 The initial temperatures of the metal and water amount to $T_{1i} = 400$ K and $T_{2i} = 294$ K. The final temperature of water and the metal immersed in it amounts to $T_f = 300$ K. The mass of the metal and water is $m_1 = 1$ kg and $m_2 = 0.3$ kg, respectively. For the change in their internal energy, we get

$$\Delta U_1 = c_1 m_1 (T_f - T_{1i}),$$

$$\Delta U_2 = c_2 m_2 (T_f - T_{2i}),$$

respectively, where c_1 and c_2 denote their specific heat. The SI derived unit of the specific hit is $\mathrm{J\,kg^{-1}\,K^{-1}}$. The whole system, i.e., the metal and water, is isolated from the surroundings, hence, $\Delta U_1 + \Delta U_2 = 0$, and we can determine the ratio

$$\frac{c_2}{c_1} = -\frac{m_1(T_f - T_{1i})}{m_2(T_f - T_{2i})}.$$

Substituting the data, we get

$$\frac{c_2}{c_1} = \frac{1\ \text{kg} \times 100\ \text{K}}{0.3\ \text{kg} \times 6\ \text{K}} \approx 55.6.$$

Thus, the specific heat of water is 55.6 times bigger than the specific heat of the metal. Substances of good cooling capabilities are those of high specific heat in the given temperature range, which means that they are difficult to warm up. Evaporation of a liquid can also be used for cooling. In the latter case, we choose substances of high heat of evaporation.

3.15 Since the internal energy is an extensive quantity, we have

$$U(T, V, n) = nu(T, V/n) = nu(T, v),$$

where u denotes the molar internal energy, and $v = V/n$ is the molar volume. For $n = 4$, we have

$$U(T, V, 4\ \text{mol}) = ATV^3 = AT(4v)^3\ \text{mol}^3 = (4\ \text{mol})u(T, v),$$

hence, $u(T, v) = BTv^3$, where $B = 16A\ \text{mol}^2 = 160\ \text{J}\,\text{mol}^2\,\text{K}^{-1}\,\text{cm}^{-9}$, and

$$U(T, V, n) = nu(T, V/n) = BT\frac{V^3}{n^2}.$$

3.16 This exercise is solved in a similar way as the previous one

$$U(T, V, 2\ \text{mol}) = aVT^4 = (2\ \text{mol})avT^4 = (2\ \text{mol})u(T, v),$$

hence $u(T, v) = avT^4$, where a is independent of n. Thus, the expression $U = aVT^4$ holds for any n.

3.17 The total energy transferred to the substance in the form of heat, i.e.,

$$Q = 12\ \text{V} \times 1\ \text{A} \times 3000\ \text{s} = 36\ \text{kJ},$$

caused the substance temperature to increase by $\Delta T = 5.5$ K. If we assume that in this range of temperature, the heat capacity of the substance, C, is independent of temperature, then we can calculate it from the formula: $C = Q/\Delta T \approx 6545\ \text{J}\,\text{K}^{-1}$.

3.18 Infinitesimal work performed by the gas is equal to $đW^* = -đW = p\text{d}V$, hence, in the process of isothermal expansion from the volume V_i to $V_f > V_i$, we get

$$W^* = nRT \int_{V_i}^{V_f} \left(\frac{1}{V} + \frac{n^2 B(T)}{V^2}\right) \text{d}V = nRT \ln \frac{V_f}{V_i} - n^2 RT B(T) \left(\frac{1}{V_f} - \frac{1}{V_i}\right).$$

The work done by the ideal gas amounts to $W^* = nRT \ln(V_f/V_i)$. At high temperatures $B(T) > 0$, and the work done by the real gas is greater than in the case of the ideal gas.

3.19 To calculate the work done by the gas, we integrate $đW^* = p\text{d}V$ from the initial volume V_i to the final volume $V_f > V_i$. For the van der Waals equation of state, we get

$$W^* = \int_{V_i}^{V_f} \left(\frac{nRT}{V - nb} - \frac{an^2}{V^2}\right) \text{d}V = nRT \ln\left(\frac{V_f - nb}{V_i - nb}\right) + an^2 \left(\frac{1}{V_f} - \frac{1}{V_i}\right).$$

The constants a and b are positive. The presence of a lowers the gas pressure in relation to the ideal gas pressure, and the presence of b increases the gas pressure. The same tendency holds for the work done by the gas. Then we expand the term $1/(V - nb)$ in a power series of nb/V:

$$\frac{1}{V - nb} = \frac{1}{V(1 - nb/V)} = \frac{1}{V}\left[1 + \frac{nb}{V} + \left(\frac{nb}{V}\right)^2 + \cdots\right].$$

For $V \gg nb$, we have

$$\frac{1}{V - nb} = \frac{1}{V(1 - nb/V)} \approx \frac{1}{V} + \frac{nb}{V^2}.$$

Substituting this approximation into the van der Waals equation of state, we get the expression for the function $B(T)$ introduced in the previous exercise:

$$B(T) = b - \frac{a}{RT}.$$

The above approximation can be used when the molar volume of the gas, $v = V/n$, is large compared to the parameter b. We notice also that $B(T)$ is negative at low temperatures and positive at high temperatures.

3.20 We use the form of the internal energy differential at constant volume: $dU = nc_v dT$. Integrating this relation from the temperature T_i to T_f, we find the change in the molar internal energy:

$$\Delta u = u_f - u_i = \int_{T_i}^{T_f} c_v dT = A(T_f - T_i) + \frac{1}{2}B\left(T_f^2 - T_i^2\right).$$

3.21 We proceed in the same way as in the previous exercise, hence

$$\Delta u = u_f - u_i = \int_{T_f}^{T_i} c_v dT = A(T_f - T_i) + \frac{1}{2}B\left(T_f^2 - T_i^2\right) + C\left(\frac{1}{T_f} - \frac{1}{T_i}\right).$$

3.22 In this case, the heat capacity per unit volume is given, thus, $dU = Vc_v dT$, and

$$\Delta U = U_f - U_i = V\int_{T_f}^{T_i} c_v dT = V\gamma\left(T_f^4 - T_i^4\right).$$

3.23 To avoid damaging of the vessel, the maximum pressure obtained by heating of the gas cannot exceed 100 atm, i.e., $p_f < 100$ atm. From the equation of state, we get

$$\frac{p_i}{T_i} = \frac{p_f}{T_f},$$

hence, the final temperature must satisfy the inequality

$$T_f = \frac{p_f T_i}{p_i} < 100 T_i = 29800\ \text{K}.$$

The molar heat capacity at constant volume amounts to $c_v = 5R/2$, and the mole number $n = p_i V/RT_i$. The heat to be delivered to warm up the gas from $T_i = 298$ K to $T_f = 29800$ K amounts to

$$Q = \frac{5}{2}nR(T_f - T_i) = \frac{5}{2}p_i V\left(\frac{T_f}{T_i} - 1\right) \approx 601.87 \text{ kJ}.$$

Note that the maximum final temperature is much higher than the melting point of any material on the earth. Therefore, it is not possible, in practice, to warm up the gas in the vessel, to achieve the pressure of 100 atm.

3.24 In the initial state, $U_i = 3n_i RT/2$, and in the final state, $U_f = 3n_f RT/2$. The change in the internal energy caused by the reduction of the mole number $\Delta n = n_f - n_i$ amounts to

$$\Delta U = U_f - U_i = \frac{3}{2}RT\Delta n.$$

3.25 As in the previous exercise, the change in the internal energy amounts to

$$\Delta U = \frac{3}{2}RT\Delta n.$$

The process occurs at constant pressure, hence, the work done in the process amounts to

$$W = -p\Delta V = -RT\Delta n,$$

where we have used the equation of state $pV = nRT$. According to the first law of thermodynamics, $\Delta U = Q + W + Z$. The change in the internal energy at constant temperature is related to the change in the mole number of the gas, hence, $\Delta U = Z$, from which it follows that

$$Q = -W = RT\Delta n.$$

Since $\Delta n < 0$, we have $W > 0$ and $Q < 0$, i.e., the work is performed on the system, and the system gives off the heat.

Exercises of Chapter 4

4.1 The process considered is shown in Fig. S4.1. At the first stage, $V = V_1$, hence $W_1 = 0$. The change in the temperature amounts to $\Delta T = V_1(p_2 - p_1)/R$, and the internal energy changes by

$$\Delta U_1 = \frac{3}{2}V_1(p_2 - p_1),$$

hence

$$Q_1 = \Delta U_1 - W_1 = \frac{3}{2}V_1(p_2 - p_1).$$

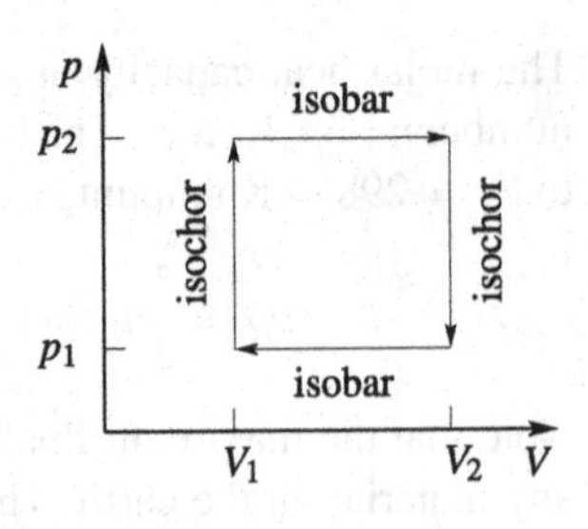

Fig. S4.1 Isochoric–isobaric cycle

At the second stage, $p = p_2$, the work amounts to

$$W_2 = -p_2(V_2 - V_1),$$

and the temperature changes by $\Delta T = p_2(V_2 - V_1)/R$, hence

$$\Delta U_2 = \frac{3}{2}p_2(V_2 - V_1),$$
$$Q_2 = \Delta U_2 - W_2 = \frac{5}{2}p_2(V_2 - V_1).$$

At the third stage, $V = V_3$, $W_3 = 0$ and

$$\Delta U_3 = \frac{3}{2}V_2(p_1 - p_2),$$
$$Q_3 = \Delta U_3 - W_3 = \frac{3}{2}V_2(p_1 - p_2).$$

At the fourth stage, we have $p = p_1$,

$$W_4 = -p_1(V_1 - V_2),$$
$$\Delta U_4 = \frac{3}{2}p_1(V_1 - V_2),$$
$$Q_4 = \Delta U_4 - W_4 = \frac{5}{2}p_1(V_1 - V_2).$$

The process is a cycle, thus

$$\Delta U = \Delta U_1 + \Delta U_2 + \Delta U_3 + \Delta U_4 = 0.$$

The total work

$$W = W_1 + W_2 + W_3 + W_4 = -(p_2 - p_1)(V_2 - V_1),$$

and the total heat

$$Q = Q_1 + Q_2 + Q_3 + Q_4 = \Delta U - W = (p_2 - p_1)(V_2 - V_1).$$

Thus, the system performs work ($W < 0$) due to the heat absorbed.

4.2 The engine efficiency η_e is equal to the ratio of the work done by the engine, $W^* = -W$, to the heat absorbed. In the cycle considered, the engine absorbs the

Fig. S4.2 Carnot cycle

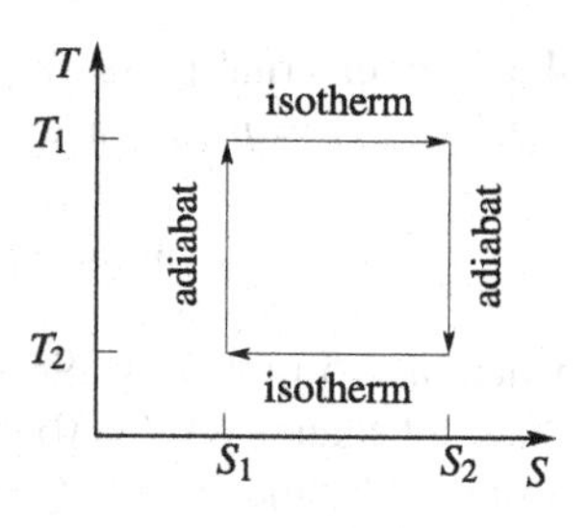

heat $Q_1 = (3/2)V_1(p_2 - p_1) > 0$ during the isochoric warming up of the gas, and the heat $Q_2 = (5/2)p_2(V_2 - V_1) > 0$ during the isobaric expansion, hence

$$\eta_e = \frac{2(p_2 - p_1)(V_2 - V_1)}{3V_1(p_2 - p_1) + 5p_2(V_2 - V_1)}.$$

As we can see, the efficiency depends on the working substance used in the engine, because the heat absorbed depends on the molar heat capacity. For instance, in the case of a two-atomic ideal gas, the engine efficiency is smaller than in the case of a monatomic gas.

4.3 The cycle can be a combination of isothermal, isobaric, adiabatic and isochoric processes. Most common is the combination of isochoric or isobaric processes with adiabatic processes, for instance, an adiabatic–isochoric cycle (Otto cycle). An interesting combination is used in the jet engine, in which air flows through a combustion chamber, and the cycle is not closed. It has two adiabatic stages and the stage of combustion, which occurs at constant pressure.

4.4 The Carnot cycle is shown in Fig. S4.2. The changes in the entropy in the isothermal processes at the temperatures T_1 and T_2 amount to $\Delta S = S_2 - S_1$ and $-\Delta S = S_1 - S_2$, respectively. The heat absorbed by the system at $T = T_1$ is equal to $Q_1 = T_1(S_2 - S_1)$, and the heat given off to the thermostat at $T = T_2$ is equal to $Q_2 = T_2(S_1 - S_2)$. Since $\Delta U = 0$ in a cyclic process, we have

$$\Delta U = Q_1 + Q_2 - W^* = 0,$$

where $W^* = -W$ is the work done by the system, hence

$$W^* = T_1(S_2 - S_1) + T_2(S_1 - S_2) = (T_1 - T_2)(S_2 - S_1), \tag{S4.1}$$

which is the area of the rectangle shown in Fig. S4.2.

4.5 The change in the entropy of the ideal gas in the isothermal process amounts to

$$\Delta S_{\text{sys}} = nR\ln\frac{V_f}{V_i} = nR\ln\frac{p_i}{p_f},$$

hence, for $n = 5$ mol, $p_i = 2$ atm, $p_f = 1$ atm, we get $\Delta S_{\text{sys}} = (5\,\text{mol})R\ln 2$. Since the process is reversible, we have

$$\Delta S_{\text{sys}} + \Delta S_{\text{sur}} = 0, \tag{S4.2}$$

hence $\Delta S_{\text{sur}} = -\Delta S_{\text{sys}}$.

4.6 The external pressure $p_{\text{ext}} = 1$ atm is constant, and the final pressure of the ideal gas $p_f = p_{\text{ext}}$, hence

$$W = -p_{\text{ext}}(V_f - V_i) = -nRT\left(1 - \frac{p_f}{p_i}\right),$$

where $n = 5$ mol, $T = 298$ K and $p_i = 2$ atm. The internal energy of the ideal gas does not change ($\Delta U = 0$), because the temperature does not change, thus, the heat absorbed by the system $Q = -W$. The change in the entropy of the system is the same as in the previous exercise, i.e.,

$$\Delta S_{\text{sys}} = nR\ln\frac{V_f}{V_i} = nR\ln\frac{p_i}{p_f} = (5\text{ mol})R\ln 2,$$

because the initial and final states of the system are also the same. However, the change in the entropy of the surroundings, which supply the heat Q to the system at constant temperature T, is different. Since the surroundings are treated as a reservoir of heat and volume, it can be assumed that they are in thermodynamic equilibrium all the time, and the change in their entropy amounts to

$$\Delta S_{\text{sur}} = -\frac{Q}{T} = -nR\left(1 - \frac{p_f}{p_i}\right) = -\left(\frac{5}{2}\text{ mol}\right)R. \tag{S4.3}$$

The total change in the entropy of the system and surroundings amounts to

$$\Delta S = \Delta S_{\text{sys}} + \Delta S_{\text{sur}} = (5\text{ mol})R\left(\ln 2 - \frac{1}{2}\right) > 0.$$

Thus, the process is irreversible, since $\Delta S > 0$.

4.7 So far we have considered the Carnot engine working between two thermostats of infinite heat capacity. Here, we assume that the heat capacity of the system at higher temperature, i.e., the system we want to cool down, is finite. This means that the temperature of the system decreases with each cycle of the engine. The engine operates until the temperature of the system becomes equal to the temperature of the radiator. Then the engine efficiency η_e reaches zero. We assume that the work done in one cycle is small compared to the total work needed to lower the system temperature to the value T_2. This assumption allows us to treat the work done by the engine in one cycle, W^*, and the heat drawn from the system by the working substance in the engine, Q_1, as infinitesimal quantities. We denote by T the actual temperature of the system, thus, the engine efficiency at T amounts to $\eta_e(T) = 1 - T_2/T$. According to the definition of η_e, we have

$$đW^* = \eta_e(T)đQ_1 = -\eta_e(T)nc_v\mathrm{d}T.$$

Integrating this equality from the initial temperature of the system, T_1, to the final temperature T_2, we obtain the total work required to cool down the system, i.e.

$$W^* = -nc_v\int_{T_1}^{T_2}\eta_e(T)\mathrm{d}T = nc_v\int_{T_2}^{T_1}\left(1 - \frac{T_2}{T}\right)\mathrm{d}T = nc_v\left(T_1 - T_2 + T_2\ln\frac{T_2}{T_1}\right).$$

The work W^* is positive, since we integrate a non-negative function $\eta_e(T)$.

4.8 Both systems and the working substance in the Carnot engine form an adiabatically isolated system which performs work on the surroundings. The system at higher temperature is treated as a heat container, and the system at lower temperature is treated as a radiator. Since the engine works reversibly, the total entropy of the composite system does not change. The change in the entropy of the heat container and radiator during one cycle is denoted by $\mathrm{d}S_1$ and $\mathrm{d}S_2$, respectively. The change in the entropy of the working substance in one cycle equals zero, hence, also $\mathrm{d}S_1 + \mathrm{d}S_2 = 0$. Denoting by T' and T'' the actual temperatures of the heat container and radiator, respectively, and using the relation between the heat capacity and entropy, we get the following equation:

$$\mathrm{d}S_1 + \mathrm{d}S_2 = C_V \frac{\mathrm{d}T'}{T'} + C_V \frac{\mathrm{d}T''}{T''} = 0.$$

It is easy to verify that the solution of the above equation can be presented as follows:

$$\ln \frac{T'}{T_1} + \ln \frac{T''}{T_2} = 0,$$

because the value $T' = T_1$ corresponds to $T'' = T_2$. At the final temperature T_f, we have $T' = T'' = T_f$, hence, $T_f = \sqrt{T_1 T_2}$ is the geometric mean of the initial temperatures of the heat container and radiator. In the case of direct thermal contact of the heat container and radiator, we have

$$C_V(T_1 - T_f) = C_V(T_f - T_2),$$

hence, the final temperature $T_f = (T_1 + T_2)/2$ is the arithmetic mean of the initial temperatures. For $T_1 \neq T_2$, the inequality $\sqrt{T_1 T_2} < (T_1 + T_2)/2$ holds, thus, the final temperature obtained by means of the Carnot engine is lower than in the case of direct thermal contact. It is so because a part of the internal energy is transferred to the surroundings in the form of work.

4.9 The change in the molar entropy follows from the relation $c_v = T(\partial s/\partial T)_v$, hence

$$\Delta s = \int_{T_i}^{T_f} (A + BT - CT^{-2}) \frac{\mathrm{d}T}{T} = A \ln \frac{T_f}{T_i} + B(T_f - T_i) + \frac{C}{2}\left(\frac{1}{T_f^2} - \frac{1}{T_i^2}\right).$$

4.10 Since $đQ = V c_v \mathrm{d}T$, from the relation $đQ = T\mathrm{d}S$, we get

$$\Delta S = V \int_{T_i}^{T_f} c_v \frac{\mathrm{d}T}{T} = V 4\gamma \int_{T_i}^{T_f} T^2 \mathrm{d}T = \frac{4}{3} V\gamma (T_f^3 - T_i^3).$$

4.11 We supply the heat Q to the system, using either an electric heater or a heat pump. We assume that the whole work of the electric current in the heater, W_h, changes into heat, hence $Q = W_h$. In the case of a heat pump, the work of the electric current, W_p, is related to the heat supplied to the system, Q, by

$$Q = \eta_p W_p = \frac{T_1}{T_1 - T_2} W_p,$$

where η_p denotes the efficiency of the heat pump, $T_1 = 23$ °C is the temperature we want to maintain at home, and $T_2 = 0$ °C is the outdoor temperature. For the given temperatures, we have $\eta_p \approx 12.9$. This is a theoretical factor by which we can lower the electricity bill, using a heat pump instead of an electric heater. In practice, the profit is smaller and the installation of a heat pump is expensive.

4.12 The artesian well serves here as a radiator at the temperature $T_2 = 278.15$ K, whereas air above the Australian desert serves as a heat container at the temperature $T_1 = 293.15$ K. In these conditions, the efficiency of the Carnot engine $\eta_e \approx 0.05$. The heat Q_1 drawn by the working substance in the engine is obtained from the formula

$$Q_1 = \frac{W^*}{\eta_e}, \tag{S4.4}$$

where W^* is the work done by the engine. The heat Q_2 which flows to the radiator (the well) follows from the relation

$$Q_1 + W + Q_2 = Q_1 - W^* - Q_2^* = 0, \tag{S4.5}$$

hence

$$Q_2^* = \left(\frac{1}{\eta_e} - 1\right) W^* = \frac{T_2}{T_1 - T_2} W^*.$$

Substituting $W^* = 500$ kJ, we get $Q_2^* \approx 9272$ kJ.

4.13 We use the relations:

$$T = \left(\frac{\partial U}{\partial S}\right)_{V,n} = 3B\frac{S^2}{nV},$$
$$p = -\left(\frac{\partial U}{\partial V}\right)_{S,n} = B\frac{S^3}{nV^2}.$$

Then, from the equation of state $U = BS^3/(nV)$, we determine the entropy as a function of U, V and n, i.e.,

$$S = \left(\frac{UVn}{B}\right)^{1/3}.$$

Substituting S into the expressions for T and p, we get:

$$T = 3\left(\frac{BU^2}{nV}\right)^{1/3},$$
$$p = \frac{U}{V}.$$

4.14 From the extensiveness of entropy, we have $S(U, V) = Vs(u)$, where $s = S/V$ and $u = U/V$. Using the relation

$$\frac{1}{T} = \left(\frac{\partial S}{\partial U}\right)_V = \frac{ds}{du} = \frac{ds}{dT}\frac{dT}{du},$$

and the equation of state $u = \gamma T^4$, we get

$$\frac{ds}{dT} = \frac{1}{T}\frac{du}{dT} = 4\gamma T^2,$$

hence

$$S = Vs = \frac{4}{3}\gamma V T^3.$$

Then, expressing T as a function of U and V, we get

$$S = \frac{4}{3}(\gamma V)^{1/4} U^{3/4}.$$

The pressure of the photon gas is determined from the relation

$$\frac{p}{T} = \left(\frac{\partial S}{\partial V}\right)_U = \frac{1}{3}\gamma^{1/4}\left(\frac{U}{V}\right)^{3/4} = \frac{1}{3}\gamma T^3,$$

hence, $p = \gamma T^4/3$. This is the second of the two equations of state for the photon gas.

4.15 We use the relation

$$ds = \frac{1}{T}du + \frac{p}{T}dv.$$

From the equation of state $u = 3RT/2 - a/v$, we obtain $du = (3R/2)dT + (a/v^2)dv$, which is then substituted into the expression for ds. Using the second equation of state, we get eventually

$$ds = \frac{3R}{2T}dT + \frac{R}{v-b}dv.$$

Integrating both sides of this equation, we obtain the molar entropy as a function of T and v:

$$s(T, v) = s_0 + \frac{3}{2}R\ln\left(\frac{T}{T_0}\right) + R\ln\left(\frac{v-b}{v_0-b}\right), \tag{S4.6}$$

where the constants T_0, v_0 and $s_0 = s(T_0, v_0)$ define the reference state. To express the molar entropy as a function of u and v, we determine $T = 2(u + a/v)/(3R)$, hence

$$s(u, v) = s_0 + \frac{3}{2}R\ln\left(\frac{u + a/v}{u_0 + a/v_0}\right) + R\ln\left(\frac{v-b}{v_0-b}\right),$$

where $u_0 = 3RT_0/2 - a/v_0$. It is easy to verify that differentiating $s(u, v)$ with respect to u and v, and using the relations $(\partial s/\partial u)_v = 1/T$ and $(\partial s/\partial v)_u = p/T$, we recover both equations of state.

4.16 First, we introduce the molar quantities: $s_A = S_A/n_A$, $u_A = U_A/n_A$ and $v_A = V_A/n_A$, and analogously for the system B. Then we have

$$s_A = \left(\frac{u_A v_A}{D}\right)^{1/3}, \qquad s_B = \left(\frac{u_B v_B}{D}\right)^{1/3},$$

hence

$$\left(\frac{\partial s_A}{\partial u_A}\right)_{v_A} = \frac{1}{3}\left(\frac{v_A}{Du_A^2}\right)^{1/3} = \frac{1}{T_A},$$
$$\left(\frac{\partial s_A}{\partial v_A}\right)_{u_A} = \frac{1}{3}\left(\frac{u_A}{Dv_A^2}\right)^{1/3} = \frac{p_A}{T_A},$$

and analogous expressions hold for the system B. From the equality of pressures and temperatures, the following equations result:

$$\frac{v_A}{u_A^2} = \frac{v_B}{u_B^2}, \qquad \frac{u_A}{v_A^2} = \frac{u_B}{v_B^2},$$

hence

$$\left(\frac{u_A}{u_B}\right)^2 = \frac{v_A}{v_B}, \qquad \left(\frac{v_A}{v_B}\right)^2 = \frac{u_A}{u_B}.$$

They are satisfied only if $u_A = u_B$ and $v_A = v_B$, hence, also $s_A = s_B$. Since we have $U = U_A + U_B = n_A u_A + n_B u_B = n u_A$ for the total internal energy, thus, $u = U/n = u_A = u_B$. In a similar way we show that $v = V/n = v_A = v_B$. The molar entropy of the fused system, $s = S/n$, must also satisfy the equilibrium conditions:

$$\left(\frac{\partial s}{\partial u}\right)_v = \frac{1}{T_A} = \frac{1}{T_B} = \frac{1}{3}\left(\frac{v}{Du^2}\right)^{1/3},$$
$$\left(\frac{\partial s}{\partial v}\right)_u = \frac{p_A}{T_A} = \frac{p_B}{T_B} = \frac{1}{3}\left(\frac{u}{Dv^2}\right)^{1/3},$$

where we have used the equalities $u_A = u_B = u$ and $v_A = v_B = v$. Therefore, we conclude that the function $s(u, v)$ has the same form as $s_A(u_A, v_A)$ and $s_B(u_B, v_B)$, i.e.,

$$s = \left(\frac{uv}{D}\right)^{1/3}.$$

Since $s = s_A = s_B$, we show eventually that

$$S = ns = (n_A + n_B)s = n_A s_A + n_B s_B = S_A + S_B.$$

4.17 Since no external lateral force acts on the vessel, the total momentum of the system is conserved and equal to zero. In Fig. S4.3, the gas occupies initially the left half of the vessel. When the internal wall is removed, the mass center of the gas shifts to the right. To maintain the zero value of the total momentum, the vessel shifts to the left, that is, in the opposite direction to the gas motion in the vessel. The gas motion causes the vessel to oscillate on the surface without friction. The oscillatory motion decays because according to the fundamental law of thermodynamics every isolated system reaches eventually an equilibrium state. In equilibrium, there is no macroscopic flow, therefore, the mass center of the gas must be at rest. Since the total momentum is conserved, the mass center of the vessel must also be at rest. Thus, the motion of the gas and vessel stops eventually.

Fig. S4.3 Broken line shows the mass center position of the vessel filled with a gas

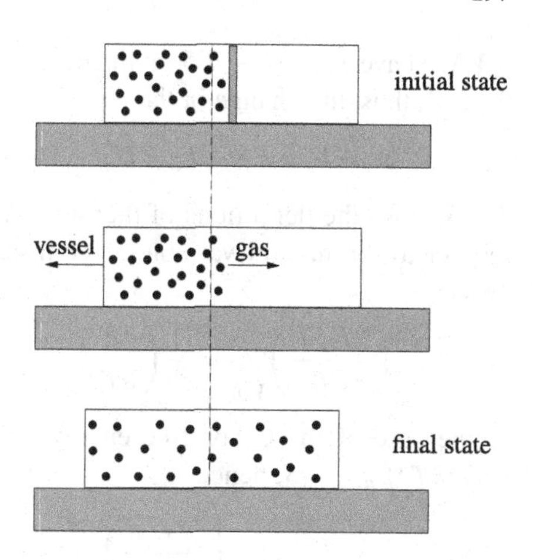

Exercises of Chapter 5

5.1 The derivative of the function $Y = Ax^2 + Bx + C$:

$$\frac{dY}{dx} = 2Ax + B,$$

is a monotonic function of the variable x, thus, the relation $dY/dx = z$ can be inverted, i.e.,

$$x = \frac{z - B}{2A}.$$

The Legendre transform of $Y(x)$ is the function

$$\Psi(z) = Y(x) - zx.$$

Substituting x expressed by z, we get

$$\Psi(z) = -\frac{z^2}{4A} + \frac{Bz}{2A} - \frac{B^2}{2A} + C,$$

hence, $d\Psi/dz = -(z - B)/(2A) = -x$. In the case of the function $Y = x + ae^x$, we have $z = dY/dx = 1 + ae^x$, hence

$$x = \ln\frac{z-1}{a}$$

and

$$\Psi(z) = z - 1 - (z-1)\ln\frac{z-1}{a}, \qquad \frac{d\Psi}{dz} = -\ln\frac{z-1}{a} = -x.$$

5.2 Since the process occurs at constant pressure, we have $H_i = U_i + pV_i$ at the beginning of the process and $H_f = U_f + pV_f$ at the end. Thus, the change in the enthalpy of the system is equal to

$$\Delta H = H_f - H_i = U_f - U_i + p(V_f - V_i) = \Delta U + p\Delta V.$$

5.3 We have $F_i = U_i - TS_i$ at the beginning of the process and $F_f = U_f - TS_f$ at the end, thus, the change in the Helmholtz free energy is equal to

$$\Delta F = F_f - F_i = U_f - U_i - T(S_f - S_i) = \Delta U - T\Delta S.$$

5.4 We use the definitions of thermodynamic potentials and their derivatives with respect to the natural variables. For instance, $F = U - TS$, $(\partial F/\partial T)_{V,n} = -S$, hence

$$\left(\frac{\partial F/T}{\partial T}\right)_{V,n} = \frac{1}{T}\left(\frac{\partial F}{\partial T}\right)_{V,n} - \frac{F}{T^2} = -\frac{1}{T^2}(F + TS) = -\frac{U}{T^2}.$$

In the case of the Gibbs free energy, we have $G = U - TS + pV = H - TS$, $(\partial G/\partial T)_{p,n} = -S$, hence

$$\left(\frac{\partial G/T}{\partial T}\right)_{p,n} = -\frac{S}{T} - \frac{G}{T^2} = -\frac{H}{T^2}.$$

The last relation is called the *Gibbs–Helmholtz relation*. In a similar way, we prove the remaining relations.

5.5 We make use of the Gibbs–Helmholtz relation derived in the previous exercise. Integrating both sides of the relation over T, from T_i to T_f, we get

$$\frac{G_f}{T_f} - \frac{G_i}{T_i} = -\int_{T_i}^{T_f} \frac{H}{T^2}\mathrm{d}T = -H\int_{T_i}^{T_f} \frac{1}{T^2}\mathrm{d}T = H\left(\frac{1}{T_f} - \frac{1}{T_i}\right),$$

hence

$$G_f = G_i\frac{T_f}{T_i} + H\left(1 - \frac{T_f}{T_i}\right).$$

5.6 Differentiating U with respect to S and V, we get

$$T = \left(\frac{\partial U}{\partial S}\right)_{V,n} = 3B\frac{S^2}{nV},$$

$$p = -\left(\frac{\partial U}{\partial V}\right)_{S,n} = B\frac{S^3}{nV^2}.$$

The Gibbs free energy is defined as $G = U - TS + pV$, and its natural variables are T, p and n. Therefore, we have to express S and V in terms of T, p and n, i.e.,

$$S = \frac{T^2 n}{9Bp},$$

$$V = \frac{T^3 n}{27Bp^2}.$$

We notice that $U = pV$, hence

$$G = U - TS + pV = 2pV - TS = \frac{2T^3 n}{27Bp} - \frac{T^3 n}{9Bp} = -\frac{T^3 n}{27Bp}.$$

Alternatively, we obtain the chemical potential from the relation $\mu = (\partial U/\partial n)_{S,V}$, and then express μ as a function of T and p, and make use of the relation $G = \mu n$.

5.7 The internal energy of the photon gas per unit volume is equal to $u = U/V = \gamma T^4$. From the definition of temperature, we have $T\mathrm{d}s = \mathrm{d}u = 4\gamma T^3 \mathrm{d}T$, where s is the entropy of the system per unit volume, hence

$$\mathrm{d}s = 4\gamma T^2 \mathrm{d}T,$$

which gives $S = (4/3)V\gamma T^3$. Then we determine the Helmholtz free energy

$$F(T, V) = U - TS = -\frac{1}{3}\gamma V T^4,$$

and the pressure as a function of T, i.e.,

$$p = -\left(\frac{\partial F}{\partial V}\right)_T = \frac{1}{3}\gamma T^4.$$

Now we can calculate the Gibbs free energy

$$G = U - TS + pV = \gamma V T^4 - \frac{4}{3}\gamma V T^4 + \frac{1}{3}\gamma V T^4 = 0.$$

The Gibbs free energy is equal to zero because the number of photons is not a thermodynamic parameter. Finally, we calculate the enthalpy

$$H = U + pV = \frac{4}{3}\gamma V T^4 = TS.$$

The natural variables of the enthalpy are S and p, therefore, we have to express T as a function of pressure, i.e.,

$$T = \left(\frac{3p}{\gamma}\right)^{1/4},$$

hence

$$H = S\left(\frac{3p}{\gamma}\right)^{1/4}.$$

5.8 We use the relation

$$p = -\left(\frac{\partial F}{\partial V}\right)_{T,n} = -\left(\frac{\partial \phi}{\partial v}\right)_T,$$

where $\phi = F/n$ denotes the molar Helmholtz free energy. Since

$$p = \frac{RT}{v - b} - \frac{a}{v^2},$$

we obtain

$$\phi(T, v) = \phi_0(T) + \int_{v_0}^{v} p(T, v')\mathrm{d}v' = \phi_0(T) + RT \ln \frac{v - b}{v_0 - b} + a\left(\frac{1}{v} - \frac{1}{v_0}\right),$$

where v_0 is the molar volume of the reference state, and $\phi_0(T) = \phi(T, v_0)$. The Helmholtz free energy $F(T, V, n) = n\phi(T, v)$. To determine the function $\phi_0(T)$, we have to know the dependence of the molar internal energy on T and v.

5.9 We use the relation (see (5.132))

$$\left(\frac{\partial U}{\partial V}\right)_{T,n} = -p + T\left(\frac{\partial p}{\partial T}\right)_{V,n},$$

into which we substitute the van der Waals equation of state

$$p = \frac{nRT}{V - nb} - \frac{an^2}{v^2},$$

which gives

$$\left(\frac{\partial U}{\partial V}\right)_{T,n} = \frac{an^2}{V^2}.$$

Integrating the last relation over V and making use of extensiveness of the internal energy, we get

$$U(T, V, n) = nf(T) + \int \frac{an^2}{V^2}\mathrm{d}V = nf(T) - \frac{an^2}{V},$$

where $f(T)$ is a function of temperature to be determined. Since

$$\left(\frac{\partial U}{\partial T}\right)_{n,V} = nc_v,$$

we obtain $f(T) = 3RT/2$, hence, the internal energy is given by the formula

$$U(T, V, n) = \frac{3}{2}nRT - \frac{an^2}{V}.$$

5.10 For a system heated at constant pressure, we have

$$\Delta H = \int_{T_i}^{T_f} c_p \mathrm{d}T = a(T_f - T_i) + \frac{b}{2}(T_f^2 - T_i^2)$$

and

$$\Delta S = \int_{T_i}^{T_f} \frac{c_p}{T}\mathrm{d}T = a\ln\frac{T_f}{T_i} + b(T_f - T_i).$$

5.11 We have

$$C_p = T\left(\frac{\partial S}{\partial T}\right)_{p,n},$$

hence

$$\left(\frac{\partial C_p}{\partial p}\right)_{T,n} = T\left(\frac{\partial^2 S}{\partial T\partial p}\right)_n = -T\left(\frac{\partial^2 V}{\partial T^2}\right)_{p,n},$$

where we have used the Maxwell relation

$$\left(\frac{\partial S}{\partial p}\right)_{T,n} = -\left(\frac{\partial V}{\partial T}\right)_{p,n}.$$

5.12 We write the differential of $V(T, p)$ at constant mole number:

$$\mathrm{d}V = \left(\frac{\partial V}{\partial T}\right)_p \mathrm{d}T + \left(\frac{\partial V}{\partial p}\right)_T \mathrm{d}p = V\alpha \mathrm{d}T - V\kappa_T \mathrm{d}p.$$

Since κ_T and α are assumed constant, we can easily integrate $\mathrm{d}V$. Dividing both sides by V and then integrating, we get

$$\ln\frac{V}{V_0} = \alpha(T - T_0) - \kappa_T(p - p_0),$$

hence

$$V = V_0 \mathrm{e}^{[\alpha(T-T_0)-\kappa_T(p-p_0)]},$$

where V_0, T_0 and p_0 define the reference state.

5.13 We use the differential of the potential Ψ, i.e.,

$$\mathrm{d}\Psi = T\mathrm{d}S - p\mathrm{d}V - n\mathrm{d}\mu.$$

From the equality of mixed second order partial derivatives, we derive the following Maxwell relations:

$$\left(\frac{\partial T}{\partial V}\right)_{S,\mu} = -\left(\frac{\partial p}{\partial S}\right)_{V,\mu},$$
$$\left(\frac{\partial T}{\partial \mu}\right)_{S,V} = -\left(\frac{\partial n}{\partial S}\right)_{V,\mu},$$
$$\left(\frac{\partial p}{\partial \mu}\right)_{S,V} = \left(\frac{\partial n}{\partial V}\right)_{S,\mu}.$$

In the case of the potential Θ, we have

$$\mathrm{d}\Theta = T\mathrm{d}S + V\mathrm{d}p - n\mathrm{d}\mu,$$

hence, the Maxwell relations follow:

$$\left(\frac{\partial T}{\partial p}\right)_{S,\mu} = \left(\frac{\partial V}{\partial S}\right)_{p,\mu},$$
$$\left(\frac{\partial T}{\partial \mu}\right)_{S,p} = -\left(\frac{\partial n}{\partial S}\right)_{p,\mu},$$
$$\left(\frac{\partial V}{\partial \mu}\right)_{S,p} = -\left(\frac{\partial n}{\partial p}\right)_{S,\mu}.$$

5.14 The potential $\Psi(S, V, \mu)$ is an extensive quantity, thus, $\psi = \Psi/V = \Psi(s, 1, \mu)$, where $s = S/V$. The natural variables are the entropy per unit volume and chemical potential. At constant volume, we have

$$\mathrm{d}\Psi = T\mathrm{d}S - n\mathrm{d}\mu.$$

Dividing both sides by V, we get

$$d\psi = T ds - \rho d\mu,$$

where $\rho = n/V = 1/v$.

5.15 It follows from the Euler relation $U = TS - pV + \mu n$ that

$$\Theta = U + pV - \mu n = TS,$$

hence, $T = \Theta/S$. At constant entropy, we have

$$d\Theta = V dp - n d\mu,$$

hence, dividing both sides by S, we get

$$dT = \frac{V}{S} dp - \frac{n}{S} d\mu.$$

We notice that it is simply a different form of the Gibbs–Duhem equation: $d\mu = -s dT + v dp$, with $s = S/n$ and $v = V/n$.

5.16 We have

$$\left(\frac{\partial v}{\partial \mu}\right)_T = \left(\frac{\partial v}{\partial p}\right)_T \left(\frac{\partial p}{\partial \mu}\right)_T = \frac{1}{v}\left(\frac{\partial v}{\partial p}\right)_T = -\kappa_T,$$

where we have used the Gibbs–Duhem equation. The second derivative is calculated in a similar way, i.e.,

$$\left(\frac{\partial s}{\partial \mu}\right)_T = \left(\frac{\partial s}{\partial p}\right)_T \left(\frac{\partial p}{\partial \mu}\right)_T = \frac{1}{v}\left(\frac{\partial s}{\partial p}\right)_T.$$

It follows from the Gibbs–Duhem equation that

$$\left(\frac{\partial s}{\partial p}\right)_T = -\left(\frac{\partial v}{\partial T}\right)_p,$$

hence

$$\left(\frac{\partial s}{\partial \mu}\right)_T = -\alpha.$$

5.17 The inversion temperature in the Joule–Thomson process satisfies the condition $T\alpha = 1$, therefore, we have to calculate the thermal expansion coefficient:

$$\alpha = \frac{1}{v}\left(\frac{\partial v}{\partial T}\right)_p,$$

for

$$pv = RT\big[1 + b(T)p\big].$$

Since we differentiate at constant pressure, we get

$$T\alpha = \frac{RT}{pv}\left(\frac{\partial pv/R}{\partial T}\right)_p = \frac{1}{1+bp}\left(\frac{\partial T(1+bp)}{\partial T}\right)_p = 1 + \frac{Tb'(T)p}{1+b(T)p},$$

where $b'(T) = db/dT$. Thus, the inversion temperature satisfies the equation

$$b'(T) = 0.$$

Exercises of Chapter 6

6.1 Since T and p are treated as independent variables, we have $\mu = \mu(T, v)$ and $p = p(T, v)$. Chemical potentials and pressures of coexisting phases are equal, i.e.,

$$\mu^\alpha(T, v^\alpha) = \mu^\beta(T, v^\beta) = \dots,$$
$$p^\alpha(T, v^\alpha) = p^\beta(T, v^\beta) = \dots,$$

where $\alpha, \beta, \dots$ correspond to different phases. Thus, we have $2(P-1)$ equations and $P+1$ variables, since the molar volume is different for different phases, in general. Subtracting the number of equations from the number of variables, we get

$$f = P + 1 - 2(P-1) = 3 - P.$$

Since $f \geq 0$, the number of coexisting phases P cannot exceed three. For instance, when two phases coexist, we can change their temperature, and the molar volumes v^α and v^β are functions of T.

6.2 It follows from the Gibbs-Duhem equation that pressure can be treated as a function of T and μ. Therefore, for the phases α and β, we have $p^\alpha(T, \mu)$ and $p^\beta(T, \mu)$, respectively. The pressures are equal on the coexistence line, hence

$$p^\alpha(T, \mu_{\text{coex}}(T)) = p^\beta(T, \mu_{\text{coex}}(T)).$$

Differentiating both sides of this equality with respect to T, we get

$$\left(\frac{\partial p^\alpha}{\partial T}\right)_{\mu_{\text{coex}}} + \left(\frac{\partial p^\alpha}{\partial \mu_{\text{coex}}}\right)_T \frac{d\mu_{\text{coex}}}{dT} = \left(\frac{\partial p^\beta}{\partial T}\right)_{\mu_{\text{coex}}} + \left(\frac{\partial p^\beta}{\partial \mu_{\text{coex}}}\right)_T \frac{d\mu_{\text{coex}}}{dT}.$$

From the Gibbs-Duhem equation: $d\mu = -s dT + v dp$, the relations $(\partial p/\partial T)_\mu = s\rho$ and $(\partial p/\partial \mu)_T = \rho$ follow, where $\rho = n/V = 1/v$, hence

$$\frac{d\mu_{\text{coex}}}{dT} = -\frac{s^\alpha \rho^\alpha - s^\beta \rho^\beta}{\rho^\alpha - \rho^\beta} = -\frac{\Delta(s\rho)}{\Delta\rho}.$$

6.3 Chemical potentials of two coexisting phases are equal. We know the chemical potential of the monatomic ideal gas (see (5.52)):

$$\mu^g(T, p) = \mu_0 \frac{T}{T_0} - \frac{5}{2} RT \ln \frac{T}{T_0} + RT \ln \frac{p}{p_0},$$

thus, $\mu_{\text{coex}}(T) = \mu^g(T, p_{\text{coex}}(T))$, where $p_{\text{coex}}(T)$ denotes the pressure on the liquid–vapour coexistence line. Since the molar volume of the vapour is much bigger than the molar volume of the liquid, we have $\Delta v \approx v^g$, where $v^g = RT/p$ for the ideal gas. Then, the pressure $p_{\text{coex}}(T)$ satisfies the Clausius-Clapeyron equation (6.41), which can be integrated if the enthalpy of transition does not depend on temperature, and we obtain see (6.42)):

$$p_{\text{coex}}(T) = p_{\text{coex}}(T_0) \exp\left[\frac{\Delta h}{R}\left(\frac{1}{T_0} - \frac{1}{T}\right)\right].$$

Substituting $p_{\mathrm{coex}}(T)$ into the expression for the chemical potential μ^g, we obtain

$$\mu_{\mathrm{coex}}(T) = \mu_0 \frac{T}{T_0} - \frac{5}{2} RT \ln \frac{T}{T_0} + \Delta h \left(\frac{T}{T_0} - 1 \right).$$

6.4 Before we start to supply heat to the system only the phase α is present, therefore, we determine the mole number of the phase β from the ratio

$$n^\beta = Q/\Delta h.$$

Hence, the ratio of the mole numbers of the two phases amounts to

$$\frac{n^\beta}{n^\alpha} = \frac{n^\beta}{n - n^\beta} = \frac{Q}{n\Delta h - Q},$$

which can also be written in the form of a lever rule:

$$n^\beta (n\Delta h - Q) = n^\alpha Q.$$

It follows from this expression that if $Q = n\Delta h$ then $n^\alpha = 0$, which means that the phase α has changed completely into the phase β.

6.5 The average molar volume is defined as

$$v = x^\alpha v^\alpha + x^\beta v^\beta,$$

where $x^\alpha = n^\alpha/n$, $x^\beta = n^\beta/n$, $n^\alpha + n^\beta = n$, hence

$$v = x^\alpha v^\alpha + (1 - x^\alpha) v^\beta = v^\beta - x^\alpha \Delta v,$$

where $\Delta v = v^\beta - v^\alpha$. The molar fraction of the phase α at the end of the process, $x_f^\alpha = n_f^\alpha/n$, is given, whereas to calculate the average molar volume v_i at the beginning of the process, we have to find x_i^α first. It follows from the context that the amount of the phase α at the beginning of the process, n_i^α, was grater than the amount of α at the end, n_f^α, thus

$$Q = (n_i^\alpha - n_f^\alpha) \Delta h,$$

hence

$$Q/n = q = (x_i^\alpha - x_f^\alpha) \Delta h.$$

Substituting x_i^α from this equation into the expression for v, we get eventually

$$v_i = v^\beta - (x_f^\alpha + q/\Delta h) \Delta v = v_f - q \Delta v/\Delta h,$$

where $v_f = v^\beta - x_f^\alpha \Delta v$.

6.6 We use the equation (see (6.39))

$$p_{\mathrm{coex}}(T) = p_{\mathrm{coex}}(T_0) + \frac{\Delta h}{\Delta v} \ln \frac{T}{T_0}.$$

Substituting the numbers, we get

$$\Delta h/\Delta v = -3.54 \times 10^9 \,\mathrm{N\,m^{-2}} = -3.54 \times 10^4 \,\mathrm{bar}$$

and $T/T_0 = 263.15/273.15$, hence $\ln(T/T_0) = -0.0373$. Finally, we get

$$p_{\text{coex}}(T) = p_{\text{coex}}(T_0) + 1320\,\text{bar}.$$

Thus, to lower the melting point of ice by 10 °C, we have to apply a pressure approximately 1300 times bigger than atmospheric pressure.

6.7 Since we want to determine Δh, we only need to transform the expression used in Exercise 6.6, i.e.,

$$\Delta h = \frac{[p_{\text{coex}}(T) - p_{\text{coex}}(T_0)]\Delta v}{\ln(T/T_0)}.$$

We have $\Delta v = v^l - v^s = 3\ \text{cm}^3\,\text{mol}^{-1}$, $T_0 = 350$ K, $T = 351$ K, $p_{\text{coex}}(T_0) = 1$ bar, $p_{\text{coex}}(T) = 100$ bar, $\ln(T/T_0) = 2.85 \times 10^{-3}$, hence, $\Delta h = 1.04\,\text{kJ}\,\text{mol}^{-1}$.

6.8 Since a liquid–gas transition is concerned, we use the equation (see (6.42))

$$p_{\text{coex}}(T) = p_{\text{coex}}(T_0)\exp\left[\frac{\Delta h}{RT_0}\left(1 - \frac{T_0}{T}\right)\right],$$

which we transform to the following form:

$$\frac{1}{T_0} - \frac{1}{T} = \frac{R}{\Delta h}\ln\left[\frac{p_{\text{coex}}(T)}{p_{\text{coex}}(T_0)}\right] = \frac{8.314\,\text{J}\,\text{K}^{-1}\,\text{mol}^{-1}}{14.4\,\text{kJ}\,\text{mol}^{-1}}\ln 2 = 4 \times 10^{-4}\ \text{K}^{-1},$$

hence $T = 194$ K.

6.9 We have two coexistence lines: the liquid–vapour line and solid–vapour line. The coordinates of the triple point, (T_3, p_3), are obtained from the intersection of these lines, i.e.,

$$15.16 - 3063\ \text{K}/T_3 = 18.70 - 3754\ \text{K}/T_3,$$

hence, $T_3 = 195.2$ K, $p_3 = 0.588$ atm. To find the enthalpy of evaporation and sublimation, we notice that the coefficient at $-1/T$ in the function $\ln[p_{\text{coex}}(T)/p_{\text{coex}}(T_0)]$ is equal to $\Delta h/R$ (see Exercise 6.8), hence

$$\Delta h_{\text{vap}} = 3063\ \text{K} \times 8.314\ \text{J}\,\text{K}^{-1}\,\text{mol}^{-1} = 25.466\ \text{kJ}\,\text{mol}^{-1},$$
$$\Delta h_{\text{sub}} = 3754\ \text{K} \times 8.314\ \text{J}\,\text{K}^{-1}\,\text{mol}^{-1} = 31.211\ \text{kJ}\,\text{mol}^{-1}.$$

Since enthalpy is a state function, we can obtain the solid–liquid transition as a combination of the solid–vapour and vapour–liquid transitions, hence

$$\Delta h_{\text{mel}} = \Delta h_{\text{sub}} - \Delta h_{\text{vap}} = 5.745\ \text{kJ}\,\text{mol}^{-1}.$$

6.10 First, we calculate the pressure on the sublimation line corresponding to the temperature of −5 °C ($T \approx 268$ K), using the expression (see Exercise 6.8)

$$\ln\left[\frac{p_{\text{coex}}(T)}{p_{\text{coex}}(T_3)}\right] = \frac{\Delta h_{\text{sub}}}{R}\left(\frac{1}{T_3} - \frac{1}{T}\right),$$

where for the triple point temperature we assume 273 K. Substituting $p_{\text{coex}}(273) = p_3 = 0.006$ bar, we get

$$p_{\text{coex}}(268) = p_{\text{coex}}(273)\exp(-0.419) = 0.004\ \text{bar}.$$

The pressure of water vapour equal to 2 Tr $= 0.0027$ bar is well below the pressure of sublimation at the temperature of -5 °C. Frost can remain only if the vapour pressure is close to the pressure of sublimation at a given temperature. Thus, the frost will disappear from the car glass.

6.11 First, we derive the barometric formula, i.e., the relation between the gas pressure and the altitude. The gravity of a gas column, of the height dz and the mass density ρ_m, is balanced by the pressure difference $dp = -\rho_m(p)g dz$, where g is the earth gravity. Using the relation $\rho_m = M\rho$, where ρ is the molar density and M denotes the molar mass, and assuming the ideal gas equation of state, $p = RT\rho$, we get

$$dp = -Mg\rho dz = -Mg\frac{p}{RT}dz.$$

Integrating this equation, we obtain the barometric formula:

$$p(z) = p_0 \exp\left[-\frac{Mg(z-z_0)}{RT}\right].$$

Deriving this formula, we make an assumption that the temperature T does not change with the altitude. Although it is not true, in general, such an approximation is sufficient for the present purpose.

Boiling consists in formation of vapour bubbles in the whole volume of a liquid. It occurs when the vapour pressure at the liquid–vapour coexistence equals atmospheric pressure. The boiling temperatures T_0 and $T_{\mathcal{H}}$ result from the equations $p_{\text{coex}}(T_0) = p_0$ and $p_{\text{coex}}(T_{\mathcal{H}}) = p_{\mathcal{H}}$, where p_0 and $p_{\mathcal{H}}$ denote the pressures at the foot of the mountain and at the altitude $\mathcal{H}$, respectively. Thus, we have

$$\ln\frac{p_{\text{coex}}(T_{\mathcal{H}})}{p_{\text{coex}}(T_0)} = \ln\frac{p_{\mathcal{H}}}{p_0} = -\frac{Mg\Delta\mathcal{H}}{RT},$$

where $\Delta\mathcal{H}$ is the height of the mountain. Using expression (6.42), we obtain the equation

$$\frac{\Delta h}{R}\left(\frac{1}{T_0} - \frac{1}{T_{\mathcal{H}}}\right) = -\frac{Mg}{RT}\Delta\mathcal{H},$$

from which we determine $\Delta\mathcal{H}$:

$$\Delta\mathcal{H} = \frac{\Delta h}{Mg}\left(\frac{T}{T_{\mathcal{H}}} - \frac{T}{T_0}\right).$$

Substituting $T = 293$ K (20 °C) and $M = 28\ \text{g mol}^{-1}$ (for N_2 molecule), we get

$$\Delta\mathcal{H} = \frac{45\ \text{kJ mol}^{-1}}{28\ \text{g mol}^{-1} \times 9.81\ \text{m s}^{-2}}\left(\frac{293}{368} - \frac{293}{378}\right) = 3.45\ \text{km}.$$

In principle, we should take into account that air is a mixture of gases, mainly nitrogen (78 %) and oxygen (21 %), and apply the barometric formula to each gas separately and then add the partial pressures of the gases. Since nitrogen predominates, and the molecular mass of O_2 (32) does not differ much from the molecular mass of N_2 (28), we do not make a big error, approximating air by nitrogen alone.

6.12 We calculate the derivative with respect to temperature of the function

$$\Delta s(T) = s^\beta\big(T, p_{\mathrm{coex}}(T)\big) - s^\gamma\big(T, p_{\mathrm{coex}}(T)\big),$$

where $p_{\mathrm{coex}}(T)$ is the pressure of coexisting phases at the temperature T. We have

$$\frac{\mathrm{d}s^\beta}{\mathrm{d}T} = \left(\frac{\partial s^\beta}{\partial T}\right)_{p_{\mathrm{coex}}} + \left(\frac{\partial s^\beta}{\partial p_{\mathrm{coex}}}\right)_T \frac{\mathrm{d}p_{\mathrm{coex}}}{\mathrm{d}T} = \frac{c_p^\beta}{T} - v^\beta \alpha^\beta \frac{\mathrm{d}p_{\mathrm{coex}}}{\mathrm{d}T},$$

where c_p^β and α^β denote the molar heat capacity at constant pressure and thermal expansion coefficient of the phase β, respectively. Deriving the above relation, we have used the Maxwell relation $(\partial s/\partial p)_T = -(\partial v/\partial T)_p$. An analogous expression holds for the phase γ. Since $\mathrm{d}p_{\mathrm{coex}}/\mathrm{d}T = \Delta s/\Delta v$, we get

$$\frac{\mathrm{d}\Delta s}{\mathrm{d}T} = \frac{\Delta c_p}{T} - \Delta(v\alpha)\frac{\Delta s}{\Delta v},$$

where $\Delta c_p = c_p^\beta - c_p^\gamma$, $\Delta v = v^\beta - v^\gamma$, $\Delta(v\alpha) = v^\beta\alpha^\beta - v^\gamma\alpha^\gamma$.

6.13 The enthalpy $H = U + pV = TS + \mu n$, hence, the molar enthalpy $h = Ts + \mu$. At the coexistence of phases β and γ, we have $\Delta h = T\Delta s$, since $\mu^\beta = \mu^\gamma$. Substituting $\Delta s = \Delta h/T$ in Exercise 6.12 and neglecting thermal expansion, we get

$$\frac{\mathrm{d}(\Delta h/T)}{\mathrm{d}T} = \frac{\Delta c_p}{T}.$$

Then, we integrate the above equality from T_0 to T, hence

$$\frac{\Delta h(T)}{T} - \frac{\Delta h(T_0)}{T_0} = \int_{T_0}^{T} \frac{\Delta c_p(T')}{T'}\mathrm{d}T' = \Delta a(T - T_0) + (\Delta b - T_0\Delta a)\ln\frac{T}{T_0},$$

where $\Delta a = a^\beta - a^\gamma$ and $\Delta b = b^\beta - b^\gamma$. Using this formula, we can calculate the enthalpy of transition at the temperature T if we know Δh at the reference temperature T_0.

6.14 The infinitesimal heat absorbed by the system in a reversible process equals $đQ = T\mathrm{d}S$. Since the process occurs at constant V and n, and the parameter that we control is temperature, the appropriate thermodynamic potential is the Helmholtz free energy $F(T, V, n)$. From the Euler relation, we get $F = U - TS = -pV + \mu n$. The system considered is a two-phase system, the pressure $p = p_{\mathrm{coex}}$ and the chemical potential $\mu = \mu_{\mathrm{coex}}$ are functions of temperature, thus (see (6.14))

$$F = U - TS = -p_{\mathrm{coex}}(T)V + \mu_{\mathrm{coex}}(T)n.$$

To calculate the heat Q absorbed by the system heated from the temperature T_0 to T_1, we notice that the change in the internal energy $\Delta U = Q$, since the volume is constant, thus, the work $W = 0$. Using the relation (see Exercise 5.4)

$$\left(\frac{\partial F/T}{\partial T}\right)_{V,n} = -\frac{U}{T^2},$$

we get

$$Q = \Delta U = -\Delta\left[T^2\left(\frac{\partial F/T}{\partial T}\right)_{V,n}\right] = -n\Delta\left[T^2\left(\frac{\partial \phi_{\text{coex}}/T}{\partial T}\right)_v\right],$$

where $\phi_{\text{coex}}(T, v) = -p_{\text{coex}}(T)v + \mu_{\text{coex}}(T)$ denotes the molar Helmholtz free energy of the system at the temperature T and at the average molar volume $v = V/n$. We can proceed with the calculation if we know the explicit form of the functions $p_{\text{coex}}(T)$ and $\mu_{\text{coex}}(T)$. For instance, for the liquid–vapour coexistence, and with the additional assumptions specified in this exercise, we have

$$p_{\text{coex}}(T) = p_{\text{coex}}(T_0)\exp\left[\frac{\Delta h}{R}\left(\frac{1}{T_0} - \frac{1}{T}\right)\right],$$

$$\mu_{\text{coex}}(T) = \mu_0\frac{T}{T_0} - \frac{5}{2}RT\ln\frac{T}{T_0} + \Delta h\left(\frac{T}{T_0} - 1\right).$$

6.15 We have (see (6.58))

$$T_{\text{cr}} = \frac{8a}{27Rb} = \frac{8 \times 0.15}{27 \times 8.314 \times 4 \times 10^{-5}}\ \text{K} = 133.6\ \text{K},$$

$$p_{\text{cr}} = \frac{a}{27b^2} = \frac{0.15}{27 \times 16 \times 10^{-10}}\ \text{Pa} = 34.7\ \text{bar},$$

$$v_{\text{cr}} = 3b = 3 \times 4 \times 10^{-5}\ \text{m}^3\,\text{mol}^{-1} = 0.12\ \text{L}\,\text{mol}^{-1}.$$

6.16 We use the relations:

$$v_{\text{cr}} = 3b, \qquad p_{\text{cr}} = \frac{a}{27b^2}, \qquad T_{\text{cr}} = \frac{8a}{27Rb}.$$

Since

$$u = \frac{f}{2}RT - \frac{a}{v},$$

at the critical point, we have

$$u_{\text{cr}} = \frac{f}{2}RT_{\text{cr}} - \frac{a}{v_{\text{cr}}} = \frac{a}{27b}(4f - 9).$$

Dividing u by u_{cr}, we get

$$\bar{u} = \frac{4f}{4f-9}\bar{T} - \frac{9}{4f-9}\frac{1}{\bar{v}}.$$

For $f = 3$, 5 and 6, we obtain, respectively:

$$\bar{u} = 4\bar{T} - \frac{3}{\bar{v}},$$

$$\bar{u} = \frac{20}{11}\bar{T} - \frac{9}{11\bar{v}},$$

$$\bar{u} = \frac{8}{5}\bar{T} - \frac{3}{5\bar{v}}.$$

6.17 We use expression (6.51):

$$\phi(T, v) = \left(\frac{1}{2}fR - s_0\right)T - \frac{a}{v} - \frac{1}{2}fRT\ln\frac{T}{T_0} - RT\ln\frac{v-b}{v_0-b},$$

where $\phi = F/n$ is the molar Helmholtz free energy. From the Euler relation, we get $F = U - TS = -pV + \mu n$, hence, $\mu = \phi + pv$. Assuming that $T > T_{\text{cr}}$, we can determine $v = v(T, p)$ from the van der Waals equation of state:

$$p = \frac{RT}{v-b} - \frac{a}{v^2},$$

hence

$$\mu(T, p) = pv(T, p) + \left(\frac{1}{2}fR - s_0\right)T - \frac{a}{v(T, p)} - \frac{1}{2}fRT\ln\frac{T}{T_0} - RT\ln\frac{v(T, p)-b}{v_0-b}.$$

As the limit $p \to 0$ corresponds to $v \to \infty$, we have

$$p = \frac{RT}{v}\left(\frac{1}{1-b/v} - \frac{a}{RTv}\right) \approx \frac{RT}{v},$$

i.e., the ideal gas equation of state, hence

$$\mu(T, p) - RT\ln\frac{p}{p_0} \to \left[\frac{1}{2}(f+2)R - s_0\right]T - \frac{1}{2}fRT\ln\frac{T}{T_0} - RT\ln\frac{RT}{(v_0-b)p_0}.$$

Taking $(v_0 - b)p_0 = RT_0$, we get eventually

$$\mu_0(T) = \left[\frac{1}{2}(f+2)R - s_0\right]T - \frac{1}{2}(f+2)RT\ln\frac{T}{T_0}.$$

Comparing $\mu_0(T)$ with the chemical potential of the ideal gas, μ^{id} (see (4.70)), we note that $\mu_0(T) = \mu^{\text{id}}(T, p_0)$. The chemical potential of the van der Waals gas can be presented as follows:

$$\mu(T, p) = \mu_0(T) + pv - RT - \frac{a}{v} + RT\ln\frac{RT}{p_0(v-b)},$$

where $v = v(T, p)$ is obtained from the van der Waals equation of state. Using the relation

$$pv - RT = \frac{RTb}{v-b} - \frac{a}{v},$$

we can eliminate p and express the chemical potential as a function of T and v:

$$\mu(T, v) = \mu_0(T) + \frac{RTb}{v-b} - \frac{2a}{v} + RT\ln\frac{RT}{p_0(v-b)}.$$

Exercises of Chapter 7

7.1 The volume of the solution amounts to

$$V = n_A v_A + n_B v_B = n(x_A v_A + x_B v_B).$$

As v_A, v_B and x_B are given, it remains to determine $n = n_A + n_B$. The mass of the solution is given by the relation

$$\begin{aligned} m &= n_A M_A + n_B M_B = n(x_A M_A + x_B M_B) \\ &= n(0.55 \times 58 + 0.45 \times 118)\,\mathrm{g\,mol^{-1}} = n \times 85\,\mathrm{g\,mol^{-1}}, \end{aligned}$$

and for $m = 0.85$ kg, we get $n = 10$ mol. For the volume V, we obtain

$$V = 10(0.55 \times 74 + 0.45 \times 80)\,\mathrm{cm^3} = 767\,\mathrm{cm^3}.$$

7.2 According to the general definition of a partial molar quantity, we have

$$h_i = \left(\frac{\partial H}{\partial n_i}\right)_{T,p,n_{j\neq i}}.$$

The differential of enthalpy

$$\mathrm{d}H = T\mathrm{d}S + V\mathrm{d}p + \sum_{i=1}^{C} \mu_i \mathrm{d}n_i,$$

hence

$$\left(\frac{\partial h_i}{\partial p}\right)_{T,x} = \left(\frac{\partial^2 H}{\partial n_i \partial p}\right)_{T,n_{j\neq i}}.$$

We use the relation

$$\left(\frac{\partial H}{\partial p}\right)_{T,n_j} = V + T\left(\frac{\partial S}{\partial p}\right)_{T,n_j},$$

and from the differential of the Gibbs free energy

$$\mathrm{d}G = -S\mathrm{d}T + V\mathrm{d}p + \sum_{i=1}^{C} \mu_i \mathrm{d}n_i,$$

we obtain the Maxwell relation

$$\left(\frac{\partial S}{\partial p}\right)_{T,n_j} = -\left(\frac{\partial V}{\partial T}\right)_{p,n_j},$$

thus,

$$\left(\frac{\partial H}{\partial p}\right)_{T,n_j} = V - T\left(\frac{\partial V}{\partial T}\right)_{p,n_j} = -T^2\left(\frac{\partial V/T}{\partial T}\right)_{p,n_j}.$$

Differentiating both sides of the above equality with respect to n_i at constant T, p and $n_{j\neq i}$, we get the relation

$$\left(\frac{\partial h_i}{\partial p}\right)_{T,x} = -T^2\left(\frac{\partial^2 V/T}{\partial T \partial n_i}\right)_{p,n_{j\neq i}} = -T^2\left(\frac{\partial v_i/T}{\partial T}\right)_{p,x}.$$

7.3 The Gibbs–Duhem equation at constant T and p has the following form:

$$\sum_{i=1}^{C} x_i \mathrm{d}\mu_i = 0.$$

From the form of the chemical potential: $\mu_i = \mu_i^*(T,p) + RT\ln x_i$, it follows that at constant T and p, we have $\mathrm{d}\mu_i = RT\mathrm{d}x_i/x_i$, hence

$$\sum_{i=1}^{C} x_i \mathrm{d}\mu_i = RT\sum_{i=1}^{C} \mathrm{d}x_i = RT\mathrm{d}\sum_{i=1}^{C} x_i = 0,$$

as the molar fractions sum up to unity. This shows that the Gibbs–Duhem equation is satisfied.

7.4 From the form of the chemical potentials, it follows that at $T = const$ and $p = const$, we have

$$x_A \mathrm{d}\mu_A + x_B \mathrm{d}\mu_B = RT\mathrm{d}x_A + x_A W'(x_B)\mathrm{d}x_B + RT\mathrm{d}x_B + x_B W'(x_A)\mathrm{d}x_A,$$

where $W'(x)$ denotes the derivative. Taking into account the relations $x_A + x_B = 1$ and $\mathrm{d}x_B = -\mathrm{d}x_A$, we get

$$x_A \mathrm{d}\mu_A + x_B \mathrm{d}\mu_B = \left[(1-x_A)W'(x_A) - x_A W'(1-x_A)\right]\mathrm{d}x_A.$$

Since $W(x) = \sum_{n=1}^{N} W_n x^n$, the right-hand side of the above equation vanishes if

$$\sum_{n=1}^{N} nW_n\left[(1-x_A)x_A^{n-1} - x_A(1-x_A)^{n-1}\right] = 0.$$

It is easy to show that the expression in the brackets vanishes only if $n = 2$, thus, the polynomial must be of the form $W(x) = W_2 x^2$, where W_2 depends on T and p, in general.

7.5 According to the definition of the Gibbs free energy of mixing, we have

$$\begin{aligned}\frac{\Delta_M G}{n} &= x_A\left(\mu_A - \mu_A^*\right) + x_B\left(\mu_B - \mu_B^*\right) \\ &= RT(x_A \ln x_A + x_B \ln x_B) + x_A W(x_B) + x_B W(x_A).\end{aligned}$$

As $W(x) = W_2 x^2$, and $x_A x_B^2 + x_B x_A^2 = x_A x_B(x_A + x_B) = x_A x_B$, we get

$$\frac{\Delta_M G}{n} = RT(x_A \ln x_A + x_B \ln x_B) + W_2 x_A x_B.$$

7.6 At constant T and p, the Gibbs–Duhem equation $x_A \mathrm{d}v_A + x_B \mathrm{d}v_B = 0$ must hold. Treating x_A as an independent variable, we get

$$x_A \frac{\partial v_A}{\partial x_A} + x_B \frac{\partial v_B}{\partial x_A} = 0.$$

The first term vanishes at $x_A = 0$, and since $x_B = 1 - x_A = 1$, we get $\partial v_B / \partial x_A = 0$ at $x_A = 0$. Analogously, treating x_B as an independent variable, we show that $\partial v_A / \partial x_B = 0$ at $x_B = 0$. Thus, the Taylor expansion of the function $v_A(x_B)$ around $x_B = 0$, and the function $v_B(x_A)$ around $x_A = 0$, has the following form:

$$v_A(x_B) = v_A^* + a_A x_B^2 + \cdots,$$
$$v_B(x_A) = v_B^* + a_B x_A^2 + \cdots,$$

where we have neglected higher order terms.

7.7 We calculate $x_A \mathrm{d}v_A + x_B \mathrm{d}v_B$, substituting the expressions for v_A and v_B:

$$\begin{aligned} x_A \mathrm{d}v_A + x_B \mathrm{d}v_B &= x_A \big[2a x_B - 2(a-b) x_B^2\big] \mathrm{d}x_B + x_B \big[2b x_A + 2(a-b) x_A^2\big] \mathrm{d}x_A \\ &= 2 x_A x_B \big[-a + (a-b)(1-x_A) + b + (a-b) x_A\big] \mathrm{d}x_A = 0. \end{aligned}$$

The molar volume of mixing is given by the formula

$$\Delta_M v = \frac{\Delta_M V}{n} = x_A \big(v_A - v_A^*\big) + x_B \big(v_B - v_B^*\big),$$

where $\Delta_M V$ is the volume of mixing. Substituting v_A and v_B, we get

$$\Delta_M v = x_A x_B \left[a x_B + b x_A + \frac{2}{3}(a-b)\big(x_A^2 - x_B^2\big) \right].$$

Then, we make use of the relation $x_A + x_B = 1$, which gives

$$\Delta_M v = \frac{x_A x_B}{3} \big[(2a+b) x_A + (a+2b) x_B\big].$$

7.8 Substituting the molar fractions into the formula for the entropy of mixing of an ideal mixture:

$$\Delta_M S = -nR \sum_i x_i \ln x_i,$$

we get

$$\begin{aligned} \frac{\Delta_M S}{n} &= -8.314(0.781 \ln 0.781 + 0.21 \ln 0.21 + 0.009 \ln 0.009)\ \mathrm{J\,K^{-1}\,mol^{-1}} \\ &= 4.682\ \mathrm{J\,K^{-1}\,mol^{-1}}. \end{aligned}$$

7.9 The total pressure of air results from the ideal gas equation of state

$$p = \frac{RT}{v} = \frac{8.314 \times 273.15}{0.0224}\ \mathrm{Pa} = 101383\ \mathrm{Pa}.$$

The partial pressure of nitrogen, oxygen and argon amounts to, respectively

$$p_{N_2} = 0.781p = 79180.12 \text{ Pa},$$
$$p_{O_2} = 0.210p = 21290.43 \text{ Pa},$$
$$p_{Ar} = 0.009p = 912.45 \text{ Pa}.$$

7.10 For the entropy of mixing, we have

$$\Delta_M S = -nR \sum_i x_i \ln x_i.$$

The Gibbs free energy of mixing:

$$\Delta G_M = \Delta_M H - T \Delta_M S,$$

depends only on temperature and on the mixture composition, because both terms on the right-hand side do not depend on pressure. Therefore, the volume of mixing vanishes:

$$\Delta_M V = \left(\frac{\partial \Delta_M G}{\partial p}\right)_{T,n_j} = 0,$$

and the internal energy of mixing equals the enthalpy of mixing:

$$\Delta_M U = \Delta_M H - p\Delta_M V = \Delta_M H.$$

7.11 We use the relations

$$\left(\frac{\partial \Delta_M G}{\partial p}\right)_{T,n_j} = \Delta_M V,$$
$$\left(\frac{\partial \Delta_M G}{\partial T}\right)_{p,n_j} = -\Delta_M S.$$

Using the equality of mixed second order partial derivatives of the function $\Delta_M G$, we obtain the Maxwell relation

$$\left(\frac{\partial \Delta_M S}{\partial p}\right)_{T,n_j} = -\left(\frac{\partial \Delta_M V}{\partial T}\right)_{p,n_j}.$$

If the volume of mixing does not change with temperature then the entropy of mixing does not depend on pressure.

7.12 We use the relation (see (7.102))

$$\ln \frac{f(T,p)}{p} = \frac{1}{RT} \int_0^p \left[v(T,p') - v^{\text{id}}(T,p')\right] \mathrm{d}p',$$

where $v^{\text{id}}(T,p') = RT/p'$ denotes the molar volume of the ideal gas. Substituting

$$\frac{v(T,p')}{RT} = \frac{1}{p'} + B(T) + C(T)p',$$

we get

$$\ln \frac{f(T,p)}{p} = B(T)p + \frac{1}{2}C(T)p^2,$$

hence

$$f(T,p) = p\exp\left[B(T)p + \frac{1}{2}C(T)p^2\right].$$

7.13 From the van der Waals equation of state,

$$p = \frac{RT}{v-b} - \frac{a}{v^2},$$

it follows that for temperatures higher than the critical temperature, v is a function of temperature and pressure which can be expressed in the following form:

$$v(T,p) = \frac{RT}{p} + v^E(T,p).$$

The first term on the right-hand side is the molar volume of the ideal gas, v^{id}, and the function $v^E(T,p)$ characterizes deviations from the ideal behaviour. Substituting $v(T,p)$ into the van der Waals equation of state and dividing both sides by p^2, we obtain the equation for $v^E(T,p)$:

$$\frac{v^E - b}{RT + p(v^E - b)} + \frac{a}{(RT + pv^E)^2} = 0,$$

hence

$$RT\left[(v^E - b)RT + a\right] + (v^E - b)(2RTv^E + a)p + (v^E - b){v^E}^2 p^2 = 0.$$

We assume that v^E can be expanded in a Taylor series in p, around $p = 0$, i.e.

$$\frac{v^E(T,p)}{RT} = B(T) + C(T)p + \cdots.$$

Substituting this expansion into the above equation, we can calculate the coefficients at consecutive powers of p:

$$B(T) = \frac{b}{RT} - \frac{a}{(RT)^2},$$

$$C(T) = \frac{a}{(RT)^3}\left(2b - \frac{a}{RT}\right).$$

For $a = 0.15\ \mathrm{J\,m^3\,mol^{-2}}$, $b = 4 \times 10^{-5}\ \mathrm{m^3\,mol^{-1}}$ and $T = 273.13$ K, we get

$$B = -1.147 \times 10^{-3}\ \mathrm{bar}^{-1}, \qquad C = 1.786 \times 10^{-6}\ \mathrm{bar}^{-2},$$

hence

$$Bp + \frac{1}{2}Cp^2 = -5.71 \times 10^{-3},$$

which gives

$$\Phi = \exp\left(Bp + \frac{1}{2}Cp^2\right) = 0.994.$$

Note that for the given values of the constants a and b, we have (see Exercise 6.15) $T_{cr} = 133.6$ K and $p_{cr} = 34.7$ bar. Thus, for $T = 273.15$ K and $p = 5$ bar, the gas behaves almost as the ideal gas.

7.14 From the solution of Exercise 6.17, we know that the chemical potential of the van der Waals gas has the following form:

$$\mu(T, v) = \mu^0(T) + \frac{RTb}{v-b} - \frac{2a}{v} + RT \ln \frac{RT}{p^0(v-b)}.$$

The gas fugacity is defined as

$$\mu(T, p) = \mu^0(T) + RT \ln \frac{f(T, p)}{p^0}.$$

As above the critical temperature the relation between p and v is unique, we can treat f as a function of T and v. From the comparison of the two expressions, it follows that

$$\ln \frac{f}{p^0} = \ln \frac{RT}{p^0(v-b)} + \frac{b}{v-b} - \frac{2a}{RTv},$$

hence

$$f(T, v) = \frac{RT}{v-b} \exp\left(\frac{b}{v-b} - \frac{2a}{RTv}\right).$$

In the limit $v \to \infty$, we obtain $f(T, v) \approx p$. The fugacity coefficient $\Phi = f/p$, hence, for T_{cr} and p_{cr}, we have $v = v_{cr}$ and

$$\Phi_{cr} = \frac{f(T_{cr}, v_{cr})}{p_{cr}} = \frac{RT_{cr}}{p_{cr}(v_{cr}-b)} \exp\left(\frac{b}{v_{cr}-b} - \frac{2a}{RT_{cr}v_{cr}}\right).$$

Substituting the values of the critical parameters of the van der Waals gas (see (6.58)):

$$v_{cr} = 3b, \qquad p_{cr} = \frac{a}{27b^2}, \qquad T_{cr} = \frac{8a}{27Rb},$$

we get

$$\Phi_{cr} = 4 \exp(-7/4) \approx 0.695.$$

Thus, the deviation of the van der Waals gas at the critical point from the ideal gas behaviour ($\Phi = 1$) amounts to about 30 %.

7.15 We have $P = 3$ and $C = 4$, hence, $f = C - P + 2 = 3$.

7.16 We have $P = 5$, and the minimum number of components corresponds to $f = 0$, thus, it amounts to $C_{min} = P - 2 = 3$.

Exercises of Chapter 8

8.1 We find the liquid composition x_A from the formula $p = p_B^* + (p_A^* - p_B^*)x_A$, which for $p = (p_A^* + p_B^*)/2$ gives $x_A = 0.5$. The relation between the liquid composition x_A and the vapour composition y_A follows from Raoult's law, $p_A = p_A^* x_A$, and Dalton's law, $p_A = p y_A$, hence

$$y_A = x_A \frac{p_A^*}{p} = \frac{p_A^*}{p_A^* + p_B^*}.$$

8.2 When the external pressure p is fixed, the temperature of the liquid–vapour coexistence at that pressure is called the boiling point. We want to derive relations between the boiling point of the solution $A + B$ and its composition x_A and the composition of the vapour above the solution, y_A. According to Raoult's law the vapour pressure above the solution with the composition x_A amounts to

$$p = p_B^*(T) + \left[p_A^*(T) - p_B^*(T)\right] x_A,$$

where $p_A^*(T)$ and $p_B^*(T)$ denote the pressures of the liquid–vapour coexistence for pure components at the temperature T. Hence, we determine the composition of the solution which boils at the given temperature T at constant pressure p:

$$x_A(T) = \frac{p - p_B^*(T)}{p_A^*(T) - p_B^*(T)}.$$

To find the function $x_A(T)$, we have to know the functions $p_A^*(T)$ and $p_B^*(T)$. It follows from the Clausius–Clapeyron equation that in the case of a pure substance, if the enthalpy of evaporation Δh does not depend on temperature, then the pressure on the liquid–vapour coexistence line, p_{coex}, is given by the formula

$$p_{\mathrm{coex}}(T) = p_{\mathrm{coex}}(T_0) \exp\left[(\Delta h/R)(1/T_0 - 1/T)\right],$$

where T_0 is the temperature of a reference state. First, we use this formula to the pure component A, substituting p_A^* for p_{coex} and the enthalpy of evaporation Δh_A^* for Δh. As the reference temperature, we assume the boiling point of the pure A at the pressure p, which we denote by T_A^*. It follows from the definition of the boiling point that $p = p_A^*(T_A^*)$, hence

$$p_A^*(T) = p\,\mathrm{e}^{(\Delta h_A^*/R)(1/T_A^* - 1/T)}.$$

Using analogous reasoning for the component B, we get

$$p_B^*(T) = p\,\mathrm{e}^{(\Delta h_B^*/R)(1/T_B^* - 1/T)},$$

where T_B^* denotes the boiling point of the pure B at the pressure p. Substituting $p_A^*(T)$ and $p_B^*(T)$ into the formula for $x_A(T)$, we get

$$x_A(T) = \frac{1 - \mathrm{e}^{(\Delta h_B^*/R)(1/T_B^* - 1/T)}}{\mathrm{e}^{(\Delta h_A^*/R)(1/T_A^* - 1/T)} - \mathrm{e}^{(\Delta h_B^*/R)(1/T_B^* - 1/T)}}.$$

This is the relation between the boiling point of the solution at constant pressure p and the liquid composition.

To find the composition of the vapour above the solution, y_A, we use the relation $y_A = p_A/p$ (see Dalton's law), where p_A is the partial vapour pressure of the component A, and then Raoult's law $p_A = p_A^* x_A$, hence

$$y_A(T) = \mathrm{e}^{(\Delta h_A^*/R)(1/T_A^* - 1/T)} x_A(T).$$

To obtain the inverse relations, i.e., the temperature as a function of the liquid or vapour composition, we need to expand the exponential function: $e^y \approx 1 + y$, assuming that y is sufficiently small. Using the approximation $1/T_A^* - 1/T \approx (T - T_A^*)/(T_A^*)^2$, we get, e.g., for $\Delta h_A^* = 20$ kJ and $T_A^* = 300$ K:

$$\frac{\Delta h_A^*}{R(T_A^*)^2} = 0.027\ \mathrm{K}^{-1}.$$

Thus, if the difference $T - T_A^*$ amounts to a few kelvins than the argument of the exponential function is really small. It happens when the temperatures T_A^* and T_B^* are not very different. We leave this simple exercise to the reader, to invert the relations $x_A = x_A(T)$ and $y_A = y_A(T)$ in the case when the approximation $e^y \approx 1 + y$ can be used.

8.3 We use the relations derived in Exercise 8.2. For pure substances, at the temperature $T = 350$ K, we get: $p_A^*(T)/p = 1.224$, $p_B^*(T)/p = 0.788$. For the solution $A + B$, the liquid composition results from Raoult's law, and the vapour composition, from Dalton's law, hence: $x_A = 0.486$ and $y_A = 0.595$. Thus, the vapour composition is richer in the more volatile component A, which in the pure state boils at a lower temperature.

8.4 First, we convert the concentration of O_2 in water from milligrams per litre to the molar fraction. We assume water to be the component A, and oxygen to be the component B. Since the solution is dilute, the molar fraction of oxygen in the solution, at the liquid–gas equilibrium, satisfies the inequality $x_B \ll 1$, hence, we get

$$x_B \approx n_B/n_A = v_A^* n_B/V_A,$$

where V_A denotes the volume of water. The molar volume of water $v_A^* = 0.018\ \mathrm{L\,mol^{-1}}$. The concentration of O_2 is equal to 4 $\mathrm{mg\,L^{-1}}$, which corresponds to $n_B/V_A = 1.25 \times 10^{-4}\ \mathrm{mol\,L^{-1}}$, hence

$$x_B = 0.018 \times 1.25 \times 10^{-4} = 2.25 \times 10^{-6}.$$

Multiplying x_B by the Henry constant for oxygen in water, $k_B = 3.3 \times 10^7$ torr, we get the partial pressure of oxygen above water surface:

$$p_B = 3.3 \times 10^7 \times 2.25 \times 10^{-6}\ \mathrm{torr} = 74.25\ \mathrm{torr}.$$

This is the pressure needed to sustain the assumed concentration of oxygen in water at the temperature of 25 °C.

8.5 We use expression (8.43) for the Ostwald absorption coefficient:

$$\frac{V_B}{V_A} = \frac{RT}{k_B v_A^*}.$$

Substituting the values $k_B = 3.3 \times 10^7$ torr and $v_A^* = 0.018\ \mathrm{L\,mol^{-1}}$, for $T = 298$ K, we get

$$V_B/V_A = \frac{8.314 \times 298.15\ \mathrm{J\,mol^{-1}}}{3.3 \times 10^7 \times 133.322\ \mathrm{Pa} \times 1.8 \times 10^{-5}\ \mathrm{m^3\,mol^{-1}}} = 0.031.$$

8.6 The elevation h of the column of water in the capillary corresponds to the osmotic pressure

$$\Pi = \rho g h,$$

where g is the gravity of earth, and ρ denotes the density of water. For a dilute solution, we can use formula (8.57):

$$\Pi = \frac{RTn_B}{V},$$

where B denotes the solute. The mole number of B, i.e., polyethylglicol, equals $n_B = m/M$. From the equality

$$\Pi = \frac{RTm}{VM} = \rho g h,$$

we determine the elevation of water level:

$$h = \frac{RTm}{VM\rho g}.$$

We notice that this formula can also be used to determine the molar mass of the solute

$$M = \frac{RTm}{V\rho g h}.$$

8.7 The boiling point elevation amounts to $\Delta T_b = 1.8$ K, hence, using formula (8.35):

$$\Delta T_b = K_b m_B,$$

we determine the molality of the solute

$$m_B = \frac{\Delta T_b}{K_b} = \frac{1.8}{2.53}\ \text{mol}\,\text{kg}^{-1} = 0.711\ \text{mol}\,\text{kg}^{-1}.$$

Using the definition of the molality:

$$m_B = \frac{n_B}{n_A M_A},$$

where $M_A = 78\ \text{g}\,\text{mol}^{-1}$ is the molar mass of benzene, we get $n_B/n_A = 0.055$. As the mass of the solution amounts to

$$m = M_A n_A + M_B n_B = n_A \left(M_A + M_B \frac{n_B}{n_A} \right) = n_A \times 81.52\ \text{g}\,\text{mol}^{-1},$$

for $m = 100$ g, we get $n_A = 1.227$ mol and $n_B = 0.067$ mol. Thus, the mass of the non-volatile component amounts to

$$M_B n_B = 64 \times 0.067 = 4.29\ \text{g}.$$

8.8 First, we calculate the mole number of ethanol (component B), of the molar mass $M_B = 46\ \text{g}\,\text{mol}^{-1}$:

$$n_B = \frac{24\ \text{g}}{46\ \text{g}\,\text{mol}^{-1}} = 0.522\ \text{mol}.$$

As the mass of the solvent (water) amounts to 1 kg, the molality of ethanol amounts to

$$m_B = 0.522 \text{ mol kg}^{-1}.$$

The freezing point depression amounts to $\Delta T_f = -0.97$ K, hence, for the cryscopic constant of water we get

$$K_f = -\frac{\Delta T_f}{m_B} = 1.86 \text{ K kg mol}^{-1}.$$

8.9 The freezing point depression $\Delta T_f = -0.5$ K, hence, the molality of acetone (component B) amounts to

$$m_B = -\frac{\Delta T_f}{K_f} = \frac{0.5}{3.70} \text{ mol kg}^{-1} = 0.135 \text{ mol kg}^{-1},$$

and the mole number of B in 1.5 kg of acetic acid amounts to

$$n_B = 0.135 \times 1.5 \text{ mol} = 0.203 \text{ mol}.$$

The molar mass of acetone $M_B = 58$ g/mol, thus, the mass of $n_B M_B = 11.77$ g of acetone should be dissolved in 1.5 kg of acetic acid.

8.10 The solvent (benzene) is denoted by A and the solute (naphthalene) is denoted by B. It follows from the assumptions that we can use formula (8.51):

$$\ln x_B = \frac{\Delta h_B^*}{R}\left(\frac{1}{T_B^*} - \frac{1}{T}\right),$$

where $T_B^* = 352.3$ K is the melting point of pure naphthalene, and $\Delta h_B^* = 19.0 \text{ kJ mol}^{-1}$ is its enthalpy of melting. This relation, called the solubility line, defines the molar fraction of the solute in saturated solution, above which the solute precipitates from the solution as a solid. For the temperature $T = 298$ K, we get

$$\ln x_B = \frac{19000}{8.314}\left(\frac{1}{352.3} - \frac{1}{298}\right) = -1.18.$$

Thus, the solubility of naphthalene in benzene at 298 K amounts to $x_B = 0.307$.

8.11 It follows from the assumptions that both lines of liquid–solid coexistence have the following form (see (8.52) and (8.53)):

$$\ln x_A = \frac{\Delta h_A^*}{R}\left(\frac{1}{T_A^*} - \frac{1}{T}\right),$$
$$\ln x_B = \frac{\Delta h_B^*}{R}\left(\frac{1}{T_B^*} - \frac{1}{T}\right),$$

where $x_B = 1 - x_A$. If we set x_A on one axis and T on the other axis we obtain two lines which intersect at the eutectic point. The temperature $T = T_e$ lies below the freezing points of both components in the pure state. Substituting the data: $x_A = x_e = 0.4$, $T_e/T_A^* = 0.9$, $T_e/T_B^* = 0.84$, we get

$$\ln x_A = \ln 0.4 = \frac{\Delta h_A^*}{RT_e}(0.9 - 1) = -0.1\,\frac{\Delta h_A^*}{RT_e},$$
$$\ln x_B = \ln 0.6 = \frac{\Delta h_B^*}{RT_e}(0.84 - 1) = -0.16\,\frac{\Delta h_B^*}{RT_e},$$

hence, the ratio of the enthalpy of melting amounts to

$$\frac{\Delta h_A^*}{\Delta h_B^*} = 1.6 \times \frac{\ln 0.4}{\ln 0.6} = 2.87.$$

Exercises of Chapter 9

9.1 We use the generalized form of Raoult's law (see (9.6) and (9.7)):

$$p_A = p_A^* \gamma_A x_A,$$
$$p_B = p_B^* \gamma_B x_B,$$

where γ_A and γ_B denote the activity coefficients (see Definition 7.15). We can determine γ_A and γ_B from measurements of the vapour pressure above the solution and above pure components. We have $p_A^* = 0.031$ bar, $p_B^* = 0.029$ bar, $x_A = 0.2$, $y_A = 0.44$ and $p = 0.041$ bar. Since vapour is treated as an ideal gas, we can use Dalton's law. Thus, the partial pressures amount to $p_A = y_A p = 0.018$ bar and $p_B = y_B p = 0.023$ bar, hence, we get

$$\gamma_A = \frac{p_A}{p_A^* x_A} = 2.9,$$
$$\gamma_B = \frac{p_B}{p_B^* x_B} = 0.99.$$

The activity coefficient of the component B is very close to unity, thus, Raoult's law is better satisfied for that component. Note that here B is a solvent because $x_B \gg x_A$.

9.2 From Dalton's law, $p_A = p y_A$, $p_B = p y_B$, where p is the vapour pressure above the solution, and p_A and p_B denote the partial pressures, hence

$$p y_A = p_A^* \gamma_A x_A, \qquad p y_B = p_B^* \gamma_B x_B.$$

At the azeotropic point, the vapour composition is the same as the liquid composition, i.e., $y_A = x_A = x_a$ and $y_B = x_B = 1 - x_a$, therefore

$$p = p_A^* \gamma_A = p_B^* \gamma_B,$$

hence

$$\frac{\gamma_B}{\gamma_A} = \frac{p_A^*}{p_B^*}.$$

Using expressions (9.13) and (9.14) for activity coefficients in the simple solution:

$$\gamma_A = \exp(g_{AB} x_B^2 / RT),$$
$$\gamma_B = \exp(g_{AB} x_A^2 / RT),$$

we get

$$\ln \frac{p_A^*}{p_B^*} = \ln \frac{\gamma_B}{\gamma_A} = \frac{g_{AB}(2x_a - 1)}{RT},$$

hence

$$\frac{g_{AB}}{RT} = \frac{\ln 2.5}{0.4} \approx 2.29.$$

For $T = 298$ K, we obtain $g_{AB} \approx 5675 \text{ J mol}^{-1}$. As $g_{AB} > 0$, the activity coefficients are grater than one, which means positive deviation from Raoult's law.

9.3 If the mixture is an azeotrope, then the composition at the azeotropic point, $x_A = x_a$, follows from the formula (see Exercise 9.2):

$$2x_a - 1 = \frac{RT}{g_{AB}} \ln \frac{p_A^*}{p_B^*} = \frac{\ln 3}{0.9} = 1.22.$$

This gives $x_a = 1.11$, which is in contradiction with the condition $x_a \leq 1$. Thus, the mixture considered does not form an azeotrope.

9.4 The condition for the critical temperature (see 9.2.2) has the following form:

$$g_{AB}(T)/RT = 5(T - T_0)\,\text{K}^{-1} - 2(T - T_0)^2\,\text{K}^{-2} = 2.$$

This equation has two roots: $T_1 = T_0 + 0.5$ K and $T_2 = T_0 + 2$ K, hence, $g_{AB}/RT < 2$, if either $T < T_1$ or $T > T_2$, and $g_{AB}/RT > 2$, if $T_1 < T < T_2$. Thus, for temperatures lower than T_1 or higher than T_2, the condition of intrinsic stability is satisfied, i.e., the components are completely miscible, whereas for temperatures $T_1 < T < T_2$, a miscibility gap exists. Thus, the miscibility curve is closed, T_1 is the lower critical temperature and T_2 is the upper critical temperature.

9.5 According to the definition of the Henry constant we have (see Exercise 9.2)

$$k_B = \lim_{x_B \to 0} \frac{p_B}{x_B},$$

where p_B is the partial pressure of B in the vapour above the solution, and x_B is the molar fraction of B in the solution. Since $p_B = p_B^* \gamma_B x_B$, where γ_B denotes the activity coefficient of B in the solution, in the case of the simple solution, we get

$$k_B = p_B^* \gamma_B(T, x_A = 1) = p_B^* \exp\left[\frac{g_{AB}(T)}{RT}\right].$$

At the composition of the azeotropic point, we have (see Exercise 9.2)

$$\frac{g_{AB}(T)}{RT} = \frac{1}{2x_a - 1} \ln \frac{p_A^*}{p_B^*},$$

hence

$$k_B = p_B^* \left(\frac{p_A^*}{p_B^*} \right)^{1/(2x_a-1)}.$$

9.6 Since the solution is ideal, Raoult's law holds. We also assume that vapour is an ideal gas. From Raoult's and Dalton's laws, the relation between the solution composition, x_A, and the vapour pressure above the solution follows:

$$p = p_A + p_B = p_B^* + (p_A^* - p_B^*)x_A,$$

hence

$$x_A(p) = \frac{p - p_B^*}{p_A^* - p_B^*}.$$

The vapour composition results from the relation $p_A = p y_A$:

$$y_A(p) = \frac{p_A^* x_A(p)}{p}.$$

Here, we treat pressure as an independent parameter and express the bubble point and dew point isotherms accordingly, i.e.,

$$\left(\frac{\partial x_A}{\partial p} \right)_T = \frac{y_A \Delta v_A + y_B \Delta v_B}{(y_A - x_A) g_{xx}},$$

$$\left(\frac{\partial y_A}{\partial p} \right)_T = \frac{x_A \Delta v_A + x_B \Delta v_B}{(y_A - x_A) g_{yy}}.$$

The molar Gibbs free energy of an ideal mixture is given by (see (9.43))

$$g = g^{\mathrm{id}} = \mu_A^* x_A + \mu_B^* x_B + RT(x_A \ln x_A + x_B \ln x_B),$$

hence

$$g_{xx} = \left(\frac{\partial^2 g}{\partial x_A^2} \right)_{T,p} = \frac{RT}{x_A x_B},$$

$$g_{yy} = \left(\frac{\partial^2 g}{\partial y_A^2} \right)_{T,p} = \frac{RT}{y_A y_B}.$$

Substituting these expressions into the equations of the bubble point and dew point isotherms, we get

$$\left(\frac{\partial \ln x_A}{\partial p} \right)_T = \frac{x_B (y_A \Delta v_A + y_B \Delta v_B)}{RT(y_A - x_A)},$$

$$\left(\frac{\partial \ln y_A}{\partial p} \right)_T = \frac{y_B (x_A \Delta v_A + x_B \Delta v_B)}{RT(y_A - x_A)}.$$

As $\ln y_A = \ln x_A - \ln(p/p_A^*)$, we have

$$\left(\frac{\partial \ln x_A}{\partial p} \right)_T - \left(\frac{\partial \ln y_A}{\partial p} \right)_T = \frac{1}{p},$$

and after simple transformations we get

$$\frac{\Delta v_A}{RT} = \frac{1}{p}, \qquad \frac{\Delta v_B}{RT} = \frac{1}{p};$$

to obtain the second relation we have replaced A with B in the equations of the isotherms. Note that the above relations are in accord with the assumption that vapour is an ideal gas and that the molar volumes of the liquids A and B can be neglected in Δv_A and Δv_B.

Exercises of Chapter 10

10.1 To prove the stability condition, we use the form of the chemical potential for the ideal gas:

$$\mu_i(T, p, x_i) = \mu_i^0(T) + RT\ln(p/p^0) + RT\ln x_i.$$

From the definition of the affinity of reaction, we get

$$\left(\frac{\partial^2 G}{\partial \xi^2}\right)_{T,p} = -\left(\frac{\partial A}{\partial \xi}\right)_{T,p} = \sum_i \nu_i \left(\frac{\partial \mu_i}{\partial \xi}\right)_{T,p}.$$

The molar fractions as functions of ξ are defined by

$$x_i(\xi) = \frac{n_i(\xi)}{n(\xi)} = \frac{n_i(0) + \nu_i \xi}{n(\xi)},$$

where $n(\xi) = \sum_i n_i(\xi) = n(0) + \xi \Delta n$, and $\Delta n = \sum_i \nu_i$. Since

$$\sum_i \nu_i \left(\frac{\partial \mu_i}{\partial \xi}\right)_{T,p} = RT \sum_i \nu_i \frac{\partial \ln x_i}{\partial \xi} = RT \sum_i \nu_i \left[\frac{\nu_i}{n_i(0) + \nu_i \xi} - \frac{\Delta n}{n(\xi)}\right],$$

we have

$$\left(\frac{\partial^2 G}{\partial \xi^2}\right)_{T,p} = \frac{RT}{n(\xi)}\left[\sum_i \frac{\nu_i^2}{x_i(\xi)} - (\Delta n)^2\right].$$

To transform the sum on the right-hand side, we use the identity $\sum_j x_j = 1$, hence

$$\sum_i \frac{\nu_i^2}{x_i} = \sum_{i,j} \frac{\nu_i^2 x_j}{x_i} = \frac{1}{2}\sum_{i,j}\left(\frac{\nu_i^2 x_j}{x_i} + \frac{\nu_j^2 x_i}{x_j}\right) = \frac{1}{2}\sum_{i,j} \frac{(\nu_i x_j - \nu_j x_i)^2}{x_i x_j} + \sum_{i,j} \nu_i \nu_j,$$

where

$$\sum_{i,j} \nu_i \nu_j = \left(\sum_i \nu_i\right)^2 = (\Delta n)^2,$$

which gives

$$\sum_i \frac{\nu_i^2}{x_i} - (\Delta n)^2 = \frac{1}{2}\sum_{i,j} \frac{(\nu_i x_j - \nu_j x_i)^2}{x_i x_j} \geq 0.$$

Although the equality holds for $x_i = const \times \nu_i$, this case is unphysical, however, because the condition $x_i \geq 0$ is not satisfied either for reactants or products. Thus, in the case of ideal gases, the stability condition holds for all acceptable values of the molar fractions, which means that G is a convex function of ξ at constant T and p.

10.2 We have $T = const$ and $pV = nRT$, hence

$$\Delta H = \Delta U + \Delta(pV) = \Delta U + RT\Delta n,$$

where $\Delta n = \sum_i \nu_i$ denotes the change in the total mole number of gases in the given reaction.

10.3 From the definition of the constant $K_x(T, p)$, we obtain

$$\frac{K_x(T, p_2)}{K_x(T, p_1)} = \left(\frac{p_2}{p_1}\right)^{-\Delta n}.$$

In the first reaction, $\Delta n = 0$, hence, K_x does not change. In the second reaction, $\Delta n = -1$, thus, for $p_2/p_1 = 2$, K_x at p_2 is twice as big as K_x at p_1.

10.4 The initial amounts of the reactants and products, expressed in moles, are: $n_A(0) = 2$, $n_B(0) = 1/3$, $n_C(0) = 1$, $n_D(0) = 1/2$, hence (see (10.5))

$$n_A = 2 - 3\xi, \qquad n_B = \frac{1}{3} - \xi, \qquad n_C = 1 + \xi, \qquad n_D = \frac{1}{2} + 2\xi.$$

Since all mole numbers must be positive, we obtain the condition for ξ:

$$-\frac{1}{4} \leq \xi \leq \frac{1}{3},$$

which specifies the minimum and maximum value of the extent of reaction. Substituting $\xi_{eq} = 1/4$ mol, we get (in moles): $n_A = 5/4$, $n_B = 1/12$, $n_C = 5/4$, $n_D = 1$, hence, the molar fractions in the state of chemical equilibrium amount to: $x_A = 15/43$, $x_B = 1/43$, $x_C = 15/43$, $x_D = 12/43$. Using the law of mass action, we get

$$K_x(T, p) = \frac{x_C x_D^2}{x_A^3 x_B} = 27.52.$$

10.5 The compounds A, B and C are ideal gases, thus, the law of mass action for this reaction has the following form:

$$\frac{x_C^2}{x_A x_B} = K_x = K^0 = 1,$$

as the pressure $p = p^0 = 1$ bar. The molar fractions depend on the extent of reaction ξ:

$$x_A(\xi) = x_A(0) - x, \qquad x_B(\xi) = x_B(0) - x, \qquad x_C(\xi) = 2x,$$

where $x = \xi/n(0)$. Thus, we obtain a quadratic equation for x, i.e.,

$$4x^2 = [x_A(0) - x][x_B(0) - x],$$

which has two roots:

$$x = \frac{1}{6}\big[-1 \pm \sqrt{1 + 12x_A(0)x_B(0)}\big].$$

The negative root is unphysical, because the molar fractions cannot be negative, and for $x_C(0) = 0$, we have $x_C \geq 0$ only if $x \geq 0$. In case (a), $x_A(0) = x_B(0) = 1/2$, hence $x_{\text{eq}} = 1/6$, and the equilibrium molar fractions amount to: $x_A = x_B = x_C = 1/3$. In case (b), $x_A(0) = 1/4$, $x_B(0) = 3/4$, hence, $x_{\text{eq}} = 0, 134$, and in chemical equilibrium we have $x_A = 0.116$, $x_B = 0.616$ and $x_C = 0.268$.

10.6 For convenience, we denote the compounds PCl_5, PCl_3 and Cl_2, by A, B and C, respectively. The molar mass of PCl_5 amounts to $M_A = 208\ \text{g mol}^{-1}$, hence, the initial mole number $n(0) = n_A(0) = 0.91 \times 10^{-2}$ mol. The mole number in the state of chemical equilibrium, $n_{\text{eq}} = n(\xi_{\text{eq}})$, results from the ideal gas equation of state:

$$n_{\text{eq}} = \frac{pV_{\text{eq}}}{RT} = \frac{0.314 \times 10^5 \times 2.4 \times 10^{-3}}{8.314 \times 593.15} = 1.53 \times 10^{-2}\ \text{mol}.$$

The changes in the mole numbers of all compounds during the reaction amount to:

$$n_A(\xi) = n(0) - \xi, \qquad n_B(\xi) = \xi, \qquad n_C(\xi) = \xi,$$

hence, $n(\xi) = n(0) + \xi$ and $\xi_{\text{eq}} = n_{\text{eq}} - n(0) = 0.62 \times 10^{-2}$ mol. The percentage of decomposed PCl_5 molecules is equal to

$$\frac{n_A(0) - n_A(\xi_{\text{eq}})}{n_A(0)} = \frac{\xi_{\text{eq}}}{n(0)} = 0.68,$$

i.e., 68 %. To calculate the equilibrium constant K^0, we have to find the molar fractions of all compounds in chemical equilibrium:

$$x_A = \frac{n_A(\xi_{\text{eq}})}{n_{\text{eq}}} = 0.190, \qquad x_B = x_C = \frac{\xi_{\text{eq}}}{n_{\text{eq}}} = 0.405.$$

As in the state of chemical equilibrium we have

$$\frac{x_B x_C}{x_A} = \left(\frac{p}{p^0}\right)^{-\Delta n} K^0(T) = \left(\frac{p}{p^0}\right)^{-1} K^0(T),$$

thus

$$K^0(320\ ^\circ\text{C}) = 0.314 \times 0.863 = 0.271.$$

10.7 According to the Le Chatelier-Braun principle, in case (a) the position of chemical equilibrium shifts in the direction of heat absorption by the system. Since the decomposition reaction is endothermic, the equilibrium shifts in the direction of higher products concentration, i.e., in the new equilibrium state, $x_A(T_1, p_0) < x_A(T_0, p_0)$. In case (b), the equilibrium shifts in the direction of decreasing volume. Due to the decomposition reaction, one mole of the gas is replaced by two moles, thus, the equilibrium shifts in the direction of higher reactant concentration, i.e., $x_A(T_0, p_1) > x_A(T_0, p_0)$.

10.8 If the temperatures T_1 and T_0 do not differ too much, then, as a first approximation, we can assume that $\Delta_r H^0$ is constant in the range $T_0 < T < T_1$. Then, from the van 't Hoff equation, we get

$$\ln K^0(T_1) = \ln K^0(T_0) + \int_{T_0}^{T_1} \frac{\Delta_r H^0}{RT^2} \mathrm{d}T = \ln K^0(T_0) + \frac{\Delta_r H^0}{R}\left(\frac{1}{T_0} - \frac{1}{T_1}\right),$$

and the same formula holds also for $T_1 < T_0$.

10.9 We assume that $\Delta_r H^0$ is independent of temperature in the temperature range of interest, thus

$$\ln \frac{K^0(T_1)}{K^0(T_0)} = \frac{32000}{8.314}\left(\frac{1}{298} - \frac{1}{310}\right) = 0.50,$$

hence, $K^0(T_1) = 1.65 K^0(T_0)$.

10.10 Since $x_B = x_C = (1 - x_A)/2$, and $\Delta n = 0$, the law of mass action for this reaction has the following form:

$$\frac{(1 - x_A)^2}{4x_A^2} = K_x(T, p) = K^0(T).$$

We have $1 - x_A = 0.40$, at $T_1 = 300$ K, and $1 - x_A = 0.42$, at $T_2 = 315$ K, hence

$$K^0(T_1) = 0.111, \qquad K^0(T_2) = 0.131.$$

We calculate the enthalpy of reaction, using the formula (see Exercise 10.8)

$$\Delta_r H^0 = \frac{R \ln[K^0(T_2)/K^0(T_1)]}{1/T_1 - 1/T_2},$$

hence, $\Delta_r H^0 = 8.68\ \mathrm{kJ\,mol^{-1}}$.

10.11 The law of mass action for this reaction has the following form:

$$\frac{(p_B/p^0)^{\nu_B}(p_C/p^0)^{\nu_C}}{p_A/p_0} = K^0(T),$$

where $p_A = p x_A$, $p_B = p x_B$ and $p_C = p x_C$ denote the partial pressures. Substituting these expressions for the partial pressures, we obtain the law of mass action expressed in terms of the molar fractions:

$$\frac{x_B^{\nu_B} x_C^{\nu_C}}{x_A} = \left(\frac{p}{p^0}\right)^{-\Delta n} K^0(T) = K_x(T, p),$$

where $\Delta n = \nu_B + \nu_C - 1$. Then, we express the molar fractions, using the degree of dissociation α. The total number of moles in the system in chemical equilibrium amounts to

$$n = n_A + n_B + n_C,$$

where, according to the definition of α, we have $n_A = n_i - n_d = (1 - \alpha)n_i$. From the reaction equation, it follows that $n_B = \nu_B n_d$, $n_C = \nu_C n_d$, hence, $n_B = \nu_B \alpha n_i$, $n_C = \nu_C \alpha n_i$ and

$$n = (1 - \alpha + \nu_B \alpha + \nu_C \alpha) n_i = (1 + \alpha \Delta n) n_i.$$

Substituting $x_A = n_A/n$, $x_B = n_B/n$, $x_C = n_C/n$, we get

$$x_A = \frac{1-\alpha}{1+\alpha\Delta n}, \qquad x_B = \frac{\nu_B\alpha}{1+\alpha\Delta n}, \qquad x_C = \frac{\nu_C\alpha}{1+\alpha\Delta n}.$$

Thus, we obtain the following relation between the degree of dissociation and the equilibrium constant

$$\frac{(\nu_B\alpha)^{\nu_B}(\nu_C\alpha)^{\nu_C}}{1-\alpha}(1+\alpha\Delta n)^{-\Delta n} = K_x(T, p).$$

When $\alpha \ll 1$, the above expression can be simplified as follows:

$$K_x(T, p) \approx \nu_B^{\nu_B}\nu_C^{\nu_C}\alpha^{\nu_B+\nu_C}.$$

10.12 In this sort of problems, we treat the symbol $\rightarrow$ in the same way as the equality sign, and add the reactions as if they were algebraic equations. The standard enthalpies of reactions are treated in the same way. It is easy to verify that adding both sides of the first and second reaction, and then subtracting the third reaction from the sum, we obtain the reaction of our interest. Therefore, $\Delta_r H^0$ for that reaction amounts to

$$\Delta_r H^0 = (-1411.3 - 285.8 + 1559.8)\ \text{kJ mol}^{-1} = -137.3\ \text{kJ mol}^{-1}.$$

A negative value of $\Delta_r H^0$ means that the reaction is exothermic, because the enthalpy of the final state is smaller than the enthalpy of the initial state, i.e., the system gives off heat to the surroundings.

10.13 The reaction of N_2O_5 formation from the elements is given by the equation

$$N_2(g) + \frac{5}{2}O_2(g) \rightarrow N_2O_5(g).$$

Multiplying the first and second reaction by 2 and then adding all three reactions, we get

$$2N_2(g) + 5O_2(g) \rightarrow 2N_2O_5(g).$$

We do the same with their standard enthalpies:

$$(-2\times 114.1 - 110.2 + 2\times 180.5)\ \text{kJ mol}^{-1} = 2\Delta_f H^0_{N_2O_5},$$

hence, $\Delta_f H^0_{N_2O_5} = 11.3\ \text{kJ mol}^{-1}$. The reaction is endothermic, since the enthalpy of the final state of the system is bigger than the enthalpy of the initial state, i.e., the system absorbs heat.

10.14 We substitute the enthalpies of formation of individual compounds, with the plus sign for products and the minus sign for reactants, and with the appropriate stoichiometric coefficient. Thus, the standard enthalpy of reaction of propane combustion amounts to

$$\Delta_r H^0 = (-3\times 393.5 - 4\times 285.8 + 103.7)\ \text{kJ mol}^{-1} = -2220\ \text{kJ mol}^{-1}.$$

Note that $\Delta_f H^0_{O_2} = 0$, since O_2 is an element. Before the reaction started there were 6 mol of gases and 0 mol of liquids, whereas after the reaction there are 3 mol of gaseous carbon dioxide and 4 mol of liquid water. The change in the volume of ideal gases amounts to

$$\Delta V = \frac{RT\Delta n}{p} = -\frac{3RT}{p} = -74.3\ \text{L},$$

where $p = 1$ bar, and $\Delta n = -3$ is the change in the mole number of gases. The amount of liquid water has increased by 4 mol, i.e., by 72 g, which occupy 72 cm^3. The volume of the liquid formed in the reaction is more than 1000 times smaller than the change in the gas volume, therefore, it can be neglected. The enthalpy $H = U + pV$, hence, the change in the internal energy at constant pressure amounts to

$$\Delta U = \Delta H - p\Delta V = \Delta H + 3RT = (-2220 + 7.4)\ \text{kJ} = -2212.6\ \text{kJ}.$$

10.15 To obtain 1 mol of ammonia from this reaction, we have to take 2 mol of hydrogen and nitrogen altogether. The change in the enthalpy amounts to $\Delta H = \Delta_f H^0_{NH_3} = -46.1\ \text{kJ mol}^{-1}$. The work done by the system per 1 mol of NH_3 amounts to

$$W = -p\Delta V = -RT\Delta n = 2.5\ \text{kJ},$$

hence

$$\Delta U = \Delta H + W = (-46.1 + 2.5)\ \text{kJ} = -43.6\ \text{kJ}.$$

10.16 We multiply the reaction of water formation by 2 and add the reaction of CO_2 formation, and then subtract the reaction of methane combustion, to get the reaction of methane formation, hence

$$\Delta_f H^0_{CH_4} = 2\Delta_f H^0_{H_2O} + \Delta_f H^0_{CO_2} - \Delta_r H^0 = -74.8\ \text{kJ mol}^{-1}.$$

For the reaction occurs at constant pressure, the heat $Q = \Delta H$, whereas for the reaction at constant volume, $Q = \Delta U$, as the system performs no work. Thus, we have

$$Q = \Delta H - \Delta(pV) \approx \Delta H - RT\Delta n,$$

where Δn is the increase in the total mole number of gases taking part in the reaction. Note that if liquids or solids also take part in a given reaction, then their contribution to pV is usually much smaller than the contribution of gases because of their much smaller molar volumes. In the reaction considered, 2 mol of hydrogen are replaced by 1 mol of methane, hence, $\Delta n = 1 - 2 = -1$ and

$$Q = \Delta_f H^0_{CH_4} + RT = -72.3\ \text{kJ mol}^{-1}.$$

10.17 We know from statistical physics (equipartition of kinetic energy) that the molar internal energy of an ideal gas is equal to the product of $RT/2$ and the number of degrees of freedom per molecule, which is treated as a solid body without internal structure. A linear molecule has 5 degrees of freedom, hence, $U_A = 5n_A RT/2$ and $U_B = 5n_B RT/2$, whereas a molecule of C has 6 degrees of freedom, hence, $U_C =$

$3n_C RT$. For all ideal gases, the equation of state has the same form: $pV = nRT$. For the molar enthalpies of individual gases, we get

$$h_A = h_B = \frac{7}{2} RT, \qquad h_C = 4RT,$$

hence, we calculate the molar heat capacity at constant pressure, c_p, differentiating the enthalpy with respect to T:

$$c_{p,A} = c_{p,B} = \frac{7}{2} R, \qquad c_{p,C} = 4R.$$

Thus, c_p depends neither on pressure nor temperature. For $\Delta_r c_p^0$, we have

$$\Delta_r c_p^0 = 3c_{p,C} - 2c_{p,A} - c_{p,B} = \frac{3}{2} R.$$

Integrating the Kirchhoff equation, we get (see (10.55))

$$\Delta_r H^0(T_2) = \Delta_r H^0(T_1) + \frac{3}{2} R(T_2 - T_1).$$

Substituting $T_1 = 298$ K and $T_2 = 340$ K, we obtain

$$\Delta_r H^0(340 \text{ K}) = -19.48 \text{ kJ mol}^{-1}.$$

10.18 The number of components in the system $C = 3$ and one chemical reaction takes place, hence, $R = 1$. The substance D is a gas, and the substances A and B are liquids. In case (1), A and B form a homogeneous mixture, thus, the number of phases $P = 2$. In case (2), A and B form two liquid phases, which are in equilibrium with the gaseous phase, thus, $P = 3$. The number of independent parameters in the system is the number of degrees of freedom, $f = C + 2 - P - R$. In case (1), $f = 3 + 2 - 2 - 1 = 2$, and in case (2), $f = 3 + 2 - 3 - 1 = 1$.

10.19 The liquids A and B form an ideal solution, thus, Raoult's law can be applied to them, i.e., $p_A = p_A^*(T)x_A$, $p_B = p_B^*(T)x_B$, where p_A and p_B denote the partial pressures of the vapour above the solution. The chemical potentials of the components of the ideal solution are given by

$$\mu_A(l) = \mu_A^*(l) + RT \ln x_A, \qquad \mu_B(l) = \mu_B^*(l) + RT \ln x_B,$$

where $\mu_A^*(l)$ and $\mu_B^*(l)$ are the chemical potentials of pure liquids A and B. Making use of Raoult's law (see Sect. 8.1.1), we get for the liquid–vapour coexistence:

$$\mu_A(l) = \mu_A(g) = \mu_A^0 + RT \ln \frac{p_A}{p^0} = \mu_A^*(T) + RT \ln x_A,$$

where $\mu_A^*(T)$ denotes the chemical potential of pure substance A on the liquid–vapour coexistence line. An analogous relation holds for the component B, i.e.,

$$\mu_B(l) = \mu_B^*(T) + RT \ln x_B.$$

The component C is an ideal gas, thus

$$\mu_C = \mu_C^0(T) + RT \ln \frac{p_C}{p^0}.$$

The condition of chemical equilibrium adopts the following form:

$$\mu_A(l) - \mu_B(l) - \mu_C(g) = \mu_A^*(T) - \mu_B^*(T) - \mu_C^0(T) + RT \ln \frac{x_A p^0}{x_B p_C} = 0.$$

Using the equilibrium constant defined by

$$\mu_A^*(T) - \mu_B^*(T) - \mu_C^0(T) = RT \ln K(T),$$

we obtain the law of mass action:

$$\frac{x_B(p_C/p_0)}{x_A} = K(T).$$

The total pressure of the gaseous phase above the solution, p, is the sum of three partial pressures:

$$p = p_A + p_B + p_C = p_A^* x_A + p_B^* x_B + p_C.$$

As the presence of the component C in the solution can be neglected, we have

$$x_A + x_B + x_C \approx x_A + x_B = 1.$$

Substituting this relation into the law of mass action, we get

$$(1 - x_A) p_C = p^0 K(T) x_A,$$

hence

$$x_A = \frac{p_C}{p^0 K + p_C}, \qquad x_B = \frac{p^0 K}{p^0 K + p_C}.$$

Finally, we obtain

$$p = \frac{p_A^* p_C + p_B^* p^0 K}{p^0 K + p_C} + p_C.$$

According to the phase rule, we have $f = C + 2 - P - R = 3 + 2 - 2 - 1 = 2$. In this case, the independent parameters are T and p_C.

10.20 First, we write the condition of chemical equilibrium (see Exercise (10.11)):

$$\mu_{CuO}(s) + \mu_{H_2}(g) - \mu_{Cu}(s) - \mu_{H_2O}(g) = 0,$$

where

$$\mu_{H_2}(g) = \mu_{H_2}^0 + RT \ln \frac{p_{H_2}}{p^0}, \qquad \mu_{H_2O}(g) = \mu_{H_2O}^0 + RT \ln \frac{p_{H_2O}}{p^0}.$$

Cu and CuO are pure substances in the solid phase. We recall that in general we can express the chemical potential of the i-the component in terms of its activity (see (7.116)), i.e.

$$\mu_i = \mu_i^0(T) + RT \ln a_i.$$

In the case of a pure substance, $\mu_i = \mu_i^*$ depends on temperature and pressure. However, liquids and solids are hardly compressible, therefore, the dependence of the chemical potential on pressure is often neglected, provided the pressure p does not differ too much from the standard pressure p^0. In the case of chemical reactions, it is usually assumed that the activity of a pure liquid or solid is equal to unity. Thus, for the reaction considered, we have $a^*_{CuO}(s) = 1$, $a^*_{Cu}(s) = 1$, and the condition of chemical equilibrium adopts the following form:

$$\mu^0_{CuO} + \mu^0_{H_2} - \mu^0_{Cu} - \mu^0_{H_2O} + RT\ln\frac{p_{H_2}}{p_{H_2O}} = 0,$$

where the standard chemical potentials depend only on temperature. According to the definition of the standard equilibrium constant, we have

$$\mu^0_{CuO} + \mu^0_{H_2} - \mu^0_{Cu} - \mu^0_{H_2O} = RT\ln K^0.$$

Finally, we get

$$\frac{p_{H_2O}}{p_{H_2}} = K^0.$$

This is the condition of chemical equilibrium for the reaction considered.

Exercises of Chapter 11

11.1 In the case of the first reaction, we have 4 components, that is, we neglect dissociation of water and assume that activity of water $a_{H_2O} = 1$. Ammonia is treated as an ideal gas, thus,

$$\mu_{NH_3}(g) = \mu^0_{NH_3} + RT\ln\frac{p_{NH_3}}{p^0},$$

where p_{NH_3} is the partial pressure of ammonia above the solution. The chemical potentials of the ions are expressed in terms of their activities:

$$\mu_{NH_4^+} = \mu^0_{NH_4^+} + RT\ln a_{NH_4^+}, \qquad \mu_{OH^-} = \mu^0_{OH^-} + RT\ln a_{OH^-}.$$

Therefore, the condition of chemical equilibrium adopts the following form:

$$\frac{(a_{NH_4^+})(a_{OH^-})p^0}{p_{NH_3}} = K^0.$$

The second reaction concerns a saturated solution of slightly soluble salt, which means that we have coexistence of practically pure crystalline salt with aqueous solution of that salt. It can be assumed that the activity of the solid is equal to unity, hence, only the activities of the ions appear in the condition of chemical equilibrium, i.e.,

$$(a_{Ba^{2+}})(a_{SO_4^{2-}}) = K^0.$$

Since the salt is slightly soluble, its concentration in the solution is small. In the case of electrolytes, such as aqueous solutions of salt for instance, the solute whose concentration in the solution is small is practically completely dissociated. Thus, the equilibrium constant K^0 is a measure of solubility of the salt in water. For this reason, it is called the solubility constant or solubility product of the salt, and is denoted by K_s. It can be easily understood if we replace the activity with the molar concentration, which is justified in the case of strong dilution. Then we have

$$[Ba^{2+}][SO_4^{2-}] = K_s,$$

and because $[Ba^{2+}] = [SO_4^{2-}]$, the molar concentration of each ion amounts to $K_s^{1/2}$. Thus, the constant K_s provides information about solubility of the salt in water.

11.2 Expressing the chemical potential in terms of the activity, we write the condition of chemical equilibrium as follows:

$$\mu^0_{A_xB_y} + RT\ln a_{A_xB_y} - x(\mu^0_A + RT\ln a_A) - y(\mu^0_B + RT\ln a_B) = 0.$$

Assuming that the activity of the pure solid $a_{A_xB_y} = 1$, we obtain the following form of the equilibrium condition:

$$a_A^x a_B^y = K_s,$$

where K_s denotes the equilibrium constant of the reaction considered. It is a generalization of the solubility product considered in Exercise 11.1. Applying this formula to aqueous solution of the salt Ag_2CO_3, which dissociates according to the equation

$$Ag_2CO_3(s) \rightleftharpoons 2Ag^+(aq) + CO_3^{2-}(aq),$$

we get

$$K_s = (a_{Ag^+})^2(a_{CO_3^{2-}}) \approx [Ag^+]^2[CO_3^{2-}].$$

For each CO_3^{2-} ion, there are two Ag^+ ions, hence, $[Ag^+] = 2[CO_3^{2-}]$ and we get

$$4[CO_3^{2-}]^3 = K_s = 6.2 \times 10^{-12},$$

i.e., $[CO_3^{2-}] = 1.16 \times 10^{-4}\ \mathrm{mol\,L^{-1}}$ and $[Ag^+] = 2.32 \times 10^{-4}\ \mathrm{mol\,L^{-1}}$.

11.3 We use relation (11.84) between the standard potential and equilibrium constant:

$$E^0 = \frac{RT}{nF}\ln K^0,$$

where $F = 9.6485 \times 10^4\ \mathrm{C\,mol^{-1}}$, and $n = 2$. The standard potential $E^0 = 0.339 - (-0.763)\ \mathrm{V} = 1.102\ \mathrm{V}$, and the coefficient RT/nF amounts to

$$\frac{RT}{nF} = \frac{8.314 \times 298}{2 \times 9.6485} \times 10^{-4}\ \mathrm{J\,C^{-1}} = 1.284 \times 10^{-2}\ \mathrm{V},$$

hence

$$K^0 = e^{E^0 nF/RT} = e^{85.8} = 1.8 \times 10^{37}.$$

Thus, the equilibrium constant of this reaction is very big.

This exercise can also be solved in a different way. We have $E^0 = E^0_R - E^0_L$, where $E^0_R = 0.339$ V, $E^0_L = -0.763$ V and

$$E^0_R = \frac{RT}{nF} \ln K^0_R, \qquad E^0_L = \frac{RT}{nF} \ln K^0_L.$$

K^0_R and K^0_L denote the standard equilibrium constants for the reactions in cells consisting of the copper and zinc electrode, respectively, in one half-cell and the standard hydrogen electrode. Thus, we have

$$\frac{RT}{nF} \ln K^0 = \frac{RT}{nF} \ln K^0_R - \frac{RT}{nF} \ln K^0_L,$$

hence, $K^0 = K^0_R / K^0_L$. Substituting the data, we get

$$K^0_R = e^{E^0_R nF/RT} = e^{26.4} = 2.9 \times 10^{11},$$
$$K^0_L = e^{E^0_L nF/RT} = e^{-59.4} = 1.6 \times 10^{-26},$$

which gives $K^0 = 1.8 \times 10^{37}$.

11.4 46 g of Na correspond to 2 mol, thus, to obtain this amount of sodium at the cathode, an electric charge of 2 mol of electrons (according to the reaction $2Na^+ + 2e^- \rightarrow 2Na$) must flow through the salt, i.e.,

$$F \times 2 \text{ mol} = 1.9297 \times 10^5 \text{ C}.$$

Since $1 \text{ A} = 1 \text{ C s}^{-1}$, the current of 10 A must flow during

$$1.9297 \times 10^4 \text{ C A}^{-1} = 19297 \text{ s} \approx 5.36 \text{ h}.$$

During this time 1 mol of gaseous chlorine is produced at the anode, because each Cl_2 molecule provides 2 electrons, donated by 2 Cl^- ions, and the charge $F \times 2$ mol has flowed. Therefore, we obtain 70.9 g of chlorine at the anode.

11.5 To solve this problem, the relation between the enthalpy of reaction and the potential difference of the cell is to be found. To do this, we use the equality (see (11.77))

$$\Delta G = -EnF\Delta\xi,$$

which holds if E does not change during the cell reaction. The change in the enthalpy is obtained from the relation

$$G = H - TS = H + T\frac{\partial G}{\partial T},$$

hence, for $T = const$, we get

$$\Delta H = \Delta G - T\frac{\partial \Delta G}{\partial T},$$

where we differentiate at constant p and ξ. Substituting the expression for ΔG, we get

$$\Delta H = nF\left[T\left(\frac{\partial E}{\partial T}\right)_p - E\right]\Delta\xi.$$

The same relation must hold for the differentials, i.e.,

$$\mathrm{d}H = nF\left[T\left(\frac{\partial E}{\partial T}\right)_p - E\right]\mathrm{d}\xi.$$

The enthalpy of reaction at constant pressure is equal to the change in the enthalpy per mole of reaction, i.e., $(\partial H/\partial\xi)_{T,p}$. Thus, the relation between the enthalpy of reaction and the potential difference of the cell has the following form:

$$\left(\frac{\partial H}{\partial\xi}\right)_{T,p} = nF\left[T\left(\frac{\partial E}{\partial T}\right)_p - E\right].$$

Now we can substitute the data for the temperature of 0 °C ($T = 273.15$ K), which gives

$$\begin{aligned}\left(\frac{\partial H}{\partial\xi}\right)_{T,p} &= -2 \times 96485\left(273.15 \times 4.02 \times 10^{-4} + 1.015\right)\,\mathrm{C\,V\,mol^{-1}}\\ &= -217.05\,\mathrm{kJ\,mol^{-1}}.\end{aligned}$$

The maximum work (not related to a change in volume) which can be obtained from a reversible cell at constant temperature and pressure is equal to ΔG. The part of the enthalpy of reaction which cannot be changed into work is equal to the difference

$$\Delta H - \Delta G = T\Delta S = -T\frac{\partial\Delta G}{\partial T} = nFT\left(\frac{\partial E}{\partial T}\right)_p \Delta\xi.$$

Taking the limit of infinitesimal changes, we get, per 1 mol of reaction,

$$T\left(\frac{\partial S}{\partial\xi}\right)_{T,p} = nFT\left(\frac{\partial E}{\partial T}\right)_p = -21.19\,\mathrm{kJ\,mol^{-1}}.$$

It is about 10 % of the enthalpy of reaction. The remaining 90 % can be used to perform work, for instance, by an electric engine. This is the maximum value, which cannot be exceeded because of limitations resulting from the laws of thermodynamics. Note, however, that the electric current obtained from the cell can be used in an electric heater, instead of performing work. Then, 100 % of the energy obtained from the chemical reaction can be changed into heat, and 10 % of that heat comes from the change in the entropy of the system, which decreases (the system gives off heat). The remaining part is the heat produced during the flow of electric current through the heater.

11.6 In the reaction, gaseous hydrogen at the pressure p_{H_2} and hydrogen ions of the activity a_{H^+} take part. The expression for the potential of the half-cell can be derived in a similar way as for a metal electrode in equilibrium with its ions (see (11.89) and (11.90)). We only have to take into account different stoichiometric coefficients and replace the activity with the ratio p_{H_2}/p^0 in the case of gaseous hydrogen, which is treated as an ideal gas. For the reaction

$$H^+(aq) + e^- \rightleftharpoons \frac{1}{2}H_2(g)$$

the coefficient at gaseous hydrogen amounts to $1/2$, hence

$$E_{H_2} = E_{H_2}^0 - \frac{RT}{nF} \ln \frac{(p_{H_2}/p^0)^{1/2}}{a_{H^+}},$$

where $n = 1$. In accord with the definition of the standard hydrogen electrode, we assume that $E_{H_2}^0 = 0$, hence

$$E_{H_2} = -\frac{RT}{F} \ln \frac{(p_{H_2}/p^0)^{1/2}}{a_{H^+}}.$$

If the activity of hydrogen ions is fixed, and the pressure of gaseous hydrogen changes from $p_{H_2} = p_1$ to $p_{H_2} = p_2$, then the change in E_{H_2} amounts to

$$\Delta E_{H_2} = E_{H_2}(p_2) - E_{H_2}(p_1) = \frac{RT}{2F} \ln \frac{p_1}{p_2}.$$

If the pressure is fixed, and the activity of hydrogen ions changes from $a_{H^+} = a_1$ to $a_{H^+} = a_2$, then

$$\Delta E_{H_2} = E_{H_2}(a_2) - E_{H_2}(a_1) = \frac{RT}{F} \ln \frac{a_2}{a_1}.$$

At the temperature $T = 298.15$ K, we have $RT/F = 0.0257$ V. In case (1), we get $p_1/p_2 = 9 = 3^2$, and in case (2), we get $a_2/a_1 = 3$. Thus, in both cases, we have

$$\Delta E_{H_2} = \frac{RT}{F} \ln 3 = 0.0282 \text{ V}.$$

11.7 We use expression (11.94) for the redox potential. Since the reduced form corresponds to pure zinc in the solid phase, we have $a_{Red} = a_{Zn} = 1$, hence

$$E_{Zn} = E_{Zn}^0 + \frac{RT}{2F} \ln a_{Zn^{2+}} = \left(-0.763 + \frac{1}{2} 0.0257 \times \ln 0.1\right) \text{V}$$
$$= -(0.763 + 0.0296) \text{ V} = -0.7926 \text{ V}.$$

11.8 Adding the reactions in the half-cells, we obtain the total cell reaction

$$\frac{1}{2}H_2(g) + AgCl(s) \rightleftharpoons Ag(s) + H^+(aq) + Cl^-(aq).$$

The potentials of the half-cells are given by the following expressions (see Exercise 11.6):

$$E_{H_2} = -\frac{RT}{F} \ln \frac{(p_{H_2}/p^0)^{1/2}}{a_{H^+}}$$

and

$$E_{Ag|AgCl} = E_{Ag|AgCl}^0 - \frac{RT}{F} \ln \left(\frac{a_{Ag} a_{Cl^-}}{a_{AgCl}}\right).$$

Since the activities of pure solids, a_{Ag} and a_{AgCl}, can be omitted as equal to 1, the potential of the cell amounts to

$$E = E_{Ag|AgCl}^0 - \frac{RT}{F} \ln \frac{(a_{H^+})(a_{Cl^-})}{(p_{H_2}/p^0)^{1/2}}.$$

We know that the activities of anions and cations cannot be measured separately, but only their product can be determined (see Sect. 11.1.2). Using the molality as a measure of concentration, we have $a_{H^+} = \gamma_+ m_+ / m^0$, $a_{Cl^-} = \gamma_- m_- / m^0$, hence

$$(a_{H^+})(a_{Cl^-}) = a_{HCl} = \gamma_\pm^2 (m/m^0)^2,$$

where $\gamma_\pm^2 = \gamma_+ \gamma_-$, and $m = m_+ = m_-$, as the molality of both ions is the same. For the pressure $p_{H_2} = p^0$, we can write

$$E + \frac{2RT}{F} \ln \frac{m}{m^0} = E^0_{Ag|AgCl} - \frac{2RT}{F} \ln \gamma_\pm .$$

The left-hand side of this equation contains quantities which are determined experimentally: the cell potential E and the molality m. It means that we can calculate the average activity coefficient of ions for the electrolyte, $\gamma_\pm$, provided that we know the standard potential $E^0_{Ag|AgCl}$. On the other hand, the standard potential can be determined from the measurement of the cell potential for dilute solutions. Since $\gamma_\pm \to 1$ when $m/m^0 \to 0$, the left-hand side of the equation must tend to the constant $E^0_{Ag|AgCl}$ if we extrapolate to $m = 0$.

References

1. P.W. Atkins, *Physical Chemistry*, 6th edn. (Oxford University Press, Oxford, 1998)
2. J.J. Binney, N.J. Dowrick, A.J. Fisher, M.E.J. Newman, *The Theory of Critical Phenomena An Introduction to the Renormalization Group* (Clarendon Press, Oxford, 1992)
3. H.B. Callen, *Thermodynamics and an Introduction to Thermostatistics*, 2nd edn. (Wiley, New York, 1985)
4. Compendium of Chemical Terminology Gold Book, v. 2.3, International Union of Pure and Applied Chemistry (2011), http://goldbook.iupac.org/
5. R. Haase, in *Thermodynamics*, ed. by W. Jost. Physical Chemistry An Advanced Treatise, vol. 1 (Academic Press, New York, 1971), Chaps. 1 and 3
6. R. Hołyst, A. Poniewierski, A. Ciach, *Termodynamika dla chemików, fizyków i inżynierów* (Wydawnictwo Uniwersytetu Kardynała Stefana Wyszyńskiego, Warsaw, 2005), in Polish
7. R. Hołyst, A. Poniewierski, *Termodynamika w zadaniach* (Wydawnictwo Uniwersytetu Kardynała Stefana Wyszyńskiego, Warsaw, 2008), in Polish
8. K. Huang, *Statistical Mechanics*, 2nd edn. (Wiley, New York, 1987)
9. D. Kondepudi, I. Prigogine, *Modern Thermodynamics* (Wiley, New York, 1999)
10. H.C. Van Ness, M.M. Abbott, *Classical Thermodynamics of Non Electrolyte Solutions* (McGraw-Hill, New York, 1982)
11. L. Pauling, *General Chemistry* (Dover, New York, 1988)
12. J.M. Prausnitz, R.N. Lichtenthaler, E. Gomes de Azevedo, *Molecular Thermodynamics of Fluid-Phase Equilibria*, 2nd edn. (Prentice-Hall, New Jersey, 1986)
13. C.F. Prutton, S.H. Maron, *Fundamental Principles of Physical Chemistry* (Mac-Millan, New York, 1948)
14. F. Reif, *Statistical Physics*, Berkeley Physics Course, vol. 5 (McGraw-Hill, New York, 1967)
15. E.B. Smith, *Basic Chemical Thermodynamics*, 5th edn. (World Scientific, Singapore, 2004)
16. H.E. Stanley, *Introduction to Phase Transitions and Critical Phenomena* (Oxford University Press, Oxford, 1987)

R. Hołyst, A. Poniewierski, *Thermodynamics for Chemists, Physicists and Engineers*,
DOI 10.1007/978-94-007-2999-5,

Index

A
Absolute zero, 26
Acceptor, 250
Acid, 250
  dissociation constant, 253
Activity, 170
  coefficient, 172, 173
    in simple solution, 204
    of ions, 248
Adiabatic
  process, 11
    irreversible, 45
    reversible, 44, 111
  wall, 9
Affinity of reaction, 226
Amount of substance, 18
Amphoteric compound, 250
Anion, 245
Anode, 255
Atmosphere, 22
Autodissociation of water, 250
Avogadro constant, 18
Azeotrope, 201, 207
Azeotropic
  mixture, 205, 207
  point, 207

B
Barometer, 22
Base, 250
  dissociation constant, 253
Blackbody radiation, 31
Boiling point, 134
  elevation, 188, 197
Boltzmann constant, 64
Brønsted–Lowry theory, 250
Bubble point
  curve, 204
  isobar, 205, 206, 208
  isotherm, 205–208

C
Carnot engine, 76
  efficiency, 77
Cathode, 255
Cation, 245
Celsius scale, 24
Chemical equilibrium, 227
Chemical potential, 27, 66, 153, 161
  of component in ideal mixture, 166
  of ideal gas, 71
  of solute in dilute solution, 172
  of solvent, 174
Chemical reaction, 49, 85
  equation, 225, 226, 234
Clapeyron equation, 138
Clausius–Clapeyron equation, 140
Closed system, 9
Colligative properties, 197
Component, 151, 239
Compound, 239
Compressibility
  adiabatic, 110
  isothermal, 108
Condensation line, 138
Conjugate acid–base pair, 250
Continuous phase transition, 124, 125
Counter ions, 248
Critical
  opalescence, 142
  phenomena, 126
  point, 126, 132, 146

R. Hołyst, A. Poniewierski, *Thermodynamics for Chemists, Physicists and Engineers*,
DOI 10.1007/978-94-007-2999-5,

Critical (*cont.*)
pressure, 132, 147
temperature, 132, 147
of solution, 210
Cryoscopic constant, 192
Curie temperature, 125

D
Dalton's law, 165
Daniell cell, 253
Debye screening length, 249
Debye–Hückel limiting law, 249
Degrees of freedom
of molecule, 29
of system, 176
Deposition line, 138
Dew point
curve, 204
isobar, 205, 206, 208
isotherm, 205–208
Diathermal wall, 9
Diesel cycle, 81
Differential, 15
form, 17
Dilute solution, 152, 172
Dissociation, 245
degree, 246
Distillation, 186
of azeotropic mixture, 207
Donor, 250

E
Ebullioscopic constant, 189
Ehrenfest classification, 127
Electric
neutrality, 245
potential difference, 256
Electrochemical
cell, 253
series, 260
Electrolysis, 255
Electrolyte, 245
Electrolytic cell, 255
Electronegative metal, 261
Electropositive metal, 260
Endothermic
process, 211
reaction, 227
Enthalpy, 43, 93, 109, 154
natural variables, 98
of evaporation, 140
of ideal gas, 99
of melting, 139
of mixing, 164
of reaction, 227
of sublimation, 140
Entropy, 6, 57
maximum principle, 61
mechanical-statistical definition, 63
of ideal gas, 70
of mixing, 164
of van der Waals gas, 144
Equation of state, 29
Dieterici's, 148
of ideal gas, 29, 32
of photon gas, 31, 32
van der Waals', 31, 32, 143
Equilibrium
constant, 232
state, 7
with respect to matter flow, 65
Euler relation, 69, 153
Eutectic
point, 194
system, 220
Evaporation, 48, 74, 123
line, 138
Excess
chemical potential, 203
functions, 175
Gibbs free energy, 204
Exothermic
process, 212
reaction, 227
Extensive parameter, 11
Extent of reaction, 226

F
Fahrenheit scale, 24
Faraday constant, 249
Ferromagnetic phase, 125
Fractional crystallization, 219
Free gas expansion, 84, 112
Freezing, 123
line, 138
point
curve, 219
depression, 192, 197
Fugacity
coefficient, 170
of gas in mixture, 169
of pure gas, 168
Functions of mixing, 163
for ideal gases, 166
Fundamental
equations of chemical thermodynamics, 154

Fundamental (*cont.*)
postulate of thermodynamics, 8
relation, 67, 89, 153

G
Galvanic cell, 255
Gas constant, 26
Gibbs free energy, 93, 107, 154
minimum principle, 104
natural variables, 98
of ideal gas, 99
of mixing, 164
Gibbs–Duhem equation, 94, 155
at constant temperature and pressure, 156
Gibbs–Helmholtz relation, 187
Grand thermodynamic potential, 95
natural variables, 98
of ideal gas, 100

H
Half-cell, 254
potential, 259
Heat, 10, 16, 40
capacity, 42
at constant pressure, 43, 108
at constant volume, 42
relation to entropy, 71
devices, 75
engine, 75
efficiency, 76
of transition, 125
pump, 80
efficiency, 80
reservoir, 75
Helmholtz free energy, 92, 108, 154
minimum principle, 101
natural variables, 97
of ideal gas, 99
of van der Waals gas, 145
Henry constant, 190
Henry's law, 190, 201
Hess' law, 234
Heteroazeotrope, 217
Heterozeotropic mixture, 217
Homoazeotrope, 218
Hydronium ion, 250
Hydroxide ion, 250

I
Ideal
dilute solution, 190
gas, 30
mixture, 167, 181
solubility, 193
solution, 152, 167
Intensive parameter, 10, 20
Internal energy, 9, 19, 39, 110, 153
differential, 41
natural variables, 97
of ideal gas, 29
of mixing, 164
of photon gas, 31
of van der Waals gas, 31
Intrinsic stability
of mixture, 156, 205
of pure substance, 115
Inversion temperature, 114
Ion product of water, 251
Ionic strength, 249
Irreversible process, 13
change in entropy, 82
Isobaric process, 11, 43, 73
Isochoric process, 11, 41, 73
Isolated system, 9
Isothermal process, 11
irreversible, 47
reversible, 47, 72

J
Joule, 43
Joule–Thomson process, 113

K
Kelvin, 25
Kelvin scale, 25
Kirchhoff equation, 237

L
Lambda transition, 136
Law of mass action, 232, 258
Laws of thermodynamics
first law, 40
second law, 58
third law, 85
zeroth law, 23
Le Chatelier–Braun principle, 230
Legendre transformation, 89
definition, 90
of entropy, 96
of internal energy, 91
Lever rule, 129, 130, 141, 185
Liquid composition line, 183
Liquid–gas coexistence, 140
Liquid–vapour
equilibrium, 181
phase diagram
at constant pressure, 185
at constant temperature, 183

M
Macroscopic system, 7
Massieu functions, 96
Maxwell
  construction, 146, 216
  relations, 107
Mechanical
  equilibrium, 65
  stability, 117
Melting, 123, 125
  line, 132, 138
  point, 134, 139
    curve, 219
Metastable states, 128
Millimeter Hg, 22
Miscibility
  curve, 210
  gap, 201, 210
    in simple solutions, 213
Mixture, 151
  ideal, 167, 181
  of ideal gases, 164
  real, 168, 201
Molality, 153
Molar
  concentration, 152
  fraction, 152, 154
Mole, 18
  number, 18

N
Natural variables, 89, 97
Nernst equation, 258

O
Order parameter, 126
Order-disorder transition, 126
Osmotic
  equilibrium, 196
  pressure, 28, 196, 198
Ostwald absorption coefficient, 191
Otto cycle, 81
Oxidation, 254

P
Paramagnetic phase, 125
Partial
  molar entropy, 162
  molar quantity, 160
  molar volume, 161
  pressure, 165
Pascal, 21
Perfect blackbody, 26
PH, 251
Phase, 123
  coexistence, 125, 127, 131
    conditions, 128, 176
    line, 128, 138
  diagram, 132
    of ^{4}He, 135
    of simple eutectic, 195
    of typical substance, 132
    of water, 134
  rule, 176
    for chemical systems, 238
  transition, 124
    of first order, 124
    of second order, 124
Photon gas, 31
Pressure, 21
Principle of corresponding states, 147
Product of reaction, 225
Pyrometer, 27

Q
Quasi-static
  heat, 17, 64
  process, 12
  work, 17, 257

R
Radiator, 75
Raoult's law, 181
  derivation, 182
  deviations from, 201, 202
Reactant, 225
Redox potential, 259
Reduced variables, 147
Reduction, 254
Refrigerator, 79
  efficiency, 80
Reversible
  cell, 256
  flow of heat, 64
  process, 13
  work source, 45

S
Salt bridge, 255
Saturated
  solution, 193
  vapour, 48, 133, 141
Screening, 248
Simple
  eutectic, 194
  solution, 203
Solid solution, 218
Solid–gas coexistence, 140

Solid–liquid coexistence, 139
Solubility line, 193
Solute, 152
Solution, 152
Solvent, 152
Standard
  cell potential, 258
  enthalpy
    of formation, 235
    of reaction, 233
  equilibrium constant, 231
  Gibbs free energy of reaction, 231
  hydrogen electrode, 260
  pressure, 165
  state, 166
    of gas, 166
    of pure substance or solvent, 170
    of solute, 174
State function, 10
State parameter, 10
Steam engine, 81
Stefan–Boltzmann law, 26
Stoichiometric coefficient, 225
Sublimation, 123, 125
  line, 138
Subsystem, 10
Super-cooled liquid, 128
Super-critical phase, 133
Super-heated liquid, 128
Super-saturated vapour, 128
Surroundings, 9
System, 9

T
Temperature
  absolute, 25
  empirical, 23
Thermal
  equilibrium, 23, 61
  expansion coefficient, 25, 108
  stability, 117
Thermodynamic
  equilibrium, 7
    conditions, 61
  potentials, 89, 154
    for ideal gas, 98
  process, 11
  temperature, 25, 63, 77
Thermometer
  mercury, 24
  platinum, 25
Thermopile, 25
Torr, 22
Triple point, 131, 132
  of CO_2, 134
  of water, 131

V
Van 't Hoff equation, 233
Vapour, 48
  composition line, 183
  pressure depression, 197
Volume, 17
  of mixing, 164

W
Wall, 9
Work, 10, 16
  chemical, 40
  mechanical, 39
Working substance, 75

Z
Zeotropic mixture, 205, 206